Contextual Approach to Quantum Formalism

Fundamental Theories of Physics

An International Book Series on The Fundamental Theories of Physics: Their Clarification, Development and Application

Volume 160

For other titles published in this series, go to www.springer.com/series/6001

Andrei Khrennikov

Contextual Approach to Quantum Formalism

Springer

Prof. Andrei Khrennikov
University of Växjö
International Center for Mathematical Modeling in Physics
and Cognitive Science
Vejdes Plats 7
35195 Växjö
Sweden
E-mail: andrei.khrennikov@vxu.se

ISBN 978-1-4020-9592-4 e-ISBN 978-1-4020-9593-1

Library of Congress Control Number: 2009920940

Printed on acid-free paper

9 8 7 6 5 4 3 2 1

springer.com

Dedication

To memory of Richard von Mises whose theory of collectives opened the way to contextual probability.

Preface

Calculus of quantum probabilities is one of the most important sectors of the mathematical formalism of quantum mechanics. This calculus is based on the representation of probabilities by *complex probability amplitudes* (or normalized complex vectors in the abstract Hilbert space formalism). This algebraic representation was created in the process of the development of quantum mechanics. Since that time it has been successfully applied to the statistical analysis of data obtained in various experiments with quantum systems.

Applications to quantum physics play an extremely stimulating role in the development of quantum probability mathematics. However, the original appearance of this mathematical apparatus in the framework of a special physical theory—quantum mechanics—created a barrier (in any event, a psychological barrier) on the way to generalizations and applications of quantum probability calculus outside quantum physics.

The quantum mechanical origin of this mathematical formalism induced the impression that the quantum probabilistic behavior can only be found in very special (often regarded as even mystical) systems, namely, quantum particles and fields.[1]

[1] "Quantum mechanics is magic," Daniel Greenberger; " Everything we call real is made of things that cannot be regarded as real," Niels Bohr; "Those who are not shocked when they first come across quantum theory cannot possibly have understood it," Niels Bohr; "If you are not completely

The aim of this book is *to demystify quantum probability.*[2] The only possibility to do this is to derive the quantum probability calculus without starting with the conventional Hilbert space formalism directly. We recall the standard scheme leading to quantum probability: *complex Hilbert state space, Born's postulate, the derivation of interference of probabilities* via the transition from one basis of eigenvectors to another (for two noncommuting operators $\hat{a}$ and $\hat{b}$), see, e.g., Dirac [84] and von Neumann [313] or for the modern presentations Holevo [125], Busch, Grabowski and Lahti [44]. The interference of probabilities is interpreted as the exhibition of quantum (and hence nonclassical) probabilistic laws (as at least Dirac [84] and von Neumann [313] as well as Feynman [91] believed).

We would like to invert the scheme. We are going to start with a very general scheme of probabilistic description of experimental statistical data. We call this scheme the contextual probabilistic model (*Växjö model*) [180]. The origin of the data (i.e., the context of observation) does not play any role. It could be statistical data from quantum mechanics as well as from classical statistical mechanics, biology, sociology, economics, or meteorology. Then we will show that the interference of probabilities (even more general than in quantum mechanics) can be found for any kind of data [145, 159–163, 167–170, 173, 181–183].

---

confused by quantum mechanics, you do not understand it," John Wheeler; "It is safe to say that nobody understands quantum mechanics," Richard Feynman; "If [quantum theory] is correct, it signifies the end of physics as a science," Albert Einstein; "I do not like [quantum mechanics], and I am sorry I ever had anything to do with it;" Erwin Schrödinger; " Quantum mechanics makes absolutely no sense," Roger Penrose.

[2] Cf. Mackey [239], Accardi [2–6], Aerts et al. [12], Bohm and Hiley [38], Ballentine [29–31], Boyer [41], Cole [61], Davidson [64], De Baere [69], De la Pena and Cetto [71–75], De Muynck [79, 80], Gudder [106–108], Helland [115], Ludwig [237], Khrennikov [139–142, 144–210], Klyshko [220], Kirkpatrick [213], Manko et al. [240–243], Nelson, [252], Nieuwenhuizen [253], Santos [276, 277], Svozil [294, 295], 't Hooft [297–299], Adenier et al. [11–178], Haven [111, 112], Choustova [52–58], Busemeyer [46, 47].

Moreover, one can easily obtain the linear space representation of probabilities from the interference of probabilities and then recover Born's rule (which will not be a postulate anymore). Thus, in our approach the quantum probabilistic calculus is just a special linear space representation of given probabilistic data [145].

One of our main contributions to this sphere is the creation of a special representation algorithm. Since our algorithm can be applied not only to physical statistical data, we will speak about *quantum-like (QL) representation* rather than about a quantum one. The latter is reserved for quantum physics. We call our algorithm for the representation of probabilistic data by complex amplitudes *quantum-like representation algorithm* or QLRA. We repeat again that QLRA can work in any domain of science.

One of the advantages of the QL representation of probabilistic data is an essential simplification of calculation of probabilities. We emphasize that linear algebra in complex Hilbert space is essentially simpler than the theory of Lebesgue integral and measurable functions! We stress this point, because quantum mechanics is commonly considered as much more mathematically complicated than classical statistical mechanics. However, it is not the case.

The basic notion in our approach is that of *context*—a complex of conditions under which the measurement is performed. We could consider contexts of different kinds: physical, biological or even political. Our approach to the subject of probability is contextual. It is meaningless to consider a probability not specifying the context of consideration (we remark that this point was already discussed by Kolmogorov [222]).

We will make a few remarks regarding the terminology in this book. The notion of context can be related to the notion of the *preparation procedure* which is widely used in quantum measurement theory [44, 125, 237, 257]. Of course, preparation procedures—devices preparing systems for subsequent measurements—give a wide class of contexts. However, the context is a more general concept. For example, we can develop models operating with social, psychological or financial contexts, [46, 47, 52–58, 90, 102–105, 111, 112, 162, 166, 167, 175, 199, 261–266].

The notion of *contextual probability* can be coupled to the notion of *conditional probability* which is used in classical (e.g., Kolmogorovian) probability models. However, once again the direct identification can be rather misleading, since the conventional meaning of the conditional probability $\mathbf{P}(B|A)$ is the probability that *event* B occurs under the condition that *event* A has occurred [222]. Thus, the conventional conditioning is the *event-conditioning.* Our conditioning is a *context-conditioning:* $\mathbf{P}(b = \beta|C)$ is the probability that observable b takes the value β in the process of measurement under context C. In principle, we are not against the term "conditional probability" if it is used in the contextual sense. We also remark that typically the definition of conditional probability is associated with the one given by Bayes' formula. In particular, such an approach preassum the possibility of operating with the joint probability distribution. However, in the general contextual probabilistic model the latter need not be defined.

The main terminological problem is related to the notion of *contextuality.* The use of the term "contextual" is characterized by a huge diversity of meanings, see Bell [34], Svozil [296] or Beltrametti and Cassinelli [35] for the notion of contextuality in quantum physics as well as Light and Butterworth [235], Bernasconi and Gustafson [36] for the notion of contextuality in cognitive science and AI. In quantum physics the contextuality is typically reduced to a rather specific contextuality—"Bell contextuality." J. Bell invented this notion in the framework of the EPR-Bohm experiment [34, 35]. We recall that such *quantum contextuality* ("Bell's contextuality") is defined as follows.

The result of the measurement of an observable a depends on another measurement of observable b, although these two observables commute with each other.

It should be emphasized that the *nonlocality* in the framework of the EPR-Bohm experiment is a special case of quantum contexuality.

Our contextuality is essentially more general than Bell's. In a very special case one can determine the context for the a-measurement by fixing an observable b which is compatible with a. However, in the general case there is nothing about a mutual dependence or compatibility of observables. The context is simply a complex of conditions (e.g., physical or biological). In particular, our description of the EPR-Bohm experiment is contextual, but there is no direct coupling with nonlocality.

Our approach to contextuality is closer to the one used in cognitive science and AI, cf. [36, 235].

The basis of linear representations of probabilities is a generalization of the well-known formula of *total probability* [286]. We recall that in the case of two dichotomous variables $a = \alpha_1, \alpha_2$ and $b = \beta_1, \beta_2$ this basic formula has the form

$$\mathbf{P}(b = \beta) = \mathbf{P}(a = \alpha_1)\mathbf{P}(b = \beta|a = \alpha_1) + \mathbf{P}(a = \alpha_2)\mathbf{P}(b = \beta|a = \alpha_2), \quad (0.1)$$

where $b = \beta_1$ or $b = \beta_2$.

Starting with a general *contextual probabilistic model M* (Växjö model) we shall obtain a generalization of the conventional formula, (0.1), which is characterized by the appearance of an additional term, an *interference term*

$$\mathbf{P}(b = \beta) = \mathbf{P}(a = \alpha_1)\mathbf{P}(b = \beta|a = \alpha_1) + \mathbf{P}(a = \alpha_2)\mathbf{P}(b = \beta|a = \alpha_2) \\ + \delta(b = \beta|a). \quad (0.2)$$

Depending on the magnitude of this term (relatively to magnitudes of probabilities in the right-hand side of (0.1)), we obtain either the conventional *trigonometric interference* which is well known in classical wave mechanics as well as in quantum mechanics or a *hyperbolic interference* which was not predicted by conventional physical theories, neither by the classical wave theory nor by quantum mechanics. Such a new type of interference arose naturally in the Växjö model, see Part II and especially Part V.

We recall that all probabilities in (0.2) are contextual. They depend on a complex of conditions, context C, for measurements of a and b. Starting with the formula of total probability with an interference term and applying QLRA we obtain two types of representations of contexts, $C \to \psi_C$, in linear spaces:

(a) representation of some special collection of contexts $\mathscr{C}^{\mathrm{tr}}$ ("trigonometric contexts")[3] in complex Hilbert space, see Part II;

(b) representation of some special collection of contexts $\mathscr{C}^{\mathrm{hyp}}$ ("hyperbolic contexts")[4] in the so-called hyperbolic Hilbert space, see Part V.

We emphasize that in general the collections of trigonometric and hyperbolic contexts, $\mathscr{C}^{\mathrm{tr}}$ and $\mathscr{C}^{\mathrm{hyp}}$, are just proper subsets of the complete collection of contexts $\mathscr{C}$ of a Växjö model M (contextual probabilistic model). Depending on model M there can exist contexts which can not be represented algebraically: neither in complex nor in hyperbolic Hilbert spaces.

Thus quantum probabilities are demystified. In the Växjö-approach the quantum probabilistic calculus appears via the application of QLRA: the linear space representation of a special collection of contexts. The origin of those contexts is not important. In any event they need not have any relation to the microworld. In our approach the quantum probabilistic model is incomplete in the following sense: there can exist prequantum models M having essentially larger collections of contexts $\mathscr{C}$ than those that can be represented in complex Hilbert space. Of course, quantum contexts are special. Such "speciality" is characterized not by the properties of systems comprising the contexts, but by the mutual relations between probabilities which are used for the linear space representation of contexts.

Through the demystification of quantum probabilities we have closed the gap between the classical Kolmogorov and the quantum probabilistic models. These

---

[3] They produce the ordinary cos-interference.

[4] They produce hyperbolic cosh-interference.

models are simply different partial representations of the general Växjö model, cf. Mackey [239]. The classical Kolmogorov model is a representation of the Växjö model such that contexts are given by elements of a σ-algebra $\mathscr{F}$ of subsets of some set Ω (sample set, or space of elementary events). The quantum probabilistic model is a representation of the Växjö model such that contexts are given by normalized vectors of complex Hilbert space (or equivalent classes of such vectors). These mentioned representations are not the only possible representations of the Växjö model. We have discovered new representations: hyperbolic and hyper-trigonometric.

Any Kolmogorov model can be considered as a contextual statistical model (with the set-representation of contexts). Hence we can apply our general algorithm of the linear space representation—QLRA—to the Kolmogorov measure-theoretic model, see Part II, Chap. 6. Here trigonometric contexts $\mathscr{C}^{\mathrm{tr}}$ form a special subcollection of σ-algebra $\mathscr{F}$. In general $\mathscr{C}^{\mathrm{tr}}$ is a proper subset of σ-algebra $\mathscr{F}$. There exist Kolmogorov "classical contexts" which can not be represented by quantum states. Contrary to rather common opinion, the quantum probabilistic model does not cover the classical Kolmogorov model. Neither can the Kolmogorov model cover the quantum one. The image of the collection of Kolmogorov's contexts in complex Hilbert space does not coincide with the whole set of quantum states—the unit sphere in complex Hilbert space. Nevertheless, all basic QL effects and structures (interference of probabilities, Born's rule, complex probabilistic amplitudes, Hilbert state space, the representation of observables by operators) are present in the classical Kolmogorov model, but in a latent form. The discovery [181, 182] of those hidden QL structures in the classical probabilistic model was very important for demystification of the quantum theory.

As was already noted, the QL representation of the Kolmogorov model can not cover the conventional quantum probabilistic model. One can be curious about a possibility of finding a classical probabilistic model which would completely cover the quantum one. Such a classical probabilistic model really exists [139]. It is the von Mises frequency model [309–311] completed by the contextual interpretation of frequency probability.[5] I am well aware of a prejudice against the von Mises approach—especially among mathematicians. (On the other hand, von Neumann used von Mises' approach to probability as the basis for quantum probability [313].) Roots of such prejudice are related to the attempt of Richard von Mises to formalize the notion of randomness through the so-called "place selections." We agree that the original attempt of von Mises was rather misleading.

Still, the crucial point is not the improvement of von Mises theory, but the fact that in quantum physics people are not interested in mathematical study of randomness of sequences produced by experiments (at least at the moment). It is assumed that these sequences are intrinsically random (thus a mathematical proof is replaced by a "physical proof"). The only thing from von Mises theory that is used in applications is the *frequency definition of probability*. This definition can work really fruitfully (as we have already shown in [139] and as we are going to confirm in the present book).

We hope that after reading this book people will pay more attention to the role of context in physics as well as in biology, cognitive and social sciences, psychology, and economics. All considerations of probabilities would be with respect to the context. All known probabilistic models, the classical ones (Kolmogorov measure-theoretic and von Mises frequency models) as well as the quantum model and more general quantum-like models, would be interpreted as special partial representations of the Växjö model, the contextual probabilistic model for description of statistical data from any domain of science.

By starting with contextual probabilities one can escape such mysterious notions as irreducible quantum randomness or quantum nonlocality. Born's rule, interfer-

[5] Probability is defined as the limit of relative frequencies if it exists. Existence of the limit-probability is also known in statistics as *the principle of statistical stabilization* of relative frequencies.

ence of probabilities, origins of complex probability amplitudes and the operator representation can be realistically explained. The mathematical formalism of quantum mechanics as well as its generalizations can be applied outside physics and provide useful, and also verifiable, results.

Växjö-Sevastopol-Copenhagen *Andrei Khrennikov*

2002–2007

Acknowledgements

I discussed different parts of this book with a huge number of physicists, mathematicians and philosophers who visited the International Center for Mathematical Modeling in Physics and Cognitive Science, Växjö University, during the last 10 years in connection with two series of international conferences [8, 10, 146, 151, 156, 177, 178]: "Foundations of Probability and Physics" 1–5 and "Quantum Theory: Reconsideration of Foundations" 1–4 (as well as the conference "Bohmian Mechanics" 2000). I also would like to mention extremely fruitful discussions during the conference "Beyond the Quantum" which was organized by efforts of G. Balian, G. 't Hooft, the author of this book, Th.M. Nieuwenhuizen, L. De la Pena and A.M. Cetto, V. Spicka [254].

I also had the benefit of many comments (sometimes rather critical) during my seminars at Bonn University, University of Illinois (Urbana-Champaign), Denver University, Boulder University, Bell Labs. of Lucent Tech., Steklov Mathematical Institute, Lebedev Physical Institute and Physics-Technology Institute of Russian Academy of Science, Moscow State University, University of Rome-2, Technical University of Vienna and Atom-institute, Vienna, University of Lecce, Danish Technical University, Copenhagen, Tokyo Science University, and Oxford University.

I am especially thankful for discussions on

- foundations of classical and quantum probability to L. Accardi, S. Albeverio, M. Appleby, L.E. Ballentine, V. Belavkin, L. Hardy, I. Bengtsson, M. Bozejko, C. Fuchs, S. Gudder, K. Gustafson, I. Helland, A. Holevo, R. Hudson, S. Kozyrev, E. Loubenets, M. Manko and V. Manko, D. Petz, R. Schack, A. Caticha,
- on quantum logic and its connection with quantum probability to D. Aerts, S. Aerts, B. Coecke, A. Dvurečenskij, A. Grib, P. Lahti, O. Nánásiová, K. Svozil, S. Pulmannová,
- on quantum nonlocality and Bell's inequality to A. Aspect, D. Mermin, L. Accardi, G. Adenier, W. De Baere, C. Carola, W. De Muynck, K. Hess and W. Philipp, R. Gill, D. Greenberger, J.-A. Larsson, A. Peres, I. Pitowsky, M. Scully, I. Volovich,
- on stochastic models for quantum mechanics to S. Albeverio, T.H. Boyer, A.M. Cetto, D. Cole, M. Davidson, E. Nelsson, Th. Nieuwenhuizen, L. de la Pena, V. Spicka, W. von Waldenfels,
- on quantum information and computing to G. Jaeger, M. Ohya, Yu. Ozhygov, J. Smolin, N. Watanabe,
- on quantum macroscopic models to D. Aerts, G. 't Hooft, A. Leggett, K. Svozil,
- on quantum cognitive systems to E. Conte, A. Elitzur, P. Suppes, W. Freudenberg, J. Busemeyer, A. Ezhov, R. Franco,
- on quantum finances to O. Choustova, E. Haven, A. Hilbert, V.P. Maslov, R. Pettersson,
- on scaling in classical and quantum physics to B. Nilsson,
- on axiomatics of quantum mechanics (including philosophic questions) to S. Aaronson, H. Atmanspacher, H.C. von Baeyer, R. Balian, I. Bialynicki-Birula, E. Binz, P. Busch, V. Belavkin, C. Fuchs, G. D' Ariano, H.D. Doebner, A. Elitzur,

G.G. Emch, G. Jaroszkiewicz, L. Hardy, Y.S. Kim, A. Plotnitsky, S. Stenholm, H.J. Folse,

- on experimental quantum physics to A. Aspect, O. Benson, Y. Hasegawa, H. Rauch, E. Sjöqvist, J. Summhammer, G. Weihs, H. Takayanagi, A. Zeilinger,
- on Bohmian mechanics to S. Goldstein, B. Hiley, G. Reinisch, H. Frisk.

Finally, I would like to thank my wife Olga Choustova for her enthusiastic support of my attempts to reconsider foundations of probability and quantum mechanics as well as attempts to move beyond quantum mechanics.

Contents

Part I
Quantum and Classical Probability

This part provides short introductions to axiomatics of quantum mechanics and classical probability theory. Such introductions may be useful, e.g., for students.[1] Experts can jump directly to Part II. Nevertheless, some sections might be interesting even for experts. In Chap. 1 we present a deep analysis of the projection postulate. In particular, we emphasize the crucial difference between von Neumann's and Lüders versions of this postulate. This analysis is important for further contextual modeling. We recall that the projection postulate plays a crucial role in theory of quantum conditional probabilities. In Chap. 2 we present briefly not only the conventional measure-theoretic (Kolmogorov, 1933) model of probability, but also the frequency model (created by von Mises, 1919). We emphasize the role of conditional probabilities in both models.

[1] I hope that this book will be readable by graduate students.

Chapter 1

Quantum Mechanics: Postulates and Interpretations

We start this chapter with a brief introduction to the quantum formalism and discussion on the two main interpretations of quantum mechanics: *Copenhagen Interpretation and Ensemble Interpretation.*

We recall that any physical theory and, in particular, quantum mechanics can be considered as consisting of two connected, but at the same time sufficiently independent, parts, see, for example, Ballentine [29, 30] or Jauch [130]:

(a) A mathematical formalism consisting of a set of primitive concepts, relations between these concepts (either postulated or obtainable by given rules of deduction), and a dynamical law.

(b) Correspondence rules which relate the theoretical concepts of (a) to the world of experience.

1.1 Quantum Mechanics

1.1.1 Mathematical Basis

Everywhere below the fields of real and complex numbers are denoted by the symbols **R** and **C**, respectively.

We recall that a *complex Hilbert space* $\mathcal{H}$ is a linear space over **C** endowed with a scalar product , $\langle \cdot, \cdot \rangle$, which is *complete* with respect to the corresponding norm

$$\|\psi\| \equiv \sqrt{\langle \psi, \psi \rangle}. \tag{1.1}$$

We set

$$S = \{\psi \in \mathcal{H} : \|\psi\| = 1\}.$$

It is the unit sphere in $\mathcal{H}$.

Any finite-dimensional complex Hilbert space is isomorphic to the coordinate space $\mathcal{H} = \mathbf{C}^m$. Its elements are complex vectors $\psi = (\psi_1, \ldots, \psi_m), \psi_j \in \mathbf{C}$. Here $\langle \psi, \phi \rangle = \sum_{j=1}^m \psi_j \overline{\phi_j}$, and $\|\psi\|^2 = \sum_{j=1}^m |\psi_j|^2$.

The most important example of a Hilbert space (of infinite dimension) which is basic in quantum mechanics is the space $\mathcal{H} = L_2(\mathbf{R}^n, dq)$ of square integrable (with respect to the Lebesgue measure $dq = dq_1 \cdots dq_n$) functions $\psi : \mathbf{R}^n \to \mathbf{C}$. Here $\langle \psi, \phi \rangle = \int_{\mathbf{R}^n} \psi(q)\overline{\phi(q)}dq$, and $\|\psi\|^2 = \int_{\mathbf{R}^n} |\psi(q)|^2 dq$.

Although this space will be considered in some general discussions on quantum foundations, e.g., on Bell's inequality, the main part of this book (devoted to contextual probability) is based solely on finite-dimensional Hilbert spaces (and the majority of constructions are presented for the two-dimensional case).

For a Hilbert space $\mathcal{H}$, we denote the space of all self-adjoint operators by the symbol

$$L_s(\mathcal{H}).$$

In the finite-dimensional case the spectral decomposition of a self-adjoint operator has a simple form

Any self-adjoint (symmetric) operator $\hat{a}$ can be represented as a linear combination of projectors onto subspaces L_{α_m} corresponding to its distinct eigenvalues, $\alpha_1, \ldots, \alpha_k$:

$$\hat{a} = \sum_{m=1}^k \alpha_m P_m^a, \tag{1.2}$$

where $P_m^a \equiv P_{L_{\alpha_m}}$.

1.1.2 Postulates

The mathematical formalism of quantum mechanics is the theory of self-adjoint operators on complex Hilbert spaces.

The symbols $\mathcal{H}$ and $\langle \cdot, \cdot \rangle$ denote separable complex Hilbert space and the scalar product on it; $\|\psi\|$ denotes the norm.

The correspondence rules for quantum theory can be formulated as the following series of postulates.

Postulate 1 (The mathematical description of quantum states). *Quantum (pure) states (wave functions) are represented by normalized vectors* ψ *(i.e.,* $\|\psi\|^2 = \langle \psi, \psi \rangle = 1$*) of a complex Hilbert space* $\mathcal{H}$*. Every normalized vector* $\psi \in \mathcal{H}$ *may represent a quantum state. If a vector* ψ *corresponding to a state is multiplied by any complex number* c, $|c| = 1$, *the resulting vector will correspond to the same state.*

The physical meaning of "a quantum state" is not defined by this postulate.

As a consequence of Postulate 1, we obtain the *superposition principle.*

A normalized linear combination of any sequence of quantum states is again a quantum state.

Postulate 1a (Equivalence of representations). *A unitary transformation of a Hilbert space generates an equivalent representation of quantum states.*

In fact, as any unitary operator corresponds to the change of an orthogonal base in $\mathcal{H}$, Postulate 1a means that the description of a quantum state ψ is invariant with respect to the choice of a system of orthogonal coordinates in $\mathcal{H}$.

Remark 1.1 (Position representation). L. De Broglie strongly criticized Postulate 1a [70]. He was sure that different (unitary equivalent) representations are not equivalent from the physical point of view. He emphasized the *exceptional role of the position representation.* This representation was considered by him as the only real physical representation. Even the momentum representation was considered as an ideal purely mathematical representation. D. Bohm and J. Bell had similar views (private communications of a former student of David Bohm).

Postulate 2 (The mathematical description of physical observables). *A physical observable a is represented by a self-adjoint operator $\hat{a}$ in complex Hilbert space $\mathcal{H}$. Different observables are represented by different operators.*

Some authors, e.q., J. von Neumann [313] (see also P. Dirac [84], p. 37), complete Postulate 2 by the additional assumption.

Postulate 2a. *Any self-adjoint operator $\hat{a}$ represents a physical observable a.*[1]

Postulate 3 (Spectral). *For a physical observable a which is represented by the self-adjoint operator $\hat{a}$ we can predict (together with some probabilities) values λ belonging to its spectral set* $\text{Spec}(\hat{a})$.

We restrict our further considerations to the class of self-adjoint operators for which the spectral representation is the same as in the finite-dimensional case. We recall that a self-adjoint operator $\hat{a}$ has *purely discrete spectrum* if it can be represented as

$$\hat{a} = \sum_m \alpha_m P_m^a, \quad \alpha_m \in \mathbf{R}, \tag{1.3}$$

where P_m^a are orthogonal projectors corresponding to the eigenvalues α_m.

Since $\hat{a}$ is in general unbounded, one should specify its domain of definition

$$D(\hat{a}) = \left\{ \psi \in \mathcal{H} : \sum_m \alpha_m^2 \| P_m^a \psi \|^2 < \infty \right\}. \tag{1.4}$$

On this domain of definition we have

$$\hat{a}\psi = \sum_m \alpha_m P_m^a \psi. \tag{1.5}$$

[1] We do not discuss such things as superselection rules.

Postulate 4 (Born's rule—in formalization of Dirac and von Neumann). *Let a physical observable a be represented by a self-adjoint operator* $\hat{a}$ *with purely discrete spectrum. The probability* $\mathbf{P}_\psi(a = \alpha_m)$ *to obtain the eigenvalue* α_m *of* $\hat{a}$ *for measurement of a in a state* ψ *is given by*

$$\mathbf{P}_\psi(a = \alpha_m) = \| P_m^a \psi \|^2. \tag{1.6}$$

This postulate provides the correspondence between probabilistic properties of physical observables and the operator formalism. Each quantum observable is considered as a classical random variable taking values in its spectral set with probabilities given by Postulate 4.

Theorem 1.1. *The average value of an observable a in a state* $\psi \in D(\hat{a})$ *is given by*

$$\langle a \rangle_\psi = \langle \hat{a}\psi, \psi \rangle. \tag{1.7}$$

Proof. We consider an operator with purely discrete spectrum, see (1.3). Its domain of definition is the linear subspace given by (1.4). By the definition of the average (mathematical expectation) Ea of a discrete random variable a in classical probability theory, see Chap. 2, we have $\langle a \rangle_\psi \equiv Ea = \sum_m \alpha_m \mathbf{P}_\psi(a = \alpha_m)$. We now use Born's rule and obtain $\langle a \rangle_\psi = \sum_m \alpha_m \| P_m^a \psi \|^2$. On its domain of definition $\hat{a}$ acts in the following way: $\hat{a}\psi = \sum_m \alpha_m P_m^a \psi$. Therefore

$$\begin{aligned}\langle \hat{a}\psi, \psi \rangle &= \Big\langle \sum_m \alpha_m P_m^a \psi, \psi \Big\rangle = \sum_m \alpha_m \langle P_m^a \psi, \psi \rangle \\ &= \sum_m \alpha_m \langle (P_m^a)^2 \psi, \psi \rangle = \sum_m \alpha_m \langle P_m^a \psi, P_m^a \psi \rangle.\end{aligned}$$

Here we have used that for any projector, $P^2 = P$ and $P^\star = P$.

P. Dirac [84] and J. von Neumann [313] generalized Postulate 4 to a few observables represented by commutative operators.

Postulate 4a (Dirac-von Neumann). *Let physical observables* $a_1, \ldots, a_N$ *be represented by self-adjoint operators* $\hat{a}_1, \ldots, \hat{a}_N$ *with purely discrete spectra*

$$\hat{a}_j = \sum_m \alpha_m^j P_m^{a_j}, \quad \alpha_m^j \in \mathbf{R}, \tag{1.8}$$

where $P_m^{a_j}$ are projectors corresponding to the eigenvalues α_m^j. These observables can be measured simultaneously if and only if the operators $\hat{a}_1, \ldots, \hat{a}_N$ commute. The joint probability distribution in a state ψ is given by

$$\mathbf{P}_\psi(a_1 = \alpha_k^1, \ldots, a_N = \alpha_m^N) = \| P_m^{a_N} \cdots P_k^{a_1} \psi \|^2. \tag{1.9}$$

Postulate 5 (Time evolution of wave function). *Let $\hat{H}$ be the Hamiltonian of a quantum system, i.e., the self-adjoint operator corresponding to the energy observable. The time evolution of the wave function $\psi \in \mathcal{H}$ is described by the Schrödinger equation*

$$i\hbar \frac{d}{dt} \psi(t) = \hat{H} \psi(t) \tag{1.10}$$

with the initial condition $\psi(0) = \psi$. Here $\hbar$ is the Dirac constant (the Planck constant h divided by 2π).

Now we are going to discuss one of the most important and complicated notions of quantum mechanics, the notion of a quantum state. There are two main points of view which are formulated in the following postulates.

Postulate 6 (The ensemble interpretation). *A wave function provides a description of certain statistical properties of an ensemble of similarly prepared quantum systems.*

This interpretation is upheld, for example by Einstein, Popper, Margenau, Bohm, Ballentine, Klyshko and recent years by, e.g., De Muynck, De Baere, Holevo, Santos, Khrennikov, Nieuwenhuizen, Adenier and many others.

Postulate 6a (The Copenhagen interpretation). *A wave function provides a complete description of an individual quantum system.*

This interpretation was supported by a great variety of members, from Schrödinger's original attempt to identify the electron with a wave function solution of his equation to the several versions of the Copenhagen interpretation (for example, Heisenberg, Bohr, Pauli, Dirac, von Neumann, Landau, Fock and recent years by, e.g., Greenberger, Mermin, Lahti, Peres, Summhammer[2]). Nowadays the individual interpretation is extremely popular, especially in quantum information and computing.

In [313] J. von Neumann presented the first mathematical (of course, still quite vague) discussion on Postulates 6 and 6a. He wrote about *irreducible quantum randomness*—randomness which could not be associated with an ensemble of quantum systems, but with a single quantum system. Moreover, he presented some mathematical arguments in favor of Postulate 6a. In contemporary literature on foundations of quantum mechanics those considerations are well-known as *von Neumann's NO-GO theorem*, see Part III, Chap. 8.[3]

The reading of von Neumann's book [313] induces a rather strange impression. J. von Neumann discussed in detail individual randomness which was for him the main distinguishing feature of quantum probability. This randomness is associated with an individual electron and not an ensemble of electrons. However, to find con-

---

[2] There is an interesting story about the correspondence between Bohr and Fock on the individual interpretation. This story was told to me by a former student of Fock who pointed out that one of the strongest supporters of this interpretation was Vladimir A. Fock, and that even though Bohr himself had doubts about its consistency, he, Fock, demonstrated to Bohr an inconsistency in the Einsteinian ensemble interpretation. Thus the interpretation which is commonly known as the Copenhagen interpretation might be as well called the "Leningrad interpretation."

[3] Recently I gave a talk on my contextual approach to quantum probabilities at Beckman Institute, University of Urbana-Champain, and I was curious: "Why did Albert Einstein never pay attention to the von Neumann "NO-GO" theorem?" (It is known that Einstein had von Neumann's book in his office). A. Leggett pointed out that the word "theorem" appeared only in the English translation of von Neumann's book [313]. In the original German version it was "ansatz." The term ansatz has the meaning of intention to prove something, so it is not a mathematical theorem. This might be an explanation of Einstein's ignoring of von Neumann's "no-go" argument (because Einstein did not consider von Neumann's arguments as rigorous).

nection with experiments, he would immediately switch to the ensemble approach. He understood well that there was no other possibility. It is impossible to proceed from randomness of an individual quantum system to experiment.

In general, *quantum theory does not predict the result of an individual measurement of some observable a*. However, the probability of each possible value α_m (which is considered as the *exact value* of the observable a) can be verified by repeating the measurement many times and computing the relative frequencies

$$\nu_N(\alpha_m) = n_N(\alpha_m)/N,$$

where $n_N(\alpha_m)$ is the number of realizations of the value α_m in the first N trials.

It is interesting that *this method of the verification is used in both frameworks—ensemble and Copenhagen.* To explain this coincidence of points of view for such different interpretations of a quantum state, it was pointed out (by Körner, Popper, Ballentine [29, 30]) that one should distinguish between the probability distribution $\mathbf{P}_\psi$, which is associated with a quantum state ψ, and the statistical frequency of results in an actual sequence of experiments (the experimental probability), cf. von Neumann [313]. The ensemble and Copenhagen interpretations have the same point of view on experimental probability, but they have different points of view on the meaning of the probability distribution $\mathbf{P}_\psi$ (which is the origin of experimental randomness).[4]

L. Ballentine used the terminology *"statistical interpretation."* However, Ballentine's terminology is rather misleading, because the term "statistical interpretation" was also used by J. von Neumann for individual randomness! For him "statistical in-

[4] Such a coincidence of the experimental verification procedure induces a huge problem in understanding the difference between the ensemble and Copenhagen interpretation. Many adherents of Copenhagen interpretation operate with terminology "statistical interpretation," but at the same time they have individual (irreducible) quantum randomness in mind (e.g., private discussions with V. Belavkin, see also [33]).

terpretation" had a meaning which is totally different from Ballentine's "ensemble-statistical interpretation." J. von Neumann wanted to emphasize the difference between deterministic (Newtonian) classical mechanics in that the state of a system is determined by values of two observables (position and momentum) and quantum mechanics in that the state is determined not by values of observables, but by probabilities.

We shall follow Albert Einstein and use the terminology *"ensemble interpretation."*

Finally, we point out recent papers concerning foundations and in particular various interpretations of quantum mechanics: [1–25, 27, 28, 31–33, 37, 41–43, 45, 48, 61–64, 69, 75, 79, 80, 94–98, 108, 109, 115–120, 123, 125, 131–213, 228, 232, 233, 236, 240–243, 247, 251, 253, 258, 269–272, 276–278, 292–299, 308, 316]; see also about possibilities to apply the mathematical formalism of QM outside physics (in particular, in biology, psychology and finances): [46, 47, 52–58, 90, 102–105, 111, 112, 162, 166, 167, 175, 199, 261–266, 306].

1.2 Projection Postulate, Collapse of Wave Function, Schrödinger's Cat

We did not include von Neumann's projection postulate [313] in the main list of postulates of quantum mechanics, because one might proceed rather far without it. In this section we discuss this postulate. Even its formulation may induce an interesting debate.

1.2.1 Von Neumann's Projection Postulate

Projection Postulate. *Let a be a physical observable represented by a self-adjoint operator $\hat{a}$ having a purely discrete nondegenerate spectrum. Any measurement of the observable a on the quantum state ψ induces transition from the state ψ into one of the eigenvectors e_k^a of the operator $\hat{a}$.*

See J. von Neumann [313], p. 216: "Under the above assumption on $\hat{a}$, a measurement of a has the consequence of changing each state ψ into one of the states $e_1^a, e_2^a, \ldots$ which are connected with respective results of measurement $\alpha_1, \alpha_2, \ldots$. The probabilities of these changes are therefore equal to the measurement probabilities for $\alpha_1, \alpha_2, \ldots$"

We emphasize that the use of von Neumann's projection postulate is meaningful only in the framework of the Copenhagen interpretation by which the ψ-function gives the state of an individual quantum system.

1.2.2 Collapse of Wave Function

The von Neumann projection postulate is often referred as the postulate about *collapse of the wave function.* Such a point of view was strongly motivated by the original von Neumann considerations:

"This discontinuous transition from ψ into one of the states e_1^a, e_2^a, ... (which are independent of ψ, because ψ enters only into the respective probabilities

$$\mathbf{P}(n) = |\langle \psi, e_n^a \rangle|^2, \quad n = 1, 2, \ldots$$

of this jump) is certainly not of the type described by the time dependent Schrödinger equation. This latter always results in a continuous change of ψ, in which the final result is uniquely determined and is dependent on ψ," [313], p. 217.

1.2.3 Schrödinger's Cat

To demonstrate *absurdness of the Copenhagen interpretation*, especially the Copenhagen viewpoint on the notion of superposition of states as states of an individual system) and the von Neumann projection postulate, Einstein and Schrödinger elaborated (see [280] on their correspondence) an example which nowadays is well known as Schrödinger's cat (Schrödinger just modified Einstein's example by using cat instead of bomb and poisson instead of gun), see [280] for details. Schrödinger wrote [279]:

> "One can even set up quite ridiculous cases. A cat is penned up in a steel chamber, along with the following device (which must be secured against direct interference by the cat): in a Geiger counter there is a tiny bit of radioactive substance, so small, that perhaps in the course of the hour one of the atoms decays, but also, with equal probability, perhaps none; if it happens, the counter tube discharges and through a relay releases a hammer which shatters a small flask of hydrocyanic acid. If one has left this entire system to itself for an hour, one would say that the cat still lives if meanwhile no atom has decayed. The ψ-function of the entire system would express this by having in it the living and dead cat (pardon the expression) mixed or smeared out in equal parts.
>
> It is typical of these cases that an indeterminacy originally restricted to the atomic domain becomes transformed into macroscopic indeterminacy, which can then be resolved by direct observation. That prevents us from so naively accepting as valid a "blurred model" for representing reality. In itself it would not embody anything unclear or contradictory. There is a difference between a shaky or out-of-focus photograph and a snapshot of clouds and fog banks."

In the Copenhagen approach the presence of Schrödinger's cats is often considered as a kind of paradox.

However, the ensemble interpretation suffers of its own complicated problem. This is the problem of explanation of the *interference of probabilities* (which is equivalent to explanation of Born's rule).[5] Schrödinger wrote to Einstein [280] that if it were possible to derive the interference of probabilities by using the ensemble interpretation, he (Schrödinger) would definitely support this interpretation. Unfortunately, Einstein was not able to derive the interference of probabilities in the ensemble approach.

We shall solve this problem (which was so disturbing for Einstein and Schrödinger) in Part II.

1.2.4 Lüders Projection Postulate

If the spectrum of an operator $\hat{a}$ is *degenerate* then (according to von Neumann [313]) the observation of the result $a = \alpha_k$ *does not induce the transition of the initial pure state* ψ *into a new pure state.* In such a case the resulting state is not determined. It can be determined only through a subsequent measurement of an observable d refining the original observable a. After such a refining measurement we obtain not a pure state, but a statistical mixture.

However, in the contemporary formulations of "the von Neumann projection postulate" the cases of nondegenerate and degenerate spectra are not distinguished! In fact, this was not the original von Neumann invention, but it was a new postulate proposed by G. Lüders in 1951 (see [44] for more details and applications).

Lüders Projection Postulate. *For any operator* $\hat{a}$ *with purely discrete spectrum, a measurement of the corresponding observable* a *giving the result* $a = \alpha$, *where* $\alpha \in \text{Spec}(\hat{a})$, *always produces the projection of the initial state* ψ *onto the state*

$$\psi_\alpha = \frac{P_\alpha \psi}{\|P_\alpha \psi\|}, \tag{1.11}$$

[5] We remark that the Copenhagen interpretation was also unable to explain the origin of Born's rule.

where as always P_α is the orthogonal projector onto the subspace corresponding to the eigenvalue α.

The crucial point is that by the Lüders Projection Postulate the post-measurement state is also a pure state independently of degeneration of spectrum.

However, von Neumann emphasized a few times in [313] that " if the eigenvalue α is multiple, then the state ϕ after the measurement is not uniquely determined by the knowledge of the result of the measurement," p. 218.

The post-measurement state ϕ is not determined. What does it mean? J. von Neumann pointed out that to determine ϕ one should determine a subsequent measurement procedure which corresponds to the choice of a concrete orthonormal basis in the subspace

$$\mathscr{H}_\alpha = P_\alpha^a \mathscr{H}.$$

This ambiguity in the determination of the post-measurement state is an important difficulty in foundations of quantum mechanics.

1.3 Statistical Mixtures

Let ψ_1 and ψ_2 be two pure states. Let us now consider an ensemble S of quantum systems in which the state ψ_j is realized with probability $\mathbf{p}_j : \mathbf{p}_1 + \mathbf{p}_2 = 1$. Then such an ensemble is described by the *density operator*

$$\rho = \mathbf{p}_1 P_{\psi_1} + \mathbf{p}_2 P_{\psi_2},$$

where P_{ψ_j} is the orthogonal projector onto vector ψ_j. This definition is especially natural for the ensemble interpretation of quantum mechanics. Let states ψ_j describe contexts (complexes of physical conditions) C_j and let S_j be corresponding ensembles of quantum systems (prepared under these contexts). Then the density matrix ρ

represents the statistical mixture (a new ensemble) S of ensembles S_j with weights $\mathbf{p}_1$ and $\mathbf{p}_2$, respectively. The ρ is called a *statistical mixture* of pure states ψ_j. The ρ describes statistical properties of the ensemble S.

By using the ensemble interpretation, Postulate 4 and the formula of total probability (see Chap. 2) we can easily find probabilities of realizations of values of a (discrete) observable $\hat{a} = \sum \alpha_m P_m$ for a statistical mixture ρ

$$\mathbf{P}_\rho(a = \alpha_m) = \mathbf{p}_1 \|P_m \psi_1\|^2 + \mathbf{p}_2 \|P_m \psi_2\|^2. \tag{1.12}$$

This expression can be written as

$$\mathbf{P}_\rho(a = \alpha_m) = \mathrm{Tr}\, \rho P_m. \tag{1.13}$$

Postulate 7 (von Neumann-Landau). *A statistical mixture of pure quantum states is represented by a density operator, namely, a self-adjoint positive trace-class operator ρ*

$$\rho \geq 0, \mathrm{Tr}\, \rho = 1.$$

Probability of the result $a = \alpha_m$ for a state ρ, is given by (1.13) *and hence the average of a is given by*

$$\langle a \rangle_\rho = \mathrm{Tr}\, \rho \hat{a} = \sum_m \alpha_m \mathrm{Tr}\, \rho P_m.$$

We need a few facts on the tensor-product representation of operators. Let $\psi_j \in \mathcal{H}$, $j = 1, 2$. Then an operator $A = \psi_1 \otimes \psi_2$ acts (by definition) as $A\phi = \langle \phi, \psi_2 \rangle \psi_1$. In the general case $A = \sum_{kl} c_{kl} f_k \otimes g_l$, where $f_k, g_l \in \mathcal{H}$ and $c_{kl} \in \mathbf{C}$, acts as

$$A\phi = \sum_{kl} c_{kl} \langle \phi, g_l \rangle f_k. \tag{1.14}$$

If

$$\rho_\psi = P_\psi = \psi \otimes \psi$$

for some normalized ψ, then the density operator ρ_ψ corresponds to a pure state. If ρ cannot be represented in such a form, then we have a nontrivial statistical mixture of pure states.

We remark that a density operator is a trace class operator, $\mathrm{Tr}\,\rho < \infty$. There exists the complete system of eigenvectors, $\rho\psi_n = \mathbf{p_n}\psi_n$. Moreover, $\mathbf{p_n} \geq 0$ and $\sum_n \mathbf{p_n} = 1$. Hence ρ can be expanded as

$$\rho = \sum_n \mathbf{p_n} P_{\psi_n} = \sum_n \mathbf{p_n} \psi_n \otimes \psi_n.$$

By both the Copenhagen and ensemble interpretations ρ describes the statistical mixture S of ensembles S_n corresponding to the pure states ψ_n. The difference is in the interpretations of pure states. By the Copenhagen interpretation each system $\omega \in S_n$ has the pure state ψ_n, but by the ensemble interpretation the real state of an $\omega \in S_n$ is unknown (it is a hidden variable) and the vector ψ_n is just a label denoting the ensemble S_n.

1.4 Von Neumann's and Lüders' Postulates for Mixed States

Let ψ be a pure state and let P be a projector. By Lüders' postulate after measurement of the observable represented by P that gives the result 1, the initial pure state ψ is transformed again into a pure state, namely, $\psi' = \frac{P\psi}{\|P\psi\|}$. Thus for corresponding density operators we have $\rho_{\psi'} = \psi' \otimes \psi' = \frac{P\psi \otimes P\psi}{\|P\psi\|^2}$. We remark that

$$\rho_{\psi'} = \frac{P\rho_\psi P}{\|P\psi\|^2}. \tag{1.15}$$

Finally, we see

$$\rho_{\psi'} = \frac{P\rho_\psi P}{\mathrm{Tr}\,\rho_\psi P}. \tag{1.16}$$

In this way Lüders' postulate is represented in the framework of density operators (still in the case of pure states). Gerhart Lüders generalized this formula (without any doubt) to an arbitrary state ρ. If we measure the observable represented by a projector P for the ensemble of systems described by ρ and then select a new ensemble corresponding to the result 1, we get the state

$$\rho' = \frac{P\rho P}{\operatorname{Tr}\rho P}. \tag{1.17}$$

Let now consider an arbitrary self-adjoint operator with purely discrete spectrum

$$\hat{a} = \sum_m \alpha_m P_m, \quad \alpha_m \in \mathbf{R}, \; P_m P_l = \delta_{ml} P_l. \tag{1.18}$$

G. Lüders pointed out [238] that in a measurement of $\hat{a}$ the initial state ρ is transformed into

$$\rho' = \sum_m P_m \rho P_m. \tag{1.19}$$

It is easy to see that such a ρ' is again positive and self-adjoint. We show that its trace equals 1. Let $\{e_n\}$ be an orthonormal basis. We have $\operatorname{Tr}\rho' = \sum_m \sum_n \langle P_m \rho P_m e_n, e_n \rangle$.

We now take the basis consisting of blocks e_{km}, $k = 1, \ldots, N_m$, such that $e_{km} \in P_m \mathscr{H}$. For a fixed m, e_{km} are eigenvectors corresponding to α_m. Here N_m can be equal to infinity. We continue

$$\operatorname{Tr}\rho' = \sum_m \sum_k^{N_m} \langle \rho e_{km}, e_{km} \rangle = \operatorname{Tr}\rho = 1.$$

This is Lüders' postulate for the transformation of a state ρ through a measurement of an observable represented by $\hat{a}$. The crucial Lüders' assumption is that, *for a pure state after a measurement of $\hat{a}$ and selection with respect to the value α_m, we always obtain again a pure state.*

Von Neumann had a completely different viewpoint on such a transformation [313]. As was already pointed out, even for a pure state ψ the result will not again be a pure state (if the operator has degenerated spectrum).

Let $\hat{a}$ have nondegenerate spectrum. Thus all P_m are one-dimensional projectors (onto eigenvectors $\{e_m\}$ of $\hat{a}$). Then by the von Neumann projection postulate a measurement of a giving the result $a = \alpha_m$ really induces the projection of the original pure state ψ onto e_m^a. The transformation of the density operator is given by

$$\rho_{\psi'} = \sum_m P_m \rho_\psi P_m \tag{1.20}$$

(so in the nondegenerate case Lüders' approach coincides with von Neumann's). Starting with an arbitrary initial state ρ we obtain the state

$$\rho' = \sum_m P_m \rho P_m. \tag{1.21}$$

We remark that, since all projectors are one-dimensional, we have $(P_m \rho P_m)\phi = P_m(\langle \phi, e_m \rangle \rho e_m) = \langle \rho e_m, e_m \rangle \langle \phi, e_m \rangle e_m = \langle \rho e_m, e_m \rangle P_m \phi$. Thus we can rewrite (1.20) as von Neumann wrote

$$\rho' = \sum_m \langle \rho e_m, e_m \rangle P_m. \tag{1.22}$$

Let us start with a pure state ψ. If $\hat{a}$ has degenerate (discrete) spectrum, then according to von Neumann [313] a measurement of a giving the value $a = \alpha_m$ *does not induce a projection of* ψ. The result will not be a pure state (in particular, not $\psi_m = P_m \psi$). Moreover, the resulting state is not determined. Only a subsequent measurement of an observable d such that $a = f(d)$ and d is represented by the operator $\hat{d}$ with nondegenerate spectrum will determine the final state.

Let $\hat{a} = P$ be an orthogonal projector onto a subspace $\mathscr{H}_0$ of the state space $\mathscr{H}$. Let us choose in $\mathscr{H}_0$ an orthonormal basis $\{\phi_n\}$. The basis $\{\phi_n\}$ can be completed to an orthonormal basis in $\mathscr{H} : \{\phi_n, \phi_l'\}$. Let us take two sequences of real numbers $\{\gamma_n\}$, $\{\gamma_n'\}$ such that all numbers are distinct. We define the corresponding self-adjoint operator $\hat{d}$ having eigenvectors $\{\phi_n, \phi_l'\}$ and eigenvalues $\{\gamma_n, \gamma_n'\}$

$$\hat{d} = \sum_n \gamma_n P_{\phi_n} + \sum_n \gamma_n' P_{\phi_n'}.$$

Its domain of definition is given by

$$D(\hat{d}) = \left\{ \psi \in \mathcal{H} : \psi = \sum_n \gamma_n^2 |\langle \psi, \phi_n \rangle|^2 + \sum_n (\gamma_n')^2 |\langle \psi, \phi_n' \rangle|^2 < \infty \right\}.$$

By Postulate 2a (which was widely used by J. von Neumann) one can construct a physical observable d described by the operator $\hat{d}$. Measurement of d can be considered as measurement of the observable a, because $a = f(d)$, where f is some function such that $f(\gamma_k) = 1$ and $f(\gamma_k') = 0$. But the d-measurement (without post-measurement selection with respect to eigenvalues) produces the statistical mixture

$$\bar{\rho} = \sum_n |\langle \psi, \phi_n \rangle|^2 P_{\phi_n} + \sum_n |\langle \psi, \phi_n' \rangle|^2 P_{\phi_n'}.$$

Since we can choose $\{\phi_n\}$ and $\{\phi_n'\}$ in many ways, by obtaining the result $a = \alpha_k$ we cannot determine the post-measurement state.

If we start with an arbitrary state ρ and an arbitrary self-adjoint operator $\hat{a}$ with purely discrete spectrum, then we can determine the post-measurement state only with the aid of the subsequent measurement of an observable $d, a = f(d)$, described by the operator $\hat{d}$ with nondegenerate spectrum. We denote by $\{\phi_{km}\}$ bases in subspaces $\mathcal{H}_m = P_m \mathcal{H}$. Then by von Neumann

$$\rho' = \sum_m \sum_k \langle \rho \phi_{km}, \phi_{km} \rangle P_{\phi_{km}}. \tag{1.23}$$

It seems that experimental investigations to compare von Neumann's or Lüders' laws of transformation of states have never been performed. It is amazing! It is not so complicated to check whether after measurement of an observable a (represented by a self-adjoint operator with degenerate purely discrete spectrum) the post measurement state is a pure state (Lüders' postulate) or some statistical mixture (von Neumann's viewpoint).

1.5 Conditional Probability

As in the classical Kolmogorov and von Mises probabilistic models, see Chap. 2, Postulate 4 should be completed by a definition of conditional probability. We present the contemporary definition which is conventional in quantum logic [35] and quantum information theory.[6]

Definition 1.1. *Let physical observables a and b be represented by self-adjoint operators with purely discrete (perhaps degenerate) spectra*

$$\hat{a} = \sum_m \alpha_m P_m^a, \qquad \hat{b} = \sum_m \beta_m P_m^b. \tag{1.24}$$

Let ψ be a pure state and let $P_k^a \psi \neq 0$. Then the probability of getting the value $b = \beta_m$ under the condition that the value $a = \alpha_k$ was observed in the preceding measurement of the observable a on the state ψ is given by

$$\mathbf{P}_\psi(b = \beta_m | a = \alpha_k) \equiv \frac{\| P_m^b P_k^a \psi \|^2}{\| P_k^a \psi \|^2}. \tag{1.25}$$

Sometimes the symbol $\mathbf{P}_\psi(P_m^b | P_k^a)$ is used. Set $\psi_k^a = \frac{P_k^a \psi}{\| P_k^a \psi \|}$. Then

$$\mathbf{P}_\psi(b = \beta_m | a = \alpha_k) = \| P_m^b \psi_k^a \|^2 = \mathbf{P}_{\psi_k^a}(b = \beta_m)$$

(the latter equality is a consequence of Postulate 4).

Let $\hat{a}$ have nondegenerate spectrum. We can write: $\mathbf{P}_\psi(b = \beta_m | a = \alpha_k) = \| P_m^b e_k^a \|^2$ (here $\hat{a} e_k^a = \alpha_k e_k^a$). Thus the conditional probability in this case does not

[6] This definition is based on Lüders postulate. Von Neumann's assumption that an observable should have a nondegenerate spectrum was totally ignored.

depend on the original state ψ. We can say that the memory of the original state was destroyed.

If also the operator $\hat{b}$ has nondegenerate spectrum then we have

$$\mathbf{P}_\psi(b = \beta_m | a = \alpha_k) = |\langle e_m^b, e_k^a \rangle|^2$$

and

$$\mathbf{P}_\psi(a = \alpha_k | b = \beta_m) = |\langle e_k^a, e_m^b \rangle|^2.$$

By using symmetry of the scalar product we obtain

Proposition 1.1. *Let both operators $\hat{a}$ and $\hat{b}$ have purely discrete nondegenerate spectra and let $P_k^a \psi \neq 0$ and $P_m^b \psi \neq 0$. Then conditional probability is symmetric and it does not depend on the original state ψ*

$$\mathbf{P}_\psi(b = \beta_m | a = \alpha_k) = \mathbf{P}_\psi(a = \alpha_k | b = \beta_m) = |\langle e_m^b, e_k^a \rangle|^2.$$

We now invent the notion of conditional probability for a quantum statistical state given by a density operator ρ. Let two observables be represented by operators (1.24). Then the probability of getting the value $b = \beta_m$ under the condition that the value $a = \alpha_k$ has been observed in the preceding measurement of the observable a on the state ρ is given by

$$\mathbf{P}_\rho(b = \beta_m | a = \alpha_k) = \mathrm{Tr}\, \rho_k^a P_m^b, \quad \text{where } \rho_k^a = \frac{P_k^a \rho P_k^a}{\mathrm{Tr}\, \rho P_k^a}. \tag{1.26}$$

Here (according to Lüders) the density operator ρ_k^a describes the quantum state after the result $a = \alpha_k$ has been obtained. We shall also use the notation $\mathbf{P}_\rho(P_m^b | P_k^a)$ for $\mathbf{P}_\rho(b = \beta_m | a = \alpha_k)$.

1.6 Derivation of Interference of Probabilities

Let $\mathscr{H} = \mathbf{C} \times \mathbf{C}$ be the two-dimensional complex Hilbert space and let $\psi \in \mathscr{H}$ be a pure quantum state. Let us consider two dichotomous physical observables $b = \beta_1, \beta_2$ and $a = \alpha_1, \alpha_2$ described by symmetric operators $\hat{b}$ and $\hat{a}$, respectively. Let $e^b = \{e^b_\beta\}$ and $e^a = \{e^a_\alpha\}$ be two orthonormal bases consisting of eigenvectors of the operators. The state ψ can be represented in two ways

$$\psi = c_1 e_1^a + c_2 e_2^a, \qquad c_\alpha = \langle \psi, e_\alpha^a \rangle; \tag{1.27}$$

$$\psi = d_1 e_1^b + d_2 e_2^b, \qquad d_\beta = \langle \psi, e_\beta^b \rangle. \tag{1.28}$$

By Postulate 4 we have

$$\mathbf{P}(a = \alpha) \equiv \mathbf{P}_\psi(a = \alpha) = |c_\alpha|^2; \tag{1.29}$$

$$\mathbf{P}(b = \beta) \equiv \mathbf{P}_\psi(b = \beta) = |d_\beta|^2. \tag{1.30}$$

The possibility to expand one basis with respect to another basis induces connection between the probabilities $\mathbf{P}(a = \alpha)$ and $\mathbf{P}(b = \beta)$. Let us expand the vectors e_α^a with respect to the basis e^b

$$e_1^a = u_{11} e_1^b + u_{12} e_2^b, \tag{1.31}$$

$$e_2^a = u_{21} e_1^b + u_{22} e_2^b, \tag{1.32}$$

where $u_{\alpha\beta} = \langle e_\alpha^a, e_\beta^b \rangle$. Thus

$$d_1 = c_1 u_{11} + c_2 u_{21}, \qquad d_2 = c_1 u_{12} + c_1 u_{22}.$$

We obtain the *quantum rule* for transformation of probabilities

$$\mathbf{P}(b = \beta) = |c_1 u_{1\beta} + c_2 u_{2\beta}|^2. \tag{1.33}$$

On the other hand, by the definition of quantum conditional probability, see (1.25), we obtain

$$\mathbf{P}(b=\beta|a=\alpha)\equiv\mathbf{P}_\phi(b=\beta|a=\alpha)=|\langle e^a_\alpha, e^b_\beta\rangle|^2. \tag{1.34}$$

By combining (1.29), (1.30) and (1.33), (1.34) we obtain the *quantum formula of total probability—the formula of the interference of probabilities*

$$\mathbf{P}(b=\beta)=\sum_\alpha \mathbf{P}(a=\alpha)\mathbf{P}(b=\beta|A=\alpha) + 2\cos\theta\sqrt{\mathbf{P}(a=\alpha_1)\mathbf{P}(b=\beta|a=\alpha_1)\mathbf{P}(a=\alpha_2)\mathbf{P}(b=\beta|a=\alpha_2)}. \tag{1.35}$$

In general $\cos\theta\neq 0$. Thus the quantum formula of total probability does not coincide with the classical formula of total probability (based on the Bayes' formula), see Chap. 2, Sect. 2.1

$$\mathbf{P}(b=\beta)=\sum_\alpha \mathbf{P}(a=\alpha)\mathbf{P}(b=\beta|a=\alpha). \tag{1.36}$$

In general the quantum rule (1.35) differs from the classical rule (1.36). The standard viewpoint to the contradiction between (1.35) and (1.36) is that (see, e.g., d'Espagnat [82, 83] or Feynman [92]):

This is the exhibition of the nonclassical probabilistic structure of quantum statistical ensembles.

Then one commonly proceeds in the following way (e.g., d'Espagnat, [82, 83] or Feynman [92]):

> In classical statistical mechanics we use classical probability (Kolmogorov or frequency, see Chap. 2). In quantum theory we cannot use classical probability theory. Hence quantum systems have some very special quantum features which produce nonclassical probability.[7]

[7] On the other hand, e.g., Kirkpatrick [213] strongly criticized such a viewpoint.

However, the difference between the quantum rule (1.35) and the classical rule (1.36) has nothing to do with some special "quantum features" of individual physical systems. This difference is a natural consequence of the probabilistic structure which induces the quantum formula of total probability. The crucial point is that one cannot use the same symbol **P** to denote all probabilities in (1.35). In one formula, (1.35), one combines probabilistic data obtained in four different experiments:

(a) measurement of the observable a under the complex of physical conditions (context) C which is represented by the initial state ψ;[8]

(b) measurement of the observable b under the same context C.

After performing the a-measurement one can create through selection procedures C_{α_1} and C_{α_2} (selections of systems with respect to the values $a = \alpha_1$ and $a = \alpha_2$) two new ensembles of systems S_{α_1} and S_{α_2}. In quantum mechanics (with the ensemble interpretation) these ensembles are represented by the eigenvectors e_1^a, e_2^a of the operator $\hat{a}$. Therefore we can perform the b-measurement for two new contexts:

(a1) measurement of the observable b under the complex of physical conditions (context) C_{α_1} which is represented by the state e_1^a;

(a2) measurement of the observable b under the complex of physical conditions (context) C_{α_2} which is represented by the state e_2^a.

The (a)-experiment gives probabilities $\mathbf{P}_\psi(a = \alpha)$; the (b)-experiment—$\mathbf{P}_\psi(b = \beta)$; the (a1)-experiment—$\mathbf{P}_{e_1^a}(b = \beta)$; the (a2)-experiment—$\mathbf{P}_{e_2^a}(b = \beta)$.

What can be a reason to assume that we can use a single probability measure **P** in all these experiments? I do not see any reason. We shall see in Chap. 2, Theorem 1.1, that the classical formula of total probability (1.36) can only be derived on the basis of one fixed probability measure **P** (one fixed Kolmogorov probability space or one fixed von Mises collective). It is surprising that one could expect (as, e.g., Feynman

[8] In Part II we shall present a contextual viewpoint on quantum mechanics: quantum states as representations of complexes of physical conditions, contexts, by complex probability amplitudes.

or d'Espagnat) that the classical rule (1.36) will be obtained for the multi-experiment statistics.

Chapter 2
Classical Probability Theories

In principle, those who are not interested in mathematical foundations of probability theory might jump directly to Part II. One should just know that, besides the Kolmogorov definition of classical probability as a probability measure, another classical probability theory was developed by von Mises: probability was defined as the limit of relative frequencies.

We are well aware of mathematical difficulties of the von Mises theory, cf., e.g., [305, 315, 300, 322, 234, 139]. These difficulties are consequences of von Mises' definition of randomness through so-called *place selections*. There are few ways to escape these difficulties. One way is to follow von Mises and choose a class of place selections depending on a series of experiments under consideration, compare also with Wald [315]. Another way for development was proposed by Kolmogorov. This is *complexity theory* for random sequences: [223, 224, 50, 51, 288–290, 322, 304, 234, 282]. We also mention the theory of recursive statistical tests (initiated by Kolmogorov and developed by Martin-Löf [244, 245]).

Besides these two mathematically advanced ways, it is possible to choose a very pragmatic way of modification of von Mises's theory to obtain a rigorous mathematical formalism. At the moment the problem of randomness of sequences of experimental data is not of great interest in quantum physics. The quantum experimental research is not yet so much devoted to randomness of statistical data. It is devoted merely to the study of relative frequencies (which stabilize to probabilities in long runs of experiments). Therefore we can forget (at least for a while) about von Mises' attempt to provide a mathematical formalization of the notion of randomness.

We shall proceed by considering only statistical stabilization of relative frequencies—*existence of the limit of a sequence of relative frequencies.* Thus we shall enjoy all advantages of the von Mises frequency approach to probability and at the same time we shall escape all difficulties related to a rigorous definition of randomness.

2.1 Kolmogorov Measure-Theoretic Model

The axiomatics of modern probability theory were proposed by Andrei Nikolaevich Kolmogorov [222] in 1933.

2.1.1 Formalism

We recall some notions of measure theory. Let Ω be a set. A system F of subsets of a set Ω is called an *algebra* if the sets $\emptyset$, Ω belong to F and the union, intersection and difference of two sets of F also belong to F. In particular, for any $A \in F$, *complement* $\bar{A} = \Omega \setminus A$ of A belongs to F.

Denote by F_Ω the family of all subsets of Ω. This is the simplest example of an algebra.

Let F be an algebra. A map $\mu : F \to \mathbf{R}_+$ is said to be a *measure* if $\mu(A \cup B) = \mu(A) + \mu(B)$, for $A, B \in F, A \cap B = \emptyset$. A measure μ is called σ*-additive* if, for every sequence $\{A_n\}_{n=1}^{\infty}$ of sets $A_n \in F$ such that their union $A = \bigcup_{n=1}^{\infty} A_n$ also belongs to F, we have: $\mu(A) = \sum_{n=1}^{\infty} \mu(A_n)$. An algebra, say $\mathscr{F}$, is said to be a σ-*algebra* if, for every sequence $\{A_n\}_{n=1}^{\infty}$ of sets $A_n \in \mathscr{F}$, their union $A = \bigcup_{n=1}^{\infty} A_n$ belongs to $\mathscr{F}$.

Some books on probability theory use the terminology *field* and σ-*field* of sets, instead of algebra and σ-algebra.

Let Ω_1, Ω_2 be arbitrary sets and let G_1, G_2 be some systems of subsets of Ω_1 and Ω_2, respectively. A map $\xi : \Omega_1 \to \Omega_2$ is called *measurable* or more precisely $((\Omega_1, G_1), (\Omega_2, G_2))$-measurable if, for any set $A \in G_2$, the set $\xi^{-1}(A) \in G_1$. Here $\xi^{-1}(A) = \{\omega \in \Omega_1 : \xi(\omega) \in A\}$. We shall use the notation $\xi : (\Omega_1, G_1) \to (\Omega_2, G_2)$ to indicate dependence on G_1, G_2. Typically we shall consider measurability of maps in the case such that the systems of sets G_j, $j = 1, 2$, are algebras or σ-algebras.

Let A be a set. The *characteristic function* I_A of the set A is defined as $I_A(x) = 1$, $x \in A$, and $I_A(x) = 0$, $x \in \bar{A}$.

Let $A = \{a_1, \ldots, a_n\}$ be a finite set. We shall denote the number of elements n of A by the symbol $|A|$.

By the Kolmogorov axiomatics [222], see also [286], a *probability space* is a triple

$$\mathcal{P} = (\Omega, \mathcal{F}, \mathbf{P}),$$

where Ω is an arbitrary set (points ω of Ω are said to be *elementary events*), $\mathcal{F}$ is an arbitrary σ-algebra of subsets of Ω (elements of $\mathcal{F}$ are said to be *events*), $\mathbf{P}$ is a σ-additive measure on $\mathcal{F}$ which yields values in the segment $[0, 1]$ of the real line and normalized by the condition $\mathbf{P}(\Omega) = 1$ (it is said to be *probability*).

Random variables on the Kolmogorov space $\mathcal{P}$ are by definition measurable functions $\xi : \Omega \to \mathbf{R}$, or in our notation $\xi : (\Omega, \mathcal{F}) \to (\mathbf{R}, \mathcal{B}))$, where $\mathcal{B}$ is the Borel σ-algebra on the real line.[1] We shall use the symbol $RV(\mathcal{P})$ to denote the space of random variables for the probability space $\mathcal{P}$. The probability distribution of $\xi \in RV(\mathcal{P})$ is defined by $\mathbf{P}_\xi(B) = \mathbf{P}(\xi^{-1}(B))$ for $B \in \mathcal{B}$. This is a σ-additive

[1] Thus $\xi^{-1}(B) \in \mathcal{F}$ for every $B \in \mathcal{B}$.

measure on the Borel σ-algebra. The *average* (mathematical expectation) of a random variable ξ is defined by

$$E\xi = \int_{\Omega} \xi(\omega) d\mathbf{P}(\omega) \tag{2.1}$$

In particular, if $\xi = a_1, \ldots, a_n, \ldots$ is a discrete random variable its average is given by

$$E\xi = \sum_n a_n \mathbf{P}(\xi = a_n). \tag{2.2}$$

Conditional probability will play an essential role in further quantum considerations. In Kolmogorov's probability model *conditional probability* is defined by well-known *Bayes' formula.* In many textbooks this formula is called Bayes' theorem. However, in the Kolmogorov model it is neither a theorem nor an axiom, but a *definition.* Conditional probability is introduced in the Kolmogorov model through the following definition:

$$\mathbf{P}(B|A) = \mathbf{P}(B \cap A)/\mathbf{P}(A), \qquad \mathbf{P}(A) > 0. \tag{2.3}$$

By Kolmogorov's interpretation this is the *probability that an event A occurs under the condition of an event B having occurred.*

The *Kolmogorov probability model* is given by the Kolmogorov probability space endowed with conditional probability via the Bayes formula.

We remark that $\mathbf{P}_A(B) \equiv \mathbf{P}(B|A)$ is again a probability measure on $\mathcal{F}$. The conditional expectation of a random variable ξ is defined by

$$E(\xi|A) = \int_{\Omega} \xi(\omega) d\mathbf{P}_A(\omega). \tag{2.4}$$

In the Kolmogorov model two events A and B are said to be *independent* if

$$\mathbf{P}(A \cap B) = \mathbf{P}(A) P(B) \tag{2.5}$$

or

$$\mathbf{P}(B|A) = P(B), \qquad \mathbf{P}(A) > 0. \tag{2.6}$$

In our further considerations an important role will be played by the *formula of total probability* - a theorem of Kolmogorov's model. Let us consider a countable family of sets $A_k \in \mathcal{F}, \mathbf{P}(A_k) > 0, k = 1, \ldots,$ such that

$$\bigcup_{k=1}^{\infty} A_k = \Omega, \quad \text{and} \quad A_k \cap A_l = \emptyset, \quad k \neq l.$$

Such a family is called a *measurable partition* of the space Ω or a *complete group of disjoint events*.

Theorem 2.1. *Let $\{A_k\}$ be a complete group of disjoint events. Then, for every set $B \in \mathcal{F}$, the following formula of total probability holds*:

$$\mathbf{P}(B) = \sum_{k=1}^{\infty} \mathbf{P}(A_k)\mathbf{P}(B|A_k). \tag{2.7}$$

Proof. We have

$$\mathbf{P}(B) = \mathbf{P}\left(B \cap \bigcup_{k=1}^{\infty} A_k\right) = \sum_{k=1}^{\infty} \mathbf{P}(B \cap A_k) = \sum_{k=1}^{\infty} \mathbf{P}(A_k) \frac{\mathbf{P}(B \cap A_k)}{\mathbf{P}(A_k)}.$$

Especially interesting for us is the case such that a complete group of disjoint events is induced by a discrete random variable a taking values $\{\alpha_k\}$. Here

$$A_k = \{\omega \in \Omega : a(\omega) = \alpha_k\}. \tag{2.8}$$

Let b be another random variable. It takes values $\{\beta_j\}$. Then we have:

$$\mathbf{P}(b = \beta) = \sum_{\alpha} \mathbf{P}(a = \alpha)\mathbf{P}(b = \beta|a = \alpha). \tag{2.9}$$

2.1.2 Discussion

The Kolmogorov axiomatics [222] played a crucial role in creation of the rigorous mathematical formalism of probability theory, see, e.g., [286, 285]. The main problem of this approach to probability is that by starting with an abstract Kolmogorov space $\mathscr{P} = (\Omega, \mathscr{F}, \mathbf{P})$ we completely lose information on the intrinsic structure of a physical statistical experiment. The main idea of Kolmogorov was that it is possible to proceed in a very abstract way without having to pay attention to structures of concrete ensembles:

"To outline the context of theory, it suffices to single out from probability theory those elements that bring out its intrinsic logical structure, but have nothing to do with the specific meaning of theory."

In quantum physics we cannot use one fixed Kolmogorov probability space $\mathscr{P} = (\Omega, \mathscr{F}, \mathbf{P})$—an *absolute Kolmogorov space*—to describe different experiments.[2] This impossibility was considered as a contradiction between the classical and quantum probabilistic descriptions.

Of course, it would be better if from the very beginning A. N. Kolmogorov defined a probability space as *a collection of conventional Kolmogorov probability spaces corresponding to a family of contexts (complexes of experimental physical conditions)* $\mathscr{C} = \{C\}$:

$$\{\mathscr{P}_C : C \in \mathscr{C}\}, \quad \text{where } \mathscr{P}_C = (\Omega_C, \mathscr{F}_C, \mathbf{P}_C). \tag{2.10}$$

Unfortunately, A. N. Kolmogorov did not do this, cf. S. Gudder [106–108]: theory of probabilistic manifolds. In our approach a structure which is similar to (2.10) will appear as a special case of the general contextual probability space, see Part II, Chap. 6 (Definition 6.8). It will be called the contextual multi-Kolmogorovian probability space. We remark that such a probabilistic construction does not contradict the original ideas of Kolmogorov [222] who noticed that each complex of experimental conditions generates its own (Kolmogorov) probability space. Thus operation with data collected for a few different complexes of, e.g., physical conditions

[2] Cf. with attempts to use absolute Newton space in classical physics.

should naturally induce families of (Kolmogorov) probability spaces—contextual multi-Kolmogorovian spaces. As was pointed out, although Kolmogorov emphasized this ideologically, he did not proceed in this direction mathematically.

2.2 Von Mises Frequency Model

Let us recall the main notions of a frequency theory of probability [309–311] of Richard von Mises (1919).[3]

2.2.1 Collective (Random Sequence)

Von Mises' probability theory is based on the notion of a *collective*. Consider a random experiment. Let a be some observable representing results of this random experiment. The set of all possible results of this experiment is $L = \{s_1, \ldots, s_m\}$—the label set or the set of attributes. It will be finite in this book.

Consider N observations of a and write a result x_j after each trial. We obtain the finite sample: $x = (x_1, \ldots, x_N),\ x_j \in L$. A *collective* is an infinite idealization of this finite sample:

$$x = (x_1, \ldots, x_N, \ldots), \quad x_j \in L, \tag{2.11}$$

for which two von Mises' principles are valid. Let us compute frequencies

$$\nu_N(s; x) = \frac{n_N(s; x)}{N}, \quad s \in L,$$

where $n_N(s; x)$ is the number of realizations of the attribute s in the first N trials.

[3] In fact, already in 1866 John Venn tried to define a probability explicitly in terms of relative frequencies.

Principle S (Statistical stabilization of relative frequencies). *For every label $s \in L$, the frequency $\nu_N(s; x)$ approaches a limit as N approaches infinity.*

In the frequency theory of probability the limit $\mathbf{P}(s) = \lim \nu_N(s; x)$ is the probability of the label s. Sometimes this probability will be denoted by $\mathbf{P}_x(s)$ (to show a dependence on the collective x).

Principle R (Randomness). *The limits of relative frequencies have to be stable with respect to a place selection (a choice of a subsequence) in* (2.11).

Heuristically it is evident that we cannot consider, for example, the sequence $z = (0, 1, 0, 1, \ldots, 0, 1, \ldots)$ as a random object (generated by a statistical experiment). Principle **S** holds for z and $\mathbf{P}(0) = \mathbf{P}(1) = 1/2$. But, the sequence z does not satisfy Principle **R**. If we choose only even places, then we obtain the zero sequence $z_0 = (0, 0, \ldots)$, where $\mathbf{P}(0) = 1, \mathbf{P}(1) = 0$.

The *average* of observable a is defined as the average with respect to the probability distribution $\mathbf{P}_x$:

$$E_x a = \sum_{s \in L} s \mathbf{P}_x(s). \tag{2.12}$$

Here x is a collective representing observations of a.

Finally, we recall the original von Mises thoughts about the notion of collective:

"We will say that a collective is a mass phenomenon or a repetitive event, or simply a long sequence of observations for which there are sufficient reasons to believe that the relative frequency of the observed attribute would tend to a fixed limit if the observations were infinitely continued. This limit will be called the probability of the attribute considered within the given collective," R. von Mises [311].

2.2.2 *Difficulties with Definition of Randomness*

However, this very natural notion of randomness was the hidden bomb in the foundations of von Mises' theory. The main problem was to define a class of place selections which would induce a fruitful theory. A very natural restriction is that a place selection in (2.11) cannot be based on the use of attributes of elements. For example, we cannot consider a subsequence of (2.11) constructed by choosing elements with the fixed label $s \in L$. Von Mises proposed the following definition of a place selection:

(PS) "a subsequence has been derived by a place selection if the decision to retain or reject the nth element of the original sequence depends on the number n and on label values $x_1, \ldots, x_{n-1}$ of the $(n-1)$ presiding elements, and not on the label value of the nth element or any following element", see [87, p. 9].

Thus a place selection can be defined by a set of functions

$$f_1, \quad f_2(x_1), \quad f_3(x_1, x_2), \quad f_4(x_1, x_2, x_3), \quad \ldots,$$

each function yielding the values 0 (rejecting the nth element) or 1 (retaining the nth element). There are some examples of place selections: (1) choose those x_n for which n is prime; (2) choose those x_n which follow the word 01; (3) toss a (different) coin; choose x_n if the nth toss yields heads.

The first two selection procedures may be called *lawlike*, the third random. It is more or less obvious that all of these procedures are place selections: the value of x_n is not used in determining whether to choose x_n.

The principle of randomness ensures that no strategy using a place selection rule can select a subsequence that allows different odds for gambling than a sequence that is selected by flipping a fair coin. This principle can be called the *law of excluded gambling strategy.*

The definition (PS) induced some mathematical problems. If a class of place selections is too extended, then the notion of the collective is too restricted (in fact, there are no sequences where probabilities are invariant with respect to all place selections). This was the main critical argument against von Mises' theory.

However, von Mises himself was totally satisfied by the following operational solution of this problem. He proposed [311] to fix for any collective a class of place selections which depends on the physical problem described by this collective. Thus he moved the problem outside the mathematical framework. For any concrete experiment, one should find a special class of place selections which would be appropriative for this experiment.

2.2.3 S-sequences

As probability is defined on the basis of the principle of the statistical stabilization of relative frequencies, it is possible to develop a quite fruitful probabilistic calculus *based only on this principle.* Instead of the notion of a collective, we can consider a more general notion.

Definition 2.1. *A sequence x, see* (2.11), *which satisfies the principle of the statistical stabilization of relative frequencies is said to be an S-sequence.*

Thus the limits of relative frequencies in an S-sequence x need not be invariant with respect to some class of place selections.

It seems that the machinery of randomness has no applications in quantum physics. Experimenters are only interested in the statistical stabilization of relative frequencies.

2.2.4 Operations for Collectives

On many occasions R. von Mises emphasized that frequency probability theory is not a calculus of probabilities, but is the *calculus of collectives* which generates the corresponding calculus of probabilities. We briefly discuss some of the basic operations for collectives (see [311] for the details) and S-sequences (see [139] for details).

(a) **Operation of mixing of labels in an S-sequence (or collective) and additivity of probabilities.** Let x be an S-sequence (in particular, it can be a collective) with the (finite) label space $L_x = \{s_1, \ldots, s_m\}$:

$$x = (x_1, \ldots, x_N, \ldots), \quad x_j \in L_x, \tag{2.13}$$

and let $E = \{s_{i_1}, \ldots, s_{i_d}\}$ be a subset of the set of labels L_x. The sequence (2.13) of x is transformed into a new sequence y_E by the following rule:

If x_j belongs to the set E, then we write 1; if x_j does not belong to the set E then we write 0.

Thus the label set of the sequence y_E constructed on the basis of this rule is $L_{y_E} = \{0, 1\}$. R. von Mises called this operation on sequences, $x \to y_E$, the operation of mixing of labels. We take a subset E of the label set L_x and we "mix" elements of E into a new label, $y = 1$; elements of the complement of E are mixed into the label $y = 0$.

Proposition 2.1. *If a sequence of labels x satisfies the principle of statistical stabilization (so it is an S-sequence), then, for any subset E of the label set L_x, the sequence y_E also satisfies the principle of statistical stabilization (so it is also an S-sequence).*

Proof. For example, for the label 1 we have:

$$\mathbf{P}_{y_E}(1) = \lim_{N\to\infty} \nu_N(E; x) = \lim_{N\to\infty} \sum_{k=1}^{d} \nu_N(s_{i_k}; x) = \sum_{k=1}^{d} \mathbf{P}_x(s_{i_k}), \tag{2.14}$$

where $\nu_N(E; x) \equiv \nu_N(1; y_E) = n_N(1; y_E)/N$ is the relative frequency of 1 in y_E. To obtain (2.14) we have only used the fact that the addition is a continuous operation on the field of real numbers $\mathbf{R}$.

We can also show that if a sequence x satisfies the principle of randomness (so it is a collective), then the sequence y_E also satisfies the principle of randomness (so it is also a collective), see [311, 139].

By this operation any S-sequence (in particular, any collective) x generates a probability distribution on the algebra F_{L_x} of all subsets of the label set L_x. By definition we have:

$$\mathbf{P}_x(E) = \mathbf{P}_{y_E}(1) = \sum_{s \in E} \mathbf{P}_x(s).$$

Now we find the properties of this probability. We start with a simple, but extremely important result.

Theorem 2.2. *For any S-sequence x, the frequency probability $\mathbf{P}_x$ yields values in the segment* $[0, 1]$.

Proof. As $\mathbf{P}_x(E) = \lim_{N\to\infty} \nu_N(E; x)$ and $0 \leq \nu_N(E) \leq 1$, then (by an elementary theorem of real analysis)

$$0 \leq \mathbf{P}_x(E) \leq 1.$$

We emphasize that in frequency probability theory this is a theorem and not an axiom (as it is in the Kolmogorov measure-theoretic model). Another very simple, but extremely important result is also an axiom in the Kolmogorov model, but a theorem in the von Mises model.

Theorem 2.3. *For any S-sequence x, the frequency probability $\mathbf{P}_x$ is normalized by* 1.

Proof. As the S-sequence y_{L_x} corresponding to the whole label set L_x does not contain zeros, we obtain that for any N the relative frequency $\nu_N(L_x; x) \equiv \nu_N(1; y_{L_x}) \equiv 1$ and, consequently,

$$\mathbf{P}_x(L_x) = \sum_s \mathbf{P}_x(s) = 1. \tag{2.15}$$

Finally by (2.14) we find that the set function

$$\mathbf{P}_x : F_{L_x} \to [0, 1]$$

is additive. Thus we have obtained

Theorem 2.4. *The frequency probability* $\mathbf{P}_x$ *is additive.*

Thus $\mathbf{P}_x$ is a normalized measure on the set-algebra F_{L_x} which yields values in $[0, 1]$.

(b) **Operation of partition of an S-sequence (or a collective) and conditional probabilities.** Let x be an S-sequence (or even a collective). We take a subset, say O, of the label set L_x such that $\mathbf{P}_x(O) \neq 0$. Thus $\mathbf{P}_x(s) = \lim_{N\to\infty} \nu_n(s; x) > 0$ for at least one label $s \in L_x$. We now derive a new sequence $z(O)$ by the following rule:

There are retained only those elements of x which belong to subset O and all other elements are discarded.

The label set of the sequence $z(O)$ coincides with the set O. This operation is obviously not a place selection, since the decision to retain or reject an element of x depends on the label of just this element.[4]

Proposition 2.2. *For any S-sequence x and any subset O of the label set L_x such that* $\mathbf{P}_x(O) \neq 0$, *the sequence $z(O)$ is again an S-sequence.*

Proof. Suppose that $s \in O$ and let y_O be the S-sequence generated by x with the aid of the mixing operation. Then

$$\mathbf{P}_{z(O)}(s) = \lim_{N\to\infty} \nu_N(s; z(O)) = \lim_{k\to\infty} \nu_{N_k}(s; z(O)),$$

where $N_k \to \infty$ is an arbitrary sequence. As $\mathbf{P}(O) \neq 0$, then

$$M_k = n_k(1; y_O) \to \infty.$$

[4] It is important to remark that this operation is not based on a new measurement. We just operate with data which was collected via a measurement represented by an S-sequence (collective) x; compare with considerations that will be presented in Part II, Chap. 1.

This is the number of labels belonging to O among the first k elements of x. Thus we have:

$$\begin{aligned}\mathbf{P}_{z(O)}(s) &= \lim_{k\to\infty} \nu_{M_k}(s; z(O)) = \lim_{k\to\infty} n_{M_k}(s; z(O))/M_k \\ &= \lim_{k\to\infty} [n_{M_k}(s; z(O))/k] : [M_k/k] = \mathbf{P}_x(s)/\mathbf{P}_x(O). \qquad (2.16)\end{aligned}$$

We have used the property that $n_{M_k}(s; z(O))$—the number of occurrences of the label s among first M_k elements of $z(O)$—is equal to $n_k(s; x)$—the number of occurrences of the label s among first k elements of x.

It is also possible to show that if x is a collective, then the sequence $z(O)$ is again a collective, see [309, 139].

Definition 2.2. *The probability $\mathbf{P}_{z(O)}(s)$ is called the conditional probability of the label $s \in L_x$ under the condition that it belongs to the subset O of the label set L_x.*

This probability is denoted by $\mathbf{P}_x(s|O)$. For any $B \subset O$, we define conditional probability by

$$\mathbf{P}_x(B|O) = \mathbf{P}_{z(O)}(B) = \sum_{s\in B} \mathbf{P}_x(s|O).$$

Sometimes we shall use the symbol $\mathbf{P}(B|O)$. However, if we forget that, in fact, the probability $\mathbf{P}$ depends on a collective x, then the symbol $\mathbf{P}(B|O)$ might induce misunderstanding (as it happens in applications of the Kolmogorov model).

Theorem 2.5 (Bayes formula). *For any S-sequence (collective) x and any subset O of the label set L_x, such that $\mathbf{P}_x(O) > 0$, the Bayes formula for conditional probability holds*:

$$\mathbf{P}(B|O) = \frac{\mathbf{P}_x(B \cap O)}{\mathbf{P}_x(O)}. \qquad (2.17)$$

Proof. We have

$$\mathbf{P}_{z(O)}(B) = \sum_{s\in B\cap O} \mathbf{P}_x(s|O) = \sum_{s\in B\cap O} \mathbf{P}_x(s)/\mathbf{P}_x(O).$$

Thus in the von Mises model the Bayes formula for conditional probabilities is a theorem and not a definition; compare with the Kolmogorov model. By using the Bayes' formula (2.17) we obtain:

Theorem 2.6 (Formula of total probability). *Let x be an S-sequence and let $\{O_k\}$ be a partition of the label set L_x:*

$$L_x = \bigcup_k O_k \quad \textit{and} \quad O_k \cap O_l = \emptyset,$$

and $\mathbf{P}_x(O_k) > 0$ for any k. Then, for any $C \subset L_x$, we have:

$$\mathbf{P}_x(C) = \sum_{k=1}^{m} \mathbf{P}_x(O_k)\mathbf{P}_x(C|O_k). \tag{2.18}$$

2.3 Combining and Independence of Collectives

The material presented in this section will not be used in our contextual probabilistic considerations. We present it simply to give a more complete picture of the von Mises model. Therefore it is possible to read this section later. One might jump directly to Part II.

In the two basic operations, mixing and partition, discussed in Sect. 2.2, one single S-sequence (or collective) x served each time as the point of departure for the construction of a new S-sequence (collective). We now consider the problem of *combining* of two or more given S-sequences (collectives).

Let $x = (x_j)$ and $y = (y_j)$ be two S-sequences with label sets L_x and L_y, respectively. We define a new sequence

$$z = (z_j), \ z_j = (x_j, y_j). \tag{2.19}$$

In general such a z is not an S-sequence with respect to the label set $L_z = L_x \times L_y$.

Let $\beta \in L_x$ and $\alpha \in L_y$. Among the first N elements of z there are $n_N(\alpha; z)$ elements with the second component equal to α. As $n_N(\alpha; z) = n_N(\alpha; y)$ is a number of $y_j = \alpha$ among the first N elements of y, we obtain that

$$\lim_{N\to\infty} \frac{n_N(\alpha; z)}{N} = \mathbf{P}_y(\alpha).$$

Among these $n_N(\alpha; z)$ elements, there are a number, say $n_N(\beta|\alpha; z)$, of elements whose first component is equal to β.[5] The frequency $\nu_N(\beta, \alpha; z)$ of elements of the sequence z labeled (β, α) will then be equal to

$$\frac{n_N(\beta|\alpha; z)}{N} = \frac{n_N(\beta|\alpha; z)}{n_N(\alpha; z)} \frac{n_N(\alpha; z)}{N}.$$

We set

$$\nu_N(\beta|\alpha; z) = \frac{n_N(\beta|\alpha; z)}{n_N(\alpha; z)}.$$

Let us assume that:

For each $\alpha \in L_y$, the subsequence $x(\alpha)$ of x which is obtained by choosing x_j such that $y_j = \alpha$ is an S-sequence.[6]

Then, for each $\alpha \in L_y$, $\beta \in L_x$, there exists

$$\mathbf{P}_z(\beta|\alpha) = \lim_{N\to\infty} \nu_N(\beta|\alpha; z) = \lim_{N\to\infty} \nu_N(\beta; x(\alpha)) = \mathbf{P}_{x(\alpha)}(\beta).$$

The existence of $\mathbf{P}_z(\beta|\alpha)$ implies the existence of

$$\mathbf{P}_z(\beta, \alpha) = \lim_{N\to\infty} \nu_N(\beta, \alpha; z).$$

Moreover, we have

$$\mathbf{P}_z(\beta, \alpha) = \mathbf{P}_y(\alpha)\mathbf{P}_z(\beta|\alpha) \tag{2.20}$$

or

[5] We remark again that it is assumed that all data was collected in the sequence z and we need not perform new measurements to obtain all these numbers.

[6] In general such a choice of the subsequence $x(\alpha)$ of x is not a place selection.

$$\mathbf{P}_z(\beta|\alpha) = \mathbf{P}_z(\beta, \alpha)/\mathbf{P}_y(\alpha), \tag{2.21}$$

if $\mathbf{P}_y(\alpha) \neq 0$. Thus in this case the sequence z is an S-sequence with the probability distribution $\mathbf{P}_z(\beta, \alpha)$. This is the "joint probability distribution" of S-sequences of observations x and y. We can repeat all previous considerations for collectives (i.e., take into account the principle of randomness), see R. von Mises [311].

Definition 2.3. *S-sequences (collectives) x and y are said to be combinable if the sequence $z = (x, y)$ is S-sequence (collective).*

Definition 2.4. *Let x and y be combinable. Quantities $\mathbf{P}_z(\beta|\alpha)$ are called conditional probabilities.*

This is the definition of conditioning of one S-sequence (collective) with respect to another. It differs from the conditioning with respect to a subset O of the label set of a single S-sequence (collective), Sect. 2.2.

Definition 2.5. *Let x and y be S-sequences (collectives). The x is said to be independent from y if all $x(\alpha)$, $\alpha \in L_y$, have the same probability distribution which coincides with the probability distribution $\mathbf{P}_x$ of x.*[7]

This implies that

$$\mathbf{P}_z(\beta|\alpha) = \lim_{N\to\infty} \nu_N(\beta|\alpha; z) = \lim_{N\to\infty} \nu_N(\beta, x(\alpha)) = \mathbf{P}_x(\beta),$$

hence

$$\mathbf{P}_z(\beta, \alpha) = \mathbf{P}_y(\alpha)\mathbf{P}_x(\beta). \tag{2.22}$$

Thus the independence implies the factorization of the two-dimensional probability $\mathbf{P}_z(a, b)$.

---

[7] We recall again that the choice of a subsequence $x(\alpha)$ of x based on a label α for y is not a place selection in x. Thus in general there are no grounds for coincidence of probabilities $\mathbf{P}_{x(\alpha)}$ with the probability $\mathbf{P}_x$.

Part II
Contextual Probability and Quantum-Like Models

This is the basic chapter of the book. We introduce a contextual probabilistic model (Växjö model) unifying classical (Kolmogorov and von Mises) and quantum (Born-Dirac-von Neumann) probabilistic models. The formula of total probability with an interference type term is obtained. Starting with this formula a special system of contexts (inducing the ordinary cos-interference) is represented in the complex Hilbert space—on the basis of a simple "quantum-like representation algorithm" (QLRA). Creation of such a representation completely demystifies the conventional quantum representation of probabilities by complex amplitudes. In our model Born's rule is not postulated, but derived.

Another type of interference term will be considered in Part V—so called hyperbolic interference. Corresponding quantum-like representation will be constructed.

Chapter 3

Contextual Probability and Interference

The *formula of total probability* is one of the basic laws of classical probability theory, see Part I: Chap. 2. In the case of two dichotomous random variables $a = \alpha_1, \alpha_2$ and $b = \beta_1, \beta_2$ it has the form

$$\mathbf{P}(b = \beta_i) = \mathbf{P}(a = \alpha_1)\mathbf{P}(b = \beta_i | a = \alpha_1) + \mathbf{P}(a = \alpha_2)\mathbf{P}(b = \beta_i | a = \alpha_2). \quad (3.1)$$

On the other hand, we have a quantum analogue of this formula, see Part I: Chap. 1. We call it the *formula of total probability with the interference term*

$$\begin{aligned} &\mathbf{P}(b = \beta_i) \\ &\quad = \mathbf{P}(a = \alpha_1)\mathbf{P}(b = \beta_i | a = \alpha_1) + \mathbf{P}(a = \alpha_2)\mathbf{P}(b = \beta_i | a = \alpha_2) \\ &\qquad + 2\cos\theta\sqrt{\mathbf{P}(a = \alpha_1)\mathbf{P}(b = \beta_i | a = \alpha_1)\mathbf{P}(a = \alpha_2)\mathbf{P}(b = \beta_i | a = \alpha_2)}, \end{aligned} \quad (3.2)$$

where θ is the phase angle. This formula was derived in the formalism of complex Hilbert space on the basis of Born's postulate and the definition of quantum conditional probability.

The Hilbert space derivation of the formula (3.2) might induce the impression that we deal with something rather strange and impossible from the point of view of classical probability theory. The appearance of the *interference term* has led to the use of the term "quantum probability" in contradiction to what could be called "regular" or "classical" probability, see, e.g., [31, 38, 80, 82, 83, 92, 139], for extended discussions on this problem.

In this chapter we provide contextual probabilistic analysis making a contribution to the understanding of the formula of total probability and its violation.

Our analysis begins with the contextual definition of the relevant probabilities. The probability for the value of one observable is then expressed in terms of the conditional (contextual) probabilities involving the values of a second ("supplementary") observable. In this way the interference term in the generalized formula of total probability gives a measure of *supplementarity* of information which can be obtained through measurements of observables a and b.[1]

The perturbing term in the generalized formula of total probability is then expressed in terms of a coefficient λ (probabilistic measure of supplementarity) whose absolute value can be less than 1 or equal to 1, or it can be greater than one for each of the values of the observable. This range of values for the coefficient λ then introduces three distinct types of probabilistic perturbations:

(a) *trigonometric*, (b) *hyperbolic*, (c) *hyper-trigonometric.*

Each case is then examined separately. Classical ($\lambda = 0$) and quantum cases ($|\lambda| \leq 1$) are then special cases of more general results. Later it will be shown, Chap. 4, that in the quantum case it is possible to reproduce a Hilbert space in which the probabilities are found in the usual way, but there is a case in which this is not possible, though the space is linear it is not a Hilbert space. In general it is not a complex linear space. In the case of hyperbolic probabilistic behavior we

[1] Of course, it would be better to use the terminology "complementarity of information." However, N. Bohr had already reserved the notion of complementarity in quantum physics. The crucial notion in Bohr's complementarity (Copenhagen complementarity) is *mutual exclusivity* (*incompatibility*), see A. Plotnitsky [269–272] for an extended discussion, see also [187]. And in our approach *supplemental information* which need not be based on incompatibility plays a crucial role, see Sect. 3.4 of this chapter.

have to use linear representation of probabilities over so-called hyperbolic numbers (Part V).

3.1 Växjö Model: Contextual Probability

A general probabilistic model for observations based on the contextual viewpoint on probability will be presented. It will be shown that classical (Kolmogorov and von Mises) as well as quantum (Born-Dirac-von Neumann) probabilistic models can be obtained as particular cases of our general contextual model—the *Växjö model.*

3.1.1 Contexts

In our model *context* C is any complex of conditions, e.g. physical or biological, or social.

In principle, the notion of context can be considered as a generalization of a widely used notion of *preparation procedure*, see, e.g., [29, 44]. However, identification of context with preparation procedure would essentially restrict our theory.

In applications outside of physics (e.g., in psychology and cognitive science) mental contexts are considered. Such contexts are not simply preparation procedures. The same can be said about economic, political and social contexts. In this book we shall not provide a deeper formalization of the notion of context. In our model the notion of context is basic and irreducible, see [36, 235] for attempts to formalize the notion of context.

Let us fix a set of contexts $\mathscr{C}$ which will be used in our model.

3.1.2 Observables

We now fix a set of observables $\mathscr{O}$. We suppose that any observable $a \in \mathscr{O}$ can be measured under a complex of physical conditions C.

We shall denote observables by Latin letters, $a, b, \ldots$, and their values by Greek letters, $\alpha, \beta, \ldots$. For an observable $a \in \mathcal{O}$, we denote the set of its possible values ("spectrum") by the symbol X_a. To simplify considerations, in this chapter we shall consider only discrete observables.

Corresponding interpretational problems (in particular, contextuality and realism) will be discussed in Sect. 3.8.

3.1.3 Contextual Probability Space and Model

Definition 3.1. *A contextual probability space is a triple*

$$\mathcal{P}_{\text{cont}} = (\mathcal{C}, \mathcal{O}, \pi),$$

where elements of $\mathcal{C}$ *and* $\mathcal{O}$ *are interpreted as contexts and observables and elements of* π *are corresponding probability distributions.*

Here $\pi = \{p_C^a\}, C \in \mathcal{C}, a \in \mathcal{O}$. For any $\alpha \in X_a$,

$$p_C^a(\alpha) \equiv \mathbf{P}(a = \alpha | C) \tag{3.3}$$

is the probability to obtain the value $a = \alpha$ for observation of a under the context C. We have

$$\sum_{\alpha \in X_a} \mathbf{P}(a = \alpha | C) = 1, \qquad \mathbf{P}(a = \alpha | C) \geq 0.$$

We prefer to call probabilities (3.3) *contextual probabilities.*[2] For any context $C \in \mathcal{C}$, we shall consider the set of probabilities

[2] Of course, it would be also possible to call them conditional probabilities, but the latter term was already used in other approaches (e.g., Bayes-Kolmogorov, von Mises). In contrast to the Kolmogorov model, the contextual probability (3.3) is not probability that an event, say B, occurs under the condition that another event, say C, has occurred. The contextual probability is probability to get the result $a = \alpha$ under the complex of physical conditions C.

$$W(\mathcal{O}, C) = \{\mathbf{P}(a = \alpha|C) : a \in \mathcal{O}, \alpha \in X_a\}.$$

Definition 3.2. *Contextual expectation $E[a|C]$ of an observable $a \in \mathcal{O}$ with respect to a context $C \in \mathcal{C}$ is given by*

$$E[a|C] = \sum_{\alpha \in X_a} \alpha p_C^a(\alpha).$$

Our probability model ("Växjö model") will be based on the contextual probability space having a sufficiently rich family of contexts—such that "transition probabilities" $\mathbf{P}(b = \beta|a = \alpha), a, b \in \mathcal{O}, \alpha \in X_a, \beta \in X_b$, are well defined.

Definition 3.3 (Växjö model). *A contextual probability model is a space $\mathcal{P}_{\text{cont}} = (\mathcal{C}, \mathcal{O}, \pi)$ containing a special family $\{C_\alpha^a\}_{a \in \mathcal{O}, \alpha \in X_a}$ of contexts (so called $[a = \alpha]$-selection contexts) satisfying the condition*

$$\mathbf{P}(a = \alpha|C_\alpha^a) = 1. \tag{3.4}$$

We present a physical interpretation of selection contexts. The context C_α^a is interpreted as the selection with respect to the result $a = \alpha$ of the a-measurement. The condition (3.4) implies that in a measurement of a under the complex of conditions C_α^a the value $a = \alpha$ is obtained with probability 1. We emphasize that the condition (3.4) is *only necessary:* the equality $\mathbf{P}(a = \alpha|C) = 1$ does not imply that $C = C_\alpha^a$. The presence of selection contexts in the model provides a possibility to determine probabilities for results of sequential measurements: first the a-measurement and then the b-measurement.

The most natural interpretation of selection contexts can be given in the model in which observables are considered as observables on systems, see Chap. 5. In this case the context C_α^a consists of the a-measurement procedure and the post-measurement selection of systems for which the result $a = \alpha$ was obtained. For example, a can be chosen as momentum observable. Then C_α^a is the selection procedure with respect to the fixed value α of momentum.

To simplify notation, we shall often use the symbol C_α, instead of C_α^a (when such a notation is not ambiguous).

Let $a, b \in \mathcal{O}$ and let $\alpha \in X_a, \beta \in X_b$. We consider the $[a = \alpha]$-selection context C_α. The contextual probability $p^{b|a}(\beta|\alpha) \equiv \mathbf{P}(b = \beta|C_\alpha)$ will play an important role in further considerations. It is a contextual realization of the "transition probability" $\mathbf{P}(b = \beta|a = \alpha)$. We shall use matrices of "transition probabilities", for pairs of observables $a, b \in \mathcal{O}$,

$$\mathbf{P}^{b|a} = (p^{b|a}(\beta|\alpha)).$$

Let $C \in \mathcal{C}$. We complete the probabilistic data $W(\mathcal{O}, C)$ by the data contained in the matrices $\mathbf{P}^{b|a}$ for all pairs $a, b \in \mathcal{O}$. We obtain a collection of contextual probabilities which will be denoted by the symbol $D(\mathcal{O}, C)$. We shall often take a subset $\mathcal{O}'$ of $\mathcal{O}$ and consider the collection of probabilistic data about contexts given by observables belonging to $\mathcal{O}'$. This collection is denoted by the symbol $D(\mathcal{O}', C)$.

Typically $\mathcal{O}'$ will consist of two observables, say a and b. Our aim is to create a quantum-like representation of contexts by using just a pair of observables (which are analogues of position and momentum observables). Such observables are called *reference observables.*

By collecting the probabilistic data $D(\mathcal{O}, C)$ for all contexts $C \in \mathcal{C}$ we obtain the collection of data

$$\mathcal{D}(\mathcal{O}, \mathcal{C}) = \bigcup_{C \in \mathcal{C}} D(\mathcal{O}, C)$$

which completely characterizes the contextual statistical model. Thus any model can be symbolically written as

$$M = (\mathcal{C}, \mathcal{O}, \mathcal{D}(\mathcal{O}, \mathcal{C})).$$

We shall consider mainly pairs of dichotomous reference observables: $O' = \{a, b\}$ and $a = \alpha_1, \alpha_2, b = \beta_1, \beta_2$. Here

$$D(a, b, C) = \{\mathbf{P}(a = \alpha|C), \mathbf{P}(b = \beta|C), \mathbf{P}(a = \alpha|C_\beta), \mathbf{P}(b = \beta|C_\alpha)\},$$

where $\alpha = \alpha_1, \alpha_2$ and $\beta = \beta_1, \beta_2$.

3.1.4 Växjö Models Induced by the Kolmogorov Model

Let $\mathcal{P} = (\Omega, \mathcal{F}, \mathbf{P})$ be a Kolmogorov probability space. It induces various Växjö models through various choices of collections of contexts $\mathcal{C}$ and observables $\mathcal{O}$ as well as systems of sets representing selection contexts for values of observables from $\mathcal{O}$.

The collection of contexts $\mathcal{C}$ can be chosen as some sub-family of $\mathcal{F}$ consisting of sets of positive probability: $\mathbf{P}(C) > 0, C \in \mathcal{C}$. The crucial point is that the collection of contexts need not form a σ-algebra or algebra.

The collection of observables $\mathcal{O}$ can be chosen as a subset of the space of random variables $RV(\mathcal{P})$. The crucial point is that not all random variables are chosen as observables.[3] In general $\mathcal{O}$ is only a proper subset of $RV(\mathcal{P})$.

For a discrete variable a, its essential range of values ("spectrum") is given by the set $X_a = \{\alpha\}$, where $\mathbf{P}(\omega \in \Omega : a(\omega) = \alpha) > 0$.

Contextual probabilities in Växjö models induced by the Kolmogorov model are given by the Bayes' formula (so, these are simply conditional probabilities).

For an observable (random variable) a and its value $\alpha \in X_a$ the $[a = \alpha]$-selection context C_α can be chosen as the set $A_\alpha = \{\omega \in \Omega : a(\omega) = \alpha\}$. The condition (3.4) evidently holds. As was mentioned, this condition does not uniquely determine selection contexts. In particular, we can choose any subset $\tilde{A}_\alpha$ of A_α (of strictly positive probability) and still satisfy (3.4). What system of sets should be declared as representing selection contexts depends on the problem under consideration. By

[3] One may say that the Kolmogorov model provides the ontic description and the Växjö model provides the epistemic description [27].

choosing various families of sets as representing selection contexts we create various Växjö models. We remark that the choice of a proper subset $\tilde{A}_\alpha$ of the set A_α can be justified by the physical situation—for example, by *losses of particles* in the process of selection.

Another class of contextual probabilistic models can be induced by the Kolmogorov model via consideration on $\mathcal{F}$ of the equivalence relation given by the condition[4]

$$\mathbf{P}(C_1 \Delta C_2) = 0, \tag{3.5}$$

and on the set $RV(\mathcal{P})$ the standard equivalence relation for random variables: ξ is equivalent to ξ' if

$$\mathbf{P}(\omega \in \Omega : \xi(\omega) \neq \xi'(\omega)) = 0.$$

We now choose contexts and observables as equivalent classes of sets and random variables, respectively. To determine a Växjö model, we fix families $\mathcal{C}$ and $\mathcal{O}$ of corresponding equivalence classes. We remark that contextual probabilities are well defined with the aid of the Bayes' formula:

Let ξ, ξ' be equivalent random variables. It is easy to see that their essential ranges of values coincide $X_\xi = X_{\xi'}$. Let U, U' be equivalent sets (of strictly positive probability).

We set $A_\alpha \equiv A_\alpha^\xi = \{\omega \in \Omega : \xi(\omega) = \alpha\}$ and $A'_\alpha \equiv A_\alpha^{\xi'} = \{\omega \in \Omega : \xi'(\omega) = \alpha\}$. First we remark that $\mathbf{P}(U) = \mathbf{P}(U')$. We also have

$$\begin{aligned}\mathbf{P}(A_\alpha \cap U) &= \mathbf{P}((A_\alpha \cap A'_\alpha \cup A_\alpha \setminus A'_\alpha) \cap (U \cap U' \cup U \setminus U'))\\ &= \mathbf{P}(A_\alpha \cap A'_\alpha \cap U \cap U') = \mathbf{P}(A'_\alpha \cap U').\end{aligned}$$

Hence, $\mathbf{P}(\omega \in \Omega : \xi(\omega) = \alpha | U) = \mathbf{P}(\omega \in \Omega : \xi'(\omega) = \alpha | U')$. Thus $\mathbf{P}(a = \alpha | C)$ does not depend on choices of representatives $\xi, \xi' \in a$ and $U, U' \in C$.

[4] The symmetric difference of two sets is defined by $C_1 \Delta C_2 = (C_1 \setminus C_2) \cup (C_2 \setminus C_1)$.

Let a be an observable—an equivalence class of random variables. For any $\alpha \in X_a$, the selection context can be chosen (by such a choice we fix the model) by the equivalence class of sets A_α^ξ for $\xi \in a$. To show this, it is enough to see that $\mathbf{P}(A_\alpha \Delta A'_\alpha) = 0$ for $\xi, \xi' \in a$.

3.1.5 Växjö Models Induced by the Quantum Model

The set of contexts $\mathscr{C}$ can be chosen as a subset of the unit sphere S of complex Hilbert space $\mathscr{H}$: each context $C \in \mathscr{C}$ is encoded by a vector $\psi \in S : C \equiv C_\psi$. The set of observables $\mathscr{O}$ can be chosen as a subset of the space of self-adjoint operators $L_s(\mathscr{H})$ having purely discrete[5] nondegenerate[6] spectra.

Contextual probabilities are defined by Born's rule. Let an operator $\hat{a} \in \mathscr{O}$ have the spectrum $X_a = \{\alpha_1, \ldots, \alpha_N, \ldots\}, \alpha_i \neq \alpha_j$ and let $e_\alpha^a, \alpha \in X_a$, be corresponding eigenvectors. Then

$$\mathbf{P}(a = \alpha_i | C_\psi) = |\langle \psi, e_{\alpha_i}^a \rangle|^2.$$

The $[a = \alpha]$-selection contexts C_α are represented by the eigenvectors: $C_\alpha \equiv C_{e_\alpha^a}$. We have

$$\mathbf{P}(b = \beta | C_\alpha) = \mathbf{P}(a = \alpha | C_\beta) = |\langle e_\alpha^a, e_\beta^b \rangle|^2.$$

We point out that transition probabilities in the Växjö model induced by QM are very special. They are symmetric. In general the Växjö model produces nonsymmetric transition probabilities (already the Växjö model induced by the Kolmogorov model).

---

[5] We recall that at the moment we have defined the Växjö model only for discrete observables, see Chap. 6 for the "continuous generalization."

[6] We consider nondegenerate spectra to escape the problem of nonunique choice of selection contexts.

Another possibility to describe quantum probabilities within the contextual probabilistic model is to proceed (similarly to the Kolmogorov case) by representing contexts not by single vectors from the unit sphere S, but by equivalence classes of these vectors: ψ_1 is equivalent ψ_2 if $\psi_1 = c\psi_2$, where $|c| = 1$.

3.1.6 Växjö Models Induced by the von Mises Model

Consider a Växjö model (contextual probabilistic model) $M = (\mathcal{C}, \mathcal{O}, \mathcal{D}(\mathcal{O}, \mathcal{C}))$. We would like to realize contextual probabilities of this model, $\pi = \{p_C^a\}_{C \in \mathcal{C}, a \in \mathcal{O}}$, in the frequency probabilistic framework.

Let $C \in \mathcal{C}$ be some context and let $a, b \in \mathcal{O}$ be two arbitrary observables. In a series of observations of b under this context[7] we obtain a sequence of values of b

$$x \equiv x(b|C) = (x_1, x_2, \ldots, x_N, \ldots), \quad x_j \in X_b. \tag{3.6}$$

In the same way we obtain a sequence of values of a

$$y \equiv y(a|C) = (y_1, y_2, \ldots, y_N, \ldots), \quad y_j \in X_a. \tag{3.7}$$

We suppose that these are S-sequences, see Definition 2.1, Part I: Chap. 2 (or even von Mises collectives). Thus the principle of statistical stabilization holds and the frequency probabilities are well defined

$$p^b(\beta) = \lim_{N\to\infty} \nu_N(\beta; x), \quad \beta \in X_b, \tag{3.8}$$

$$p^a(\alpha) = \lim_{N\to\infty} \nu_N(\alpha; y), \quad \alpha \in X_a, \tag{3.9}$$

where $\nu_N(\beta; x) = \frac{n(\beta;x)}{N}$, $\nu_N(\alpha; x) = \frac{n(\alpha;x)}{N}$ are relative frequencies of realizations of labels $b = \beta$ and $a = \alpha$ in sequences of observations x and y, respectively

[7] Any context should be repeatable infinitely many times.

(relative frequencies of observations of the results $b = \beta$ and $a = \alpha$ under the context C).

We define contextual probabilities of the Växjö model M as frequency probabilities $p_C^b(\beta) = p^b(\beta)$ and $p_C^a(\alpha) = p^a(\alpha)$.

We remark that it is not assumed that the observables a and b can be measured simultaneously.

Let C_α be $[a = \alpha]$-selection, $\alpha \in X_a$. By observation of b under the context C_α we obtain a sequence

$$x^\alpha \equiv x(b|C_\alpha) = (x_1, x_2, \dots, x_N, \dots), \quad x_j \in X_b. \tag{3.10}$$

It is also assumed that the x^α ($\alpha \in X_a$) are von Mises collectives or at least S-sequences. Thus the frequency probabilities with respect to the x^α are well defined

$$p^{b|a}(\beta|\alpha) = \lim_{N\to\infty} \nu_N(\beta; x^\alpha), \tag{3.11}$$

where $\nu_N(\beta; x^\alpha) = \frac{n(\beta; x^\alpha)}{N}$ are relative frequencies of realizations of the label β in x^α (relative frequencies of observations of the result $b = \beta$ under the context C_α).

One should consider a few different S-sequences (collectives) $x, y, x^\alpha, \alpha \in X_a$, which produce probability distributions

$$p_C^b(\beta), \, p_C^a(\alpha), \, p^{b|a}(\beta|\alpha). \tag{3.12}$$

3.2 Contextual Probabilistic Description of Double Slit Experiment

In this section we analyze the well-known double slit experiment from the contextual probabilistic viewpoint. Denote by G a source of light, by U_{12} a screen with two open slits, by U_j the same screen with only the jth open slit, $j = 1, 2$; by W a screen covered by photoemulsion. We consider the following contexts

$$C_{12} = \{G, U_{12}, W\}, \qquad C_j = \{G, U_j, W\}, \quad j = 1, 2,$$

We also consider observables:

b is the "momentum observable." It is given by the position of a dot appearing on the registration screen W. This position determines the direction of the photon's velocity.

a is the "position observable." It is realized as the slit observable. To measure a, we place two detectors $d_j, j = 1, 2$ directly after slits. If d_j produces a click then $a = j$. If light has very low intensity, then two detectors practically never produce clicks simultaneously.

Thus we consider the set of contexts $\mathscr{C} = \{C_{12}, C_1, C_2\}$ and the set of observables $\mathscr{O} = \{a, b\}$. The context C_j is selection with respect to the value $a = j$ of the a-observable: if we measure position a under the context C_j (i.e., only the jth slit is open) then we get the value $a = j$ with probability 1.

We consider the dichotomous version of the momentum observable. We choose some domain on the registration screen and put $b = 1$ if a dot appears inside this domain and $b = 0$ if a dot appears outside this domain. We consider the following S-sequences (collectives):

$x \equiv x(b|C_{12})$—results of the b-measurement when both slits are open;

$y \equiv y(a|C_{12})$—results of the a-measurement when both slits are open;

$x^j \equiv x(b|C_j)$— results of the b-measurement when only the jth slit is open, $j = 1, 2$.

We obtain the probabilistic data

$$\mathbf{P}(b = i|C_{12}) = \mathbf{P}_x(b = i), \qquad \mathbf{P}(a = j|C_{12}) = \mathbf{P}_y(a = j),$$

$$p^{b|a}(i|j) \equiv \mathbf{P}(b = i|C_j) = \mathbf{P}_{x^j}(b = i).$$

3.3 Formula of Total Probability and Measures of Supplementarity

Consider a Växjö model $M = (\mathscr{C}, \mathscr{O}, \mathscr{D}(\mathscr{O}, \mathscr{C}))$. Let $C \in \mathscr{C}$ be some context and let $a, b \in \mathscr{O}$ be two arbitrary observables. For simplicity we assume that they are *dichotomous.*

There are no reasons to assume the classical (Kolmogorovian) formula of total probability, Part I: Chap. 2:

$$\mathbf{P}(b = \beta) = \sum_{\alpha} \mathbf{P}(a = \alpha)\mathbf{P}(b = \beta|a = \alpha) \tag{3.13}$$

would be true for an arbitrary contextual probabilistic model. In principle, it can be violated. We explain this point in more detail. In fact, all probabilities in (3.13) are contextual. In (3.13) we omitted the indexes of contexts. However, in reality three different contexts were involved. A context C is chosen for observations of a or b. The following contextual probabilities are given: $\mathbf{P}(a = \alpha) \equiv \mathbf{P}(a = \alpha|C), \mathbf{P}(b = \beta) \equiv \mathbf{P}(b = \beta|C)$. The selection contexts $C_\alpha, \alpha \in X_a$, are also involved. The correct contextual definition of conditional probabilities in (3.13) is given by

$$\mathbf{P}(b = \beta|a = \alpha) = \mathbf{P}(b = \beta|C_\alpha).$$

In general there are no reasons to assume that probabilities with respect to the different contexts C, C_α should match the Kolmogorovian probabilistic law given by the formula of total probability.

In the von Mises framework the formula of total probability holds for a partition $\{A_k\}$ of the label set L of a fixed S-sequence (collective) u

$$\mathbf{P}_u(b = \beta) = \sum_{\alpha} \mathbf{P}_u(a = \alpha)\mathbf{P}_u(b = \beta|a = \alpha). \tag{3.14}$$

This formula can not be derived in the contextual probabilistic model with frequency probabilities, where the conditional probabilities $\mathbf{P}(b = \beta|a = \alpha)$ are defined as contextual probabilities $\mathbf{P}(b = \beta|C_\alpha)$. In this (contextual) approach different S-sequences (collectives) are involved

$$x = x(b|C), \qquad y = y(a|C), \qquad x^\alpha = x(b|C_\alpha), \quad \alpha \in X_a.$$

Thus, by using the contextual probabilistic model, in general one could not exclude the possibility that the following probabilistic coefficient

$$\delta(\beta|a, C) = \mathbf{P}_x(\beta) - \sum_\alpha \mathbf{P}_y(\alpha)\mathbf{P}_{x^\alpha}(\beta) \neq 0. \tag{3.15}$$

As was mentioned, in the Kolmogorov or von Mises models, see (3.13), (3.14), we have

$$\delta(\beta|a, C) = 0. \tag{3.16}$$

Definition 3.4. *The quantity $\delta(\beta|a, C)$ is said to be a probabilistic measure of $b|a$-supplementarity in the context C.*

We can write the equality (3.15) in a form which is similar to the classical formula of total probability

$$\mathbf{P}_x(\beta) = \sum_\alpha \mathbf{P}_y(\alpha)\mathbf{P}_{x^\alpha}(\beta) + \delta(\beta|a, C), \tag{3.17}$$

or by using shorter notation

$$p^b(\beta) = \sum_\alpha p^a(\alpha) p^{b|a}(\beta|\alpha) + \delta(\beta|a, C). \tag{3.18}$$

This formula has the same structure as the quantum formula of total probability:

[classical part] + additional term,

cf. (3.2). To write the additional term in the same form as in the quantum representation of statistical data, we perform the normalization of the probabilistic measure of supplementarity by the square root of the product of all probabilities

$$\lambda(\beta|a, C) = \frac{\delta(\beta|a, C)}{2\sqrt{\prod_\alpha p^a(\alpha) p^{b|a}(\beta|\alpha)}}. \tag{3.19}$$

The coefficient $\lambda(\beta|a, C)$ also will be called the probabilistic measure of supplementarity.

By using this coefficient we rewrite (3.18) in the QL form

$$p^b(\beta) = \sum_\alpha p^a(\alpha) p^{b|a}(\beta|\alpha) + 2\lambda(\beta|a, C)\sqrt{\prod_\alpha p^a(\alpha) p^{b|a}(\beta|\alpha)}. \tag{3.20}$$

The coefficient $\lambda(\beta|a, C)$ is well defined only in the case when all probabilities $p^a(\alpha)$, $p^{b|a}(\beta|\alpha)$ are strictly positive. We consider the matrix of transition probabilities $\mathbf{P}^{b|a} = (p^{b|a}(\beta|\alpha))$. We remark that the matrix $\mathbf{P}^{b|a}$ is always *stochastic*

$$\sum_\beta p^{b|a}(\beta|\alpha) = 1 \tag{3.21}$$

for any $\alpha \in X_a$.

Definition 3.5. *A context C is said to be a-nondegenerate if*

$$\mathbf{P}(a = \alpha|C) > 0$$

for all $\alpha \in X_a$.

We remark that the context C_α^a is b-nondegenerate iff

$$p^{b|a}(\beta|\alpha) \neq 0, \quad \beta \in X_b. \tag{3.22}$$

The representation (3.20) can be used only for nondegenerate contexts C and C_α^a.

We can repeat all previous considerations by changing $b|a$-conditioning to $a|b$-conditioning. We consider contexts C_β^b corresponding to selections with respect to values of the observable b. Probabilistic measures of supplementarity $\delta(\alpha|b, C)$ and $\lambda(\alpha|b, C), \alpha \in X_a$, can be defined. The contexts C_β^b are a-nondegenerate iff

$$p^{a|b}(\alpha|\beta) \neq 0, \quad \alpha \in X_a. \tag{3.23}$$

For nondegenerate contexts C and C_β^b, we have

$$p^a(\alpha) = \sum_\beta p^b(\beta) p^{a|b}(\alpha|\beta) + 2\lambda(\alpha|b, C) \sqrt{\prod_\beta p^b(\beta) p^{a|b}(\alpha|\beta)}. \tag{3.24}$$

Definition 3.6. *Observables a and b are called probabilistically conjugate if* (3.22) *and* (3.23) *hold.*

Theorem 3.1. *Let observables be probabilistically conjugate and let a context $C \in \mathscr{C}$ be both a- and b-nondegenerate. Then quantum-like formulas of total probability* (3.20) *and* (3.24) *hold.*

3.4 Supplementary Observables

Definition 3.7. *Observables a and b are called $b|a$-supplementary in a context C if*

$$\delta(\beta|a, C) \neq 0 \quad \textit{for some } \beta \in X_b. \tag{3.25}$$

Lemma 3.1. *For any context $C \in \mathscr{C}$, we have*

$$\sum_{\beta \in X_b} \delta(\beta|a, C) = 0. \tag{3.26}$$

Proof. We have

$$1 = \sum_{\beta \in X_b} p^b(\beta) = \sum_{\beta \in X_b} \sum_{\alpha \in X_a} p^a(\alpha) p^{b|a}(\beta|\alpha) + \sum_{\beta \in X_b} \delta(\beta|a, C).$$

Since $\mathbf{P}^{b|a}$ is always a stochastic matrix, we have for any $\alpha \in X_a$: $\sum_{\beta \in X_b} p^{b|a}(\beta|\alpha) = 1$. By using that $\sum_{\alpha \in X_a} p^a(\alpha) = 1$ we obtain (3.26).

We point out that by Lemma 3.1 the coefficient $\delta(\beta_1|a, C) = 0$ iff $\delta(\beta_2|a, C) = 0$. Thus $b|a$-supplementarity is equivalent to the condition $\delta(\beta|a, C) \neq 0$ both for β_1 and β_2.

Definition 3.8. *Observables a and b are called supplementary in a context C if they are $b|a$ or $a|b$ supplementary*

$$\delta(\beta|a, C) \neq 0 \quad \textit{or} \quad \delta(\alpha|b, C) \neq 0 \quad \textit{for some } \beta \in X_b, \ \alpha \in X_a. \tag{3.27}$$

By Lemma 3.1. observables are supplementary iff the coefficient $\delta(\beta|a, C) \neq 0$ for all $\beta \in X_b$ or the coefficient $\delta(\alpha|b, C) \neq 0$ for all $\alpha \in X_a$.

Let us consider a contextual probabilistic model with the set of contexts $\mathscr{C}$. Observables a and b are said to be supplementary in this model if there exists $C \in \mathscr{C}$ such that they are supplementary in the context C.

Observables a and b are called *nonsupplementary* in the context C if they are neither $b|a$ nor $a|b$-supplementary

$$\delta(\beta|a, C) = 0 \quad \text{and} \quad \delta(\alpha|b, C) = 0 \quad \text{for all } \beta \in X_b, \alpha \in X_a. \tag{3.28}$$

Thus in the case of $b|a$-supplementarity we have (for $\beta \in X_b$)

$$p^b(\beta) \neq \sum_\alpha p^a(\alpha) p^{b|a}(\beta|\alpha); \tag{3.29}$$

in the case of $a|b$-supplementarity we have (for $\alpha \in X_a$)

$$p^a(\alpha) \neq \sum_\beta p^b(\beta) p^{a|b}(\alpha|\beta); \tag{3.30}$$

in the case of supplementarity we have (3.29) or (3.30). In the case of nonsupplementarity we have both representations

$$p^b(\beta) = \sum_\alpha p^a(\alpha) p^{b|a}(\beta|\alpha), \quad \beta \in X_b, \tag{3.31}$$

$$p^a(\alpha) = \sum_\beta p^b(\beta) p^{a|b}(\alpha|\beta), \quad \alpha \in X_a. \tag{3.32}$$

3.5 Principle of Supplementarity

After the careful study of Bohr's views on complementarity and especially discussions with A. Plotnitsky (see also his works on Bohr's complementarity [269–272]), I found that by complementarity N. Bohr understood complementarity based on *mu-*

tual exclusivity. In the Växjö approach, mutual exclusivity of experimental conditions is not important. The crucial role is played by supplementarity of information in the sense of "additional information."

The Principle of Supplementarity:

There exist physical observables, say a and b, such that for some context C they produce supplementary statistical information. The classical formula of total probability is violated. Supplementarity of the observables a and b under the context C induces interference of probabilities $\mathbf{P}(b = \beta|C)$ and $\mathbf{P}(a = \alpha|C)$.

3.6 Supplementarity and Kolmogorovness

Let us consider a Växjö model and let $a, b \in \mathcal{O}$ be two dichotomous observables.

Definition 3.9. *Probabilistic data $D(a, b, C)$ is said to be Kolmogorovian if there exists a Kolmogorov probability space $\mathcal{P} = (\Omega, \mathcal{F}, \mathbf{P})$ and random variables ξ_a and ξ_b on $\mathcal{P}$ such that*

$$p^a(\alpha) = \mathbf{P}(\xi_a = \alpha), \qquad p^b(\beta) = \mathbf{P}(\xi_b = \beta); \tag{3.33}$$

$$p^{b|a}(\beta|\alpha) = \mathbf{P}(\xi_b = \beta|\xi_a = \alpha), \qquad p^{a|b}(\alpha|\beta) = \mathbf{P}(\xi_a = \alpha|\xi_b = \beta). \tag{3.34}$$

Here the conditional probabilities are defined by the Bayes formula.

If data $D(a, b, C)$ is Kolmogorovian, then the observables a and b can be represented by Kolmogorovian random variables ξ_a and ξ_b. We remark that Kolmogorovness of statistical data in the sense of Definition 3.9 is not so natural from the physical viewpoint. In fact, probabilities p^a, p^b, $p^{b|a}$, $p^{a|b}$ correspond to different complexes of physical conditions (contexts) C, C_α, C_β. It would be more natural to assume that each context determines its own Kolmogorov probability measure. Nevertheless, Kolmogorovian data appear in many models, e.g., in classical statistical physics.

Lemma 3.2. *Data $D(a, b, C)$ is Kolmogorovian if and only if*

$$p^a(\alpha)p^{b|a}(\beta|\alpha) = p^b(\beta)p^{a|b}(\alpha|\beta). \tag{3.35}$$

Proof. (a) If data $D(a, b, C)$ is Kolmogorovian then (3.35) is reduced to the equality $\mathbf{P}(O_1 \cap O_2) = \mathbf{P}(O_2 \cap O_1)$ for $O_1, O_2 \in \mathscr{F}$.

(b) Let (3.35) hold true. We set $\Omega = X_a \times X_b$, where $X_a = \{\alpha_1, \alpha_2\}$, $X_b = \{\beta_1, \beta_2\}$. We define the probability distribution on Ω by

$$\mathbf{P}(\alpha, \beta) = p^b(\beta)p^{a|b}(\alpha|\beta) = p^a(\alpha)p^{b|a}(\beta|\alpha);$$

and define the random variables $\xi_a(\omega) = \alpha$, $\xi_b(\omega) = \beta$ for $\omega = (\alpha, \beta)$. We have

$$\begin{aligned}\mathbf{P}(a = \alpha) &= \sum_\beta \mathbf{P}(\alpha, \beta) = \sum_\beta p^a(\alpha)p^{b|a}(\beta|\alpha) \\ &= p^a(\alpha)\sum_\beta p^{b|a}(\beta|\alpha) = p^a(\alpha).\end{aligned}$$

And in the same way $\mathbf{P}(b = \beta) = p^b(\beta)$. Thus

$$\mathbf{P}(a = \alpha|b = \beta) = \frac{\mathbf{P}(a = \alpha, b = \beta)}{\mathbf{P}(b = \beta)} = \frac{p^b(\beta)p^{a|b}(\alpha|\beta)}{p^b(\beta)} = p^{a|b}(\alpha|\beta).$$

And in the same way we prove that $p^{b|a}(\beta|\alpha) = \mathbf{P}(b = \beta|a = \alpha)$.

We now investigate the relation between Kolmogorovness and nonsupplementarity. If $D(a, b, C)$ is Kolmogorovian, then the formula of total probability holds true and we have (3.28). Thus observables a and b are nonsupplementary (in the context C). Thus:

Kolmogorovness implies nonsupplementarity

or as we also can say

Supplementarity implies non-Kolmogorovness.

However, in the general case nonsupplementarity does not imply that probabilistic data $D(a, b, C)$ is Kolmogorovian. Let us investigate in more detail the case when both matrices $\mathbf{P}^{a|b}$ and $\mathbf{P}^{b|a}$ are *double stochastic*. We recall that a matrix $\mathbf{P}^{b|a} = (p^{b|a}(\beta|\alpha))$ is double stochastic if it is stochastic (so (3.21) holds) and, moreover,

$$\sum_{\alpha} p^{b|a}(\beta|\alpha) = 1, \quad \beta = \beta_1, \beta_2. \tag{3.36}$$

3.6.1 Double Stochasticity as the Law of Probabilistic Balance

As was mentioned, the equality (3.21) holds automatically. This is a consequence of additivity and normalization by 1 of the probability distribution.

However, the equality (3.36) is an additional condition on the observables a and b. Thus by *considering double stochastic matrices we choose a very special pair of observables.*

I propose the following physical interpretation of the equality (3.21). Since

$$p^{b|a}(\beta|\alpha_2) = 1 - p^{b|a}(\beta|\alpha_1),$$

the C_{α_1} and C_{α_2} contexts compensate each other in "preparation of the property" $b = \beta$. Thus (3.36) could be interpreted as the *law of probabilistic balance* for the property $b = \beta$. If both matrices $\mathbf{P}^{b|a}$ and $\mathbf{P}^{a|b}$ are double stochastic, then we have laws of probabilistic balance for both properties: $a = \alpha$ and $b = \beta$.

3.6.2 Probabilistically Balanced Observables

Definition 3.10. *Observables a and b are said to be probabilistically balanced if both matrices $\mathbf{P}^{b|a}$ and $\mathbf{P}^{a|b}$ are double stochastic.*

It is useful to recall the following well-known result about double stochasticity for Kolmogorovian random variables:

Lemma 3.3. *Let* ξ_a *and* ξ_b *be random variables on a Kolmogorov space* $\mathscr{P} = (\Omega, \mathscr{F}, \mathbf{P})$. *Then the following conditions are equivalent:*

(1) *The matrices* $\mathbf{P}^{a|b} = (\mathbf{P}(\xi_a = \alpha|\xi_b = \beta))$, $\mathbf{P}^{b|a} = (\mathbf{P}(\xi_b = \beta|\xi_a = \alpha))$ *are double stochastic.*

(2) *Random variables are uniformly distributed*

$$\mathbf{P}(\xi_a = \alpha) = \mathbf{P}(\xi_b = \beta) = \frac{1}{2}.$$

(3) *Random variables are "symmetrically conditioned" in the following sense*

$$\mathbf{P}(\xi_a = \alpha|\xi_b = \beta) = \mathbf{P}(\xi_b = \beta|\xi_a = \alpha), \tag{3.37}$$

so $\mathbf{P}^{a|b} = \mathbf{P}^{b|a}$.

Proof. We set $A_i = \{\omega \in \Omega : \xi_a(\omega) = \alpha_i\}$ and $B_j = \{\omega \in \Omega : \xi_b(\omega) = \beta_j\}$, $j = 1, 2$ (we recall that we consider dichotomous random variables). First we prove that (3) is equivalent (2):

(a1) Let $\mathbf{P}(A_i|B_j) = \mathbf{P}(B_j|A_i)$. Then

$$\frac{\mathbf{P}(A_1B_1)}{\mathbf{P}(B_1)} = \frac{\mathbf{P}(B_1A_1)}{\mathbf{P}(A_1)}, \quad \frac{\mathbf{P}(A_2B_2)}{\mathbf{P}(B_2)} = \frac{\mathbf{P}(B_2A_2)}{\mathbf{P}(A_2)}, \quad \frac{\mathbf{P}(A_1B_2)}{\mathbf{P}(B_2)} = \frac{\mathbf{P}(B_2A_1)}{\mathbf{P}(A_1)}. \tag{3.38}$$

Thus we obtain $\mathbf{P}(B_1) = \mathbf{P}(A_1), \mathbf{P}(A_2) = \mathbf{P}(B_2), \mathbf{P}(B_2) = \mathbf{P}(A_1)$. Thus

$$\mathbf{P}(B_1) = \mathbf{P}(B_2) = 1/2 \quad \text{and} \quad \mathbf{P}(A_1) = \mathbf{P}(A_2) = 1/2. \tag{3.39}$$

(b1) Starting with (3.39) we obtain (3.38) and consequently a, b-symmetry of transition probabilities.

We now prove that (1) is equivalent (2):

(a2) Let $\mathbf{P}(B_1|A_1) = \mathbf{P}(B_2|A_2), \mathbf{P}(B_1|A_2) = \mathbf{P}(B_2|A_1)$. Then

$$\frac{\mathbf{P}(B_1A_1)}{\mathbf{P}(A_1)} = \frac{\mathbf{P}(B_2A_2)}{\mathbf{P}(A_2)}, \qquad \frac{\mathbf{P}(B_1A_2)}{\mathbf{P}(A_2)} = \frac{\mathbf{P}(B_2A_1)}{\mathbf{P}(A_1)}; \tag{3.40}$$

$$\frac{\mathbf{P}(A_1B_1)}{\mathbf{P}(B_1)} = \frac{\mathbf{P}(A_2B_2)}{\mathbf{P}(B_2)}, \qquad \frac{\mathbf{P}(A_1B_2)}{\mathbf{P}(B_2)} = \frac{\mathbf{P}(A_2B_1)}{\mathbf{P}(B_1)}. \tag{3.41}$$

By these equations we obtain

$$\frac{\mathbf{P}(B_1)}{\mathbf{P}(A_1)} = \frac{\mathbf{P}(B_2)}{\mathbf{P}(A_2)}, \qquad \frac{\mathbf{P}(B_2)}{\mathbf{P}(A_1)} = \frac{\mathbf{P}(B_1)}{\mathbf{P}(A_2)}.$$

So $\frac{\mathbf{P}(A_2)}{\mathbf{P}(A_1)} = \frac{\mathbf{P}(A_1)}{\mathbf{P}(A_2)}$. Thus we obtain (3.39).

(b2) Let (3.39) hold true. Then we have already proved that transition probabilities are symmetric. Thus

$$\mathbf{P}(B_i|A_1) + \mathbf{P}(B_i|A_2) = \mathbf{P}(A_1|B_i) + \mathbf{P}(A_2|B_i) = 1$$

(since every matrix of transition probabilities is always stochastic).

In general the Kolmogorovian characterization of probabilistically balanced random variables is not valid for observables of the Växjö model.

Proposition 3.1. *A Kolmogorov model for data $D(a, b, C)$ need not exist even in the case of nonsupplementary probabilistically balanced observables having the uniform probability distribution (for the context C).*

Proof. Let us consider probabilistic data $D(a, b, C)$ such that $p^a(\alpha) = p^b(\beta) = 1/2$ (here $p^a(\alpha) \equiv \mathbf{P}(a = \alpha|C)$, $p^b(\beta) \equiv \mathbf{P}(b = \beta|C)$) and both matrices $\mathbf{P}^{a|b}$ and $\mathbf{P}^{b|a}$ are double stochastic. Let us assume that $p^{a|b}(\alpha|\beta) \neq p^{b|a}(\beta|\alpha)$. Then by Lemma 3.2, data $D(a, b, C)$ is non-Kolmogorovian, but

$$2\delta(\alpha|\beta, C) = 1 - \sum_{\beta} p^{a|b}(\alpha|\beta) = 0, \qquad 2\delta(\beta|\alpha, C) = 1 - \sum_{\alpha} p^{b|a}(\beta|\alpha) = 0.$$

It seems to be that symmetrical conditioning plays a crucial role in these considerations.

3.6.3 Symmetrically Conditioned Observables

Take an arbitrary Växjö model.

Definition 3.11. *Observables $a, b \in \mathcal{O}$ are called symmetrically conditioned if*

$$p^{a|b}(\alpha|\beta) = p^{b|a}(\beta|\alpha). \tag{3.42}$$

Lemma 3.4. *If observables a and b are symmetrically conditioned, then they are probabilistically balanced (so the matrices $\mathbf{P}^{a|b}$ and $\mathbf{P}^{b|a}$ are double stochastic).*

Proof. We have that, e.g., $\sum_\beta p^{a|b}(\alpha|\beta) = \sum_\beta p^{b|a}(\beta|\alpha) = 1$.

As we have seen in Proposition 3.1, probabilistically balanced observables need not be symmetrically conditioned, cf. Lemma 3.3.

Proposition 3.2. *Let observables a and b be symmetrically conditioned. Probabilistic data $D(a, b, C)$ is Kolmogorovian iff the observables a and b are nonsupplementary in the context C.*

Proof. Suppose that a and b are nonsupplementary. We set

$$p^{b|a}(1|1) = p^{b|a}(2|2) = p \quad \text{and} \quad p^{b|a}(1|2) = p^{b|a}(2|1) = 1 - p$$

(we recall that by Lemma 3.4 the matrix $\mathbf{P}^{b|a}$ is double stochastic). By (3.31), (3.32) we have

$$\begin{aligned} p^a(\alpha_i) &= \sum_\beta p^b(\beta) p^{a|b}(\alpha_i|\beta) = \sum_\beta \sum_\alpha p^a(\alpha) p^{b|a}(\beta|\alpha) p^{a|b}(\alpha_i|\beta) \\ &= \sum_\alpha p^a(\alpha) \sum_\beta p^{b|a}(\beta|\alpha) p^{b|a}(\beta|\alpha_i). \end{aligned}$$

Let us consider the case $i = 1$:

$$\begin{aligned} p^a(\alpha_1) &= p^a(\alpha_1)(p^2 + (1-p)^2) + 2p^a(\alpha_2)p(1-p) \\ &= p^a(\alpha_1)(1 - 4p + 4p^2) + 2p(1-p). \end{aligned}$$

Thus $p^a(\alpha_1) = 1/2$. Hence $p^a(\alpha_2) = 1/2$. In the same way we get that $p^b(\beta_1) = p^b(\beta_2) = 1/2$. Thus the condition (3.35) holds true and there exists a Kolmogorov model $\mathcal{P} = (\Omega, \mathcal{F}, \mathbf{P})$ for probabilistic data $D(a, b, C)$.

Conclusion. In the case of symmetrical conditioning Kolmogorovness is equivalent to nonsupplementarity.

Corollary 3.1. *For symmetrically conditioned observables, probabilistic data $D(a, b, C)$ is Kolmogorovian iff the observables a and b are uniformly distributed*

$$p^a(\alpha_1) = p^a(\alpha_2) = 1/2; \qquad p^b(\beta_1) = p^b(\beta_2) = 1/2.$$

3.7 Incompatibility, Supplementarity and Existence of Joint Probability Distribution

The notions of incompatible and complementary variables are considered as synonymous in the Copenhagen quantum mechanics. Moreover, compatibility (and consequently noncomplementarity) is considered as equivalent to existence of the joint probability distribution

$$\mathbf{P}(\alpha, \beta) = \mathbf{P}(a = \alpha, b = \beta).$$

The joint probability distribution can be interpreted as the probability distribution for simultaneous measurement. The latter is always possible for compatible observables (represented by commutative operators).

We now consider similar questions for our Växjö model (as was pointed out, we use the notion of *supplementarity*, instead of *complementarity*, since the latter has already been reserved by N. Bohr).

3.7.1 Joint Probability Distribution

The frequency definition of probability is the most appropriate for considerations in this section.

By definition, observables a and b are compatible in a context C if it is possible to perform a simultaneous observation of them under C. For any instant of time t, a pair of values $z(t) = (a(t), b(t))$ can be observed. The sequence of results of observations is well defined

$$z(a, b|C) = (z_1, z_2, \ldots, z_N, \ldots), \quad z_j = (y_j, x_j),$$

where $y_j = \alpha_1$ or α_2 and $x_j = \beta_1$ or β_2. Observables a and b are incompatible in a context C if it is impossible to perform a simultaneous observation of them under C.

Definition 3.12. *Observables a and b are probabilistically compatible in a context C if they are compatible in C and collectives (or S-sequences) $y = y(a|C)$ and $x = x(b|C)$ are combinable, see Part* I: *Chap.* 2.

Observables a and b are probabilistically compatible iff the sequence $z(a, b|C)$ is a collective (or an S-sequence).[8] Thus there exists the frequency simultaneous probability distribution

$$\mathbf{P}(\alpha, \beta) \equiv \mathbf{P}_z(\alpha, \beta) = \lim_{N \to \infty} \frac{n_N(\alpha, \beta; z)}{N}. \tag{3.43}$$

Probabilistic compatibility implies Kolmogorovness of data $D(a, b, C)$ and hence nonsupplementarity.

[8] In the opposite case observables are probabilistically incompatible.

3.7.2 Compatibility and Probabilistic Compatibility

In quantum physics compatibility of observables—the possibility to perform a simultaneous observation—is typically identified with probabilistic compatibility—existence of the frequency simultaneous probability distribution (3.43). It is a natural consequence of the Kolmogorovian psychology. In the frequency probability theory we should distinguish compatibility and probabilistic compatibility.

We present an example in which observables are compatible (so they can be measured simultaneously), but the limit (3.43) does not exist. Here observables are not probabilistically compatible. Of course, this means that S-sequences (collectives) $y = y(a|C)$ and $x = x(b|C)$ are not combinable.

Differences between compatibility and probabilistic compatibility will play an important role in analysis of Bell's considerations, Part III. Bell did not differ between these two notions. Such a position induced misunderstanding of the interrelation between the possibility to use the realistic description for quantum correlations and violation of Bell's inequality.

We shall use some well-known results about the generalized probability given by the density of natural numbers, see [139]. For a subset $A \subset \mathbf{N}$, where $\mathbf{N}$ is the set of natural numbers, the quantity

$$\mathbf{P}(A) = \lim_{N \to \infty} \frac{|A \cap \{1, \ldots, N\}|}{N}$$

is called the *density* of A if the limit exists. Here the symbol $|V|$ is used to denote the number of elements in the finite set V.

Let $\mathcal{G}$ denote the collection of all subsets of $\mathbf{N}$ which admit density. It is evident that each finite $A \subset \mathbf{N}$ belongs to $\mathcal{G}$ and $\mathbf{P}(A) = 0$. It is also evident that each subset $B = \mathbf{N} \setminus A$, where A is finite, belongs to $\mathcal{G}$ and $\mathbf{P}(B) = 1$ (in particular, $\mathbf{P}(\mathbf{N}) = 1$). The reader can easily find examples of sets $A \in \mathcal{G}$ such that $0 < \mathbf{P}(A) < 1$. The "generalized probability" $\mathbf{P}$ has the following properties:

Proposition 3.3. *Let $A_1, A_2 \in \mathcal{G}$ and $A_1 \cap A_2 = \emptyset$. Then $A_1 \cup A_2 \in \mathcal{G}$ and*

$$\mathbf{P}(A_1 \cup A_2) = \mathbf{P}(A_1) + \mathbf{P}(A_2).$$

Proposition 3.4. *Let $A_1, A_2 \in \mathcal{G}$. The following conditions are equivalent*

$$(1) \quad A_1 \cup A_2 \in \mathcal{G}; \qquad (2) \quad A_1 \cap A_2 \in \mathcal{G};$$

$$(3) \quad A_1 \setminus A_2 \in \mathcal{G}; \qquad (4) \quad A_2 \setminus A_1 \in \mathcal{G}.$$

There are standard formulas

$$\mathbf{P}(A_1 \cup A_2) = \mathbf{P}(A_1) + \mathbf{P}(A_2) - \mathbf{P}(A_1 \cap A_2);$$

$$\mathbf{P}(A_1 \setminus A_2) = \mathbf{P}(A_1) - \mathbf{P}(A_1 \cap A_2).$$

It is possible to find sets $A, B \in \mathcal{G}$ such that, for example, $A \cap B \notin \mathcal{G}$. Let A be the set of even numbers. Take any subset $C \subset A$ which has no density. In fact, you can find C such that

$$\frac{1}{N}|C \cap \{1, 2, \ldots, N\}|$$

is oscillating. There happen two cases: $C \cap \{2n\} = \{2n\}$ or $= \emptyset$. Set

$$B = C \cup \{2n - 1 : C \cap \{2n\} = \emptyset\}.$$

Then, both A and B have densities $1/2$. But $A \cap B = C$ has no density. Thus $\mathcal{G}$ is not a set algebra.

We now consider a context C which produces natural numbers. We introduce two dichotomous observables

$$a(n) = I_A(n), \qquad b(n) = I_B(n),$$

where $I_O(x)$ is the characteristic function of a set O. We assume that these observables are compatible: we can, e.g., look at a number n and find both values $a(n)$ and

$b(n)$.[9] We obtain two S-sequences

$$y = y(a|C) = (y_1, \ldots, y_N, \ldots),$$

$$x = x(b|C) = (x_1, \ldots, x_N, \ldots), \quad y_j, x_j = 0, 1.$$

The frequency probability distributions are well defined

$$p^a(\alpha) \equiv \mathbf{P}_y(\alpha) = 1/2, \qquad p^b(\beta) \equiv \mathbf{P}_x(\beta) = 1/2.$$

However, the S-sequences y and x are not combinable. Thus observables a and b are not probabilistically compatible; for example, the frequency probability $\mathbf{P}(1, 1)$ does not exist.

A philosopher may say that the observables a and b are real. However, as we have pointed out, realism (compatibility) does not imply probabilistic realism (probabilistic compatibility).

3.8 Interpretational Questions

3.8.1 Contextuality

It is necessary to discuss the meaning of the term *contextuality*, as it can obviously be interpreted in many different ways, see, e.g., Karl Svozil [294, 295] and especially [296] for details. The most common meaning (in QM and quantum logic, [35], and especially in consideration of Bell's inequality [34]) is that the outcome for a measurement of an observable u under a contextual model is calculated using a different (albeit hidden) measure space, depending on whether or not compatible observables $v, w, \ldots$ were also made in the same experiment.

[9] We can consider natural numbers as systems and interpret $a(n)$ and $b(n)$ as values of observables a and b on the system n.

We remark that the well-known "no-go" theorems (of, e.g., Bell) cannot be applied to such contextual models.

In our approach the term contextuality has an essentially more general meaning. Physical context is any complex of physical conditions.

In particular, one can create a context by fixing the values of observables $v, w, \ldots$ which are compatible with u. However, in this way we can obtain only a very special class of contexts.

Thus the Växjö contextual model will cover the conventional quantum contextual model, but it can be applied for observations which could not be described by the quantum model.

3.8.2 Realism

We remark that in general the Växjö model does not contain physical systems (see Chap. 5 for a special class of models in which contexts are represented by ensembles of physical systems, see also Ballentine [29]). At the moment we do not (and need not) consider observables as observables on physical systems. It is only supposed that:

If a context C is fixed, then we can perform a measurement of any observable $a \in \mathcal{O}$ under this complex of conditions.

In general we do not try to assign objective properties to a physical system and consider observables as giving quantitative values of these objective properties. To measure $a \in \mathcal{O}$, one should first determine a context $a \in \mathcal{O}$ for measurement. If a context is not determined, then it is meaningless to speak about measurement. Thus the Växjö model is very general and it can be applied to practically all statistical measurements in any domain of science (in particular, it covers quantum measurements).

On the other hand, we would like to check how far it would be possible to proceed by keeping to realism. We show that practically all quantum probabilistic effects, e.g., the interference of probabilities and even the violation of Bell's inequality, can be described in the realistic (contextual) framework.

3.9 Historical Remark: Comparing with Mackey's Model

From the mathematical point of view our probabilistic model is quite close to the well-known Mackey's model. George Mackey [239] presented a program of huge complexity and importance:

To deduce the probabilistic formalism of quantum mechanics starting with a system of natural probabilistic axioms.

(Here "natural" has the meaning of a natural formulation in classical probabilistic terms.) Mackey tried to realize this program starting with a system of 8 axioms—Mackey axioms, see [239]. This was an important step in clarification of the probabilistic structure of quantum mechanics. However, he did not totally succeed, see [239] for details. The crucial axiom (about the complex Hilbert space) was not formulated in natural (classical) probabilistic terms.

As Mackey [239] pointed out, probabilities cannot be considered as abstract quantities defined outside any reference to a concrete complex of physical conditions C. All probabilities are conditional, or better to say contextual.[10] Mackey did a lot to unify classical and quantum probabilistic description and, in particular,

---

[10] We remark that the same point of view can be found in the works of A.N. Kolmogorov and R. von Mises. However, it seems that Mackey's book [239] was the first thorough presentation of a program of conditional probabilistic description of measurements, both in classical and quantum physics.

demystify quantum probability. One crucial step is however missing in Mackey's work. In his book Mackey [239] introduced the quantum probabilistic model (based on the complex Hilbert space) by means of a special axiom (Axiom 7, p. 71) that looked rather artificial in his general conditional probabilistic framework.

Mackey's model is based on a system of eight axioms, when our own model requires only two axioms. Let us briefly mention the content of Mackey's first axioms. The first four axioms concern conditional structure of probabilities, that is, they can be considered as axioms of a classical probabilistic model. The fifth and sixth axioms are of a logical nature (about questions). We reproduce below Mackey's "quantum axiom", and Mackey's own comments on this axiom (see [239], pp. 71–72):

Axiom 7 (Mackey). *The partially ordered set of all questions in quantum mechanics is isomorphic to the partially ordered set of all closed subsets of a separable, infinite dimensional Hilbert space.*[11]

Our activity can be considered as an attempt to find a list of physically plausible assumptions from which the Hilbert space structure can be deduced. We showed that this list can be presented in the form of a compact definition of the contextual probabilistic model, see Sect. 3.1.

---

[11] "This axiom has rather a different character from Axioms 1 through 4. These all had some degree of physical naturalness and plausibility. Axiom 7 seems entirely *ad hoc*. Why do we make it? Can we justify making it? What else might we assume? We shall discuss these questions in turn. The first is the easiest to answer. We make it because it "works", that is, it leads to a theory which explains physical phenomena and successfully predicts the results of experiments. It is conceivable that a quite different assumption would do likewise but this is a possibility that no one seems to have explored. Ideally one would like to have a list of physically plausible assumptions from which one could deduce Axiom 7."

3.10 Subjective and Contextual Probabilities in Quantum Theory

From the probabilistic viewpoint (by the Växjö interpretation) quantum mechanics is a generalization of Bayesian statistical analysis based on the formula of total probability. The classical Bayesian analysis is based on the classical formula of total probability and "quantum Bayesian analysis" is based on the interference formula of total probability.

If the coefficients of supplementarity $\lambda(\beta|a, C) = 0$, $\beta \in X_b$, then for $b|a$-conditioning under the complex of physical conditions C we can use the classical Bayesian analysis, if $|\lambda(\beta|a, C)| \leq 1$, $\beta \in X_b$, then we can use the Hilbert space variant of Bayesian analysis (which coincides with the classical one for observables which are not supplementary under the context C), see Chap. 4.

If $|\lambda(\beta|a, C)| \geq 1$, $\beta \in X_b$, then we could not use the classical Bayesian analysis nor its complex Hilbert space generalization, but it is possible to use hyperbolic Bayesian analysis based on the formula of total probability with the cosh-interference, see Part V.

From this point of view it is not surprising that quantum mechanics (its probabilistic part) induced a strong tendency toward idealism. Even the classical Bayesian analysis induced a similar tendency. In fact, the subjective interpretation of probability was developed starting with the Bayesian statistical analysis.

It should be expected that quantum mechanics would sooner or later induce the subjective interpretation of quantum probabilities. And it really has happened, see recent investigations of C. Fuchs, M. Appleby, A. Caticha, J.-A. Larsson, R. Schack, see, e.g., [24, 49, 95, 96, 278] on the subjective probabilistic approach to quantum foundations and quantum information. We remark that these investigations are totally justified from the purely mathematical viewpoint (as well as investigations on classical subjective probability), because they are based on the correct formalism.

This is nothing wrong with De Finetti's approach from a purely mathematical viewpoint. The only unfair assumption is that the initial probability distribution in the Bayesian analysis is chosen subjectively.

As was pointed out already by Kolmogorov [222] and Gnedenko [100], the choice of the initial probability distribution is based on the results of previous experiments. Statistical data which was collected in such experiments gives one the possibility to propose some form of a priori probability distribution. For example, when one uses the hypothesis that the distribution is Gaussian it is not just his purely subjective proposal. It is, in fact, the result of the frequency data from the huge number of statistical experiments.

There is nothing wrong with the use of subjective probability as the basis of an experimental statistical methodology. The main negative consequence of this approach is that it can support idealist views on physics (quantum as well as classical). Therefore Kolmogorov [222] and Gnedenko [100] criticized so strongly the use of subjective probability in science.

Chapter 4

Quantum-Like Representation of Contextual Probabilistic Model

Starting with the formula of total probability with an interference term, we shall construct the representation of a special class of contexts of the Växjö model, so-called *trigonometric contexts*, in complex Hilbert space. Then we obtain Born's rule and a representation of reference observables[1] by (noncommutative) self-adjoint operators $\hat{a}$ and $\hat{b}$. (Noncommutativity of operators is equivalent to consideration of probabilistically conjugate reference observables.) Thus quantum probabilistic formalism can be derived from the Växjö model on the basis of formula of total probability with interference term. In this chapter we shall realize this program of derivation of quantum probabilistic formalism.

We shall present a simple algorithm for transferring the probabilistic data $D(a, b, C)$ about a context C, see Sect. 3.1, Chap. 3, into a complex probabilistic amplitude: *quantum-like representation algorithm*, QLRA. The main distinguishing feature of QLRA is that classical probabilistic data is coupled with its QL-image by Born's rule.

---

[1] These are two specially chosen fundamental observables playing the role of the position and momentum observables.

4.1 Trigonometric, Hyperbolic, and Hyper Trigonometric Contexts

We consider a contextual probabilistic model (Växjö model) $M = (\mathcal{C}, \mathcal{O}, \mathcal{D}(\mathcal{O}, \mathcal{C}))$ and choose two *dichotomous* reference observables $a, b \in \mathcal{O}$. The formula of total probability with the interference term, Chap. 3, plays a fundamental role in further considerations. It was shown that for observables a, b and a context C such that

(Na) all elements of the matrix of $\mathbf{P}^{b|a}$ are strictly positive: $p^{b|a}(\beta|\alpha) > 0$,

(Nb) the context C is a-nondegenerate i.e., $p^a(\alpha) \equiv \mathbf{P}(a = \alpha|C) > 0$ for all $\alpha \in X_a$,

we have the following interfering-representation of probabilities:

$$p^b(\beta) = \sum_\alpha p^a(\alpha) p^{b|a}(\beta|\alpha) + 2\lambda(\beta|\alpha, C)\sqrt{\prod_\alpha p^a(\alpha) p^{b|a}(\beta|\alpha)}, \tag{4.1}$$

where $\lambda(\beta|\alpha, C)$ is the coefficient of $b|a$-supplementarity with respect to the context C. Depending on the magnitude of this coefficient the generalized formula of total probability (4.1) can be rewritten either in the form of the well-known trigonometric cos-interference or in the form of the so-called hyperbolic cosh-interference.

(1) Suppose that the coefficients of $b|a$-supplementarity $\lambda(\beta|\alpha, C)$ with respect to the context C are relatively small

$$|\lambda(\beta|a, C)| \leq 1, \quad \beta \in X_b. \tag{4.2}$$

In this case we can introduce new probabilistic parameters ("probabilistic angles") $\theta(\beta|\alpha, C) \in [0, 2\pi]$ and represent the coefficients in the trigonometric form

$$\lambda(\beta|a, C) = \cos\theta(\beta|a, C).$$

Parameters $\theta(\beta|\alpha, C)$ are said to be *$b|a$-relative phases* with respect to the context C. In this case we obtain the following interference formula of total probability

$$p^b(\beta) = \sum_\alpha p^a(\alpha)p^{b|a}(\beta|\alpha) + 2\cos\theta(\beta|\alpha, C)\sqrt{\prod_\alpha p^a(\alpha)p^{b|a}(\beta|\alpha)}. \quad (4.3)$$

This is nothing else than the famous *formula of interference of probabilities.*[2]

Definition 4.1. *Let C be an a-nondegenerate context and let* (4.2) *hold. Such a context is called $b|a$-trigonometric.*

For a trigonometric context C, starting from (4.3) and applying the QLRA we shall construct a complex probability amplitude ψ_C.

(2) Suppose that the coefficients of $b|a$-supplementarity $\lambda(\beta|\alpha, C)$ with respect to the context C are relatively large

$$|\lambda(\beta|a, C)| \geq 1, \quad \beta \in X_b. \quad (4.4)$$

In this case we can introduce new probabilistic parameters ("hyperbolic probabilistic angles") $\theta\ (\beta|a, C) \in (-\infty, +\infty)$ and represent the coefficients in the hyperbolic form

$$\lambda(\beta|a, C) = \pm\cosh\theta(\beta|a, C).$$

Parameters $\theta(\beta|a, C)$ are said to be (hyperbolic) *$b|a$-relative phases* with respect to the context C. In this case we obtain the following hyperbolic interference formula of total probability

---

[2] Typically this formula is derived by using the Hilbert space (unitary) transformation corresponding to the transition from one orthonormal basis to another and Born's probability postulate, see Part I: Chap. 1.

We demonstrated that, opposite to the common opinion, nontrivial interference of probabilities is not related to some special (and even mystical) "quantum features." In the Växjö approach all probabilistic considerations are purely classical (but contextual!). We shall not consider waves or appeal to wave-particle duality. Interference of probabilities for observables a and b in a context C is a consequence of the presence in these observables of some supplementary information about the context C. The coefficient λ gives the measure of this supplementarity.

$$p^b(\beta) = \sum_\alpha p^a(\alpha)p^{b|a}(\beta|\alpha) \pm 2\cosh\theta(\beta|\alpha, C)\sqrt{\prod_\alpha p^a(\alpha)p^{b|a}(\beta|\alpha)}. \quad (4.5)$$

Definition 4.2. *Let C be an a-nondegenerate context and let* (4.4) *hold. Such a context is called hyperbolic.*

For a hyperbolic context C, starting from (4.5) and applying the hyperbolic version of QLRA we shall construct a hyperbolic probability amplitude ψ_C taking values in the hyperbolic algebra **G**—the two dimensional Clifford algebra with the basis $e_1 = 1$ and $e_2 = j$, where $j^2 = +1$, see Part V.

(3) Suppose that the coefficients of $b|a$-supplementarity $\lambda(\beta|\alpha, C)$ with respect to the context C are relatively small for some β and relatively large for another. In this case we obtain the hyper-trigonometric interference of probabilities. This case has not yet been studied in detail.

4.2 Quantum-Like Representation Algorithm—QLRA

4.2.1 Probabilistic Data about Context

We denote the set of trigonometric contexts by the symbol $\mathscr{C}^{\mathrm{tr}}$. We emphasize that $\mathscr{C}^{\mathrm{tr}}$ depends on the choice of reference observables a, b

$$\mathscr{C}^{\mathrm{tr}} \equiv \mathscr{C}^{\mathrm{tr}}_{b|a,a}.$$

Here the index $b|a$ points to the definition of the coefficients of supplementarity through $b|a$-conditioning, and the index a points to consideration of contexts which are a-nondegenerate. Everywhere below we assume that the (dichotomous) reference observables a and b are *probabilistically conjugate.* Hence $p^{b|a}(\beta|\alpha) > 0$ for all $\alpha \in X_a$ and $\beta \in X_b$.

Let a context $C \in \mathscr{C}^{\text{tr}}$. We would like to notice the dependence of probabilities on the context C

$$p^a(\alpha) \equiv p_C^a(\alpha), \qquad p^b(\beta) \equiv p_C^b(\beta), \quad \alpha \in X_a, \; \beta \in X_b.$$

We rewrite the generalized formula of total probability (4.3) in the context dependent form

$$p_C^b(\beta) = \sum_\alpha p_C^a(\alpha) p^{b|a}(\beta|\alpha) + 2\cos\theta(\beta|\alpha, C)\sqrt{\prod_\alpha p_C^a(\alpha) p^{b|a}(\beta|\alpha)}. \tag{4.6}$$

4.2.2 Construction of Complex Probabilistic Amplitudes

By using the elementary formula

$$D = A + B + 2\sqrt{AB}\cos\theta = |\sqrt{A} + e^{i\theta}\sqrt{B}|^2,$$

for real numbers $A, B > 0, \theta \in [0, 2\pi]$, we can represent the probability $p_C^b(\beta)$ as the square of the complex amplitude (Born's rule)

$$p_C^b(\beta) = |\psi_C(\beta)|^2. \tag{4.7}$$

Here

$$\psi(\beta) \equiv \psi_C(\beta) = \sqrt{p_C^a(\alpha_1) p^{b|a}(\beta|\alpha_1)} + e^{i\theta_C(\beta)}\sqrt{p_C^a(\alpha_2) p^{b|a}(\beta|\alpha_2)}, \quad \beta \in X_b, \tag{4.8}$$

where $\theta_C(\beta) \equiv \theta(\beta|\alpha, C)$.

The formula (4.8) gives the QL representation algorithm—QLRA. For any trigonometric context C by starting with the probabilistic data—$p_C^b(\beta)$, $p_C^a(\alpha)$, $p^{b|a}(\beta|\alpha)$—QLRA produces the complex amplitude ψ_C. This algorithm can be used in any domain of science to create the QL-representation of probabilistic data (for a special class of contexts).[3]

We point out that QLRA contains the reference observables as parameters. Hence the complex amplitude given by (4.8) depends on $a, b : \psi_C \equiv \psi_C^{b|a}$.

Remark 4.1 (Choice of probabilistic phases). For each $\beta \in X_b$ the phase $\theta_C(\beta)$ can be chosen in two ways—by taking signs $+$ or $-$. Hence, the representation of contexts by complex amplitudes is not uniquely defined by the formula (4.8). In general each trigonometric context can be represented by four complex probability amplitudes based on the same formula (4.6). Suppose that we have chosen two fixed probabilistic angles.

Choice 1: $\theta_C(\beta_1) = \theta_1, \theta_C(\beta_2) = \theta_2$.

Then we can also make the following choices:

Choice 2: $\theta_C(\beta_1) = -\theta_1, \theta_C(\beta_2) = \theta_2$,

Choice 3: $\theta_C(\beta_1) = \theta_1, \theta_C(\beta_2) = -\theta_2$,

Choice 4: $\theta_C(\beta_1) = -\theta_1, \theta_C(\beta_2) = -\theta_2$.

For each of the complex probability amplitudes corresponding to these choices of the phase, we obtain Born's rule (4.7). Denote the corresponding amplitudes by the symbols $\psi_{C,j}, j = 1, \ldots, 4$. Then $\overline{\psi_{C,1}} = \psi_{C,4}$ as well as $\overline{\psi_{C,2}} = \psi_{C,3}$. Thus the choices one and four are conjugately-equivalent as well as the choices two and three. In future we shall see that only the choices one and four induce a natural QL-representation. But, at the moment we just fix, for any trigonometric context C, one of four possible probability amplitudes.

We denote the space of functions $\psi : X_b \to \mathbf{C}$ by the symbol

$$E_b \equiv \Phi(X_b, \mathbf{C}),$$

[3] I was even thinking about getting a patent for QLRA, but then I decided that QLRA should be free for all users.

where $\mathbf{C}$ is the field of complex numbers. Since the observable b takes only two values—$X_b = \{\beta_1, \beta_2\}$, the E_b is the two-dimensional complex linear space. The Dirac δ-functions $\{\delta(\beta - \beta_1), \delta(\beta - \beta_2)\}$ form the canonical basis in this space. Each $\psi \in E_b$ can be expanded with respect to this basis

$$\psi(\beta) = \psi(\beta_1)\delta(\beta - \beta_1) + \psi(\beta_2)\delta(\beta - \beta_2).$$

By using the representation (4.8) we construct the map

$$J^{b|a} : \mathscr{C}^{\text{tr}} \to E_b. \tag{4.9}$$

The $J^{b|a}$ maps contexts (complexes of, e.g., physical conditions) into complex amplitudes. The representation (4.7) of probability as the square of the absolute value of the complex $b|a$-amplitude is nothing else than the famous *Born rule.*

Remark 4.2 (Role of reference observables). We repeat that the complex linear space representation (4.8) of the set of contexts $\mathscr{C}^{\text{tr}}$ is based on a pair (a, b) of *probabilistically conjugate* observables. By choosing another pair we shall get a different representation.

Remark 4.3 (Origin of complex numbers). The appearance of complex numbers is one of the mysteries of QM. In our contextual approach complex numbers appeared as just a special representation of the formula of total probability with trigonometric interference term. Instead of the formula (4.6) and a collection of contextual probabilities, we preferred to work with Born's rule (4.7) and complex probability amplitudes.

Definition 4.3. *The complex amplitude ψ_C produced by QLRA is called a QL wave function (of the complex of physical conditions, context C) or a QL state.*

By the Växjö interpretation the wave function does not provide the (complete) description of the state of an individual system. The wave function is a special (in general incomplete) representation of context.

In fact, the multi-value structure of QLRA is even more complicated than was pointed out in Remark 4.1. We might represent each context $C \in \mathscr{C}^{\mathrm{tr}}$ by a family of complex amplitudes

$$\psi(x) \equiv \psi_C(\beta) = \sum_{\alpha \in X_a} \sqrt{p_C^a(\alpha) p^{b|a}(\beta|\alpha)} e^{i\xi_C(\beta|\alpha)} \tag{4.10}$$

such that

$$\xi_C(\beta|\alpha_1) - \xi_C(\beta|\alpha_2) = \theta_C(\beta).$$

4.3 Hilbert Space Representation of *b*-Observable

4.3.1 Born's Rule

We consider the following basis in the space E_b $(\beta \in X_b)$: $e_\beta^b(x) = \delta(x - \beta)$. Born's rule for complex amplitudes (4.7) can be rewritten in the following form

$$p_C^b(\beta) = |\langle \psi_C, e_\beta^b \rangle|^2, \quad \beta \in X_b. \tag{4.11}$$

Here the scalar product in the complex linear space E_b is defined by the standard formula

$$\langle \psi_1, \psi_2 \rangle = \sum_{\beta \in X_b} \psi_1(\beta) \overline{\psi_2(\beta)}. \tag{4.12}$$

The system of functions $\{e_\beta^b\}_{x \in X_b}$ is an orthonormal basis in the Hilbert space

$$\mathscr{H} \equiv \mathscr{H}_b^{b|a} = (E_b, \langle \cdot, \cdot \rangle). \tag{4.13}$$

In the symbolic notation $\mathscr{H}_b^{b|a}$ the upper index $b|a$ indicates that the representation of contexts was created through $b|a$-supplementarity, the low index b indicates that the observable b plays the fundamental role. Contexts are represented by functions defined on the range of values X_b of the observable b. We emphasize that the reference observables a and b are involved in the $\mathscr{H}_b^{b|a}$-representation in an asymmetric way, see especially Sect. 4.6.

4.3.2 Fundamental Physical Observable: Views of De Broglie and Bohm

L. De Broglie permanently emphasized the exceptional role of the *position observable* q in QM [70]. Similar views can be found in works of D. Bohm. In our approach the b-observable plays the role of the position observable q. The a-observable (which is chosen as probabilistically conjugate to b) plays the role of the momentum observable p. L. De Broglie wrote in [70], p. 55: "A consideration of foregoing examples, and others which we could imagine, necessarily leads us to the conclusion that it is the representation in space and time which is objective, and not the Fourier analysis which only exists in the mind of the theoretician. The various Fourier components can only be observed by means of devices which change completely the initial state of affairs and modify the phase relationships. In the language of the Theory of Transformations, this can be expressed by saying that the q-representation is the only objective representation, whilst the p-representation—the abstract representation in momentum space—exists only in the mind of the theoretician. This shows, contrary to what the Theory of Transformations usually asserts, that the two representations—q and p—are by no means equivalent. It is the wave function that describes the physical reality and not the coefficients c_i considered separately. Moreover, this conclusion is the consequence of the obvious fact that the three dimensional space is a physical reality and the essential framework of our experiment, whilst momentum space is only an abstract mathematical representation." Here "Theory of Transformations" is theory of unitary transformations performing transition from one orthonormal basis in complex Hilbert space to another.

4.3.3 b-Observable as Multiplication Operator

Let $X_b \subset \mathbf{R}$ (so we assume that the results of measurements of the observable b are given by real numbers). By using the Hilbert space representation (4.11) of Born's rule we obtain the Hilbert space representation of the contextual expectation of the observable b

$$E[b|C] = \sum_{\beta \in X_b} \beta p_C^b(\beta) = \sum_{\beta \in X_b} \beta |\psi_C(\beta)|^2$$

$$= \sum_{\beta \in X_b} \beta \langle \psi_C, e_\beta^b \rangle \overline{\langle \psi_C, e_\beta^b \rangle} = \langle \hat{b}\psi_C, \psi_C \rangle. \tag{4.14}$$

Here the (self-adjoint) operator $\hat{b} : \mathscr{H} \to \mathscr{H}$ is determined by its eigenvectors: $\hat{b}e_\beta^b = \beta e_\beta^b, \beta \in X_b$. This is the multiplication operator in the linear space of complex-valued functions $E_b = \Phi(X_b, \mathbf{C})$: $\hat{b}\psi(\beta) = \beta\psi(\beta)$. It is natural to represent the b-observable by the operator $\hat{b}$ in the Hilbert state space.

4.3.4 Interference

We set

$$u_j^a = \sqrt{p_C^a(\alpha_j)}, \qquad u_j^b = \sqrt{p_C^b(\beta_j)}, \qquad p_{ij}^{b|a} = p^{b|a}(\beta_j|\alpha_i),$$
$$u_{ij} = \sqrt{p_{ij}^{b|a}}, \qquad \theta_j = \theta_C(\beta_j). \tag{4.15}$$

We remark that the coefficients u_j^a, u_j^b depend on a context C; so $u_j^a = u_j^a(C), u_j^b = u_j^b(C)$. We consider the matrix $\mathbf{P}^{b|a} = (p_{ij}^{b|a})$. We have, see (4.10), that

$$\psi_C = v_1^b e_{\beta_1}^b + v_2^b e_{\beta_2}^b, \qquad \text{where } v_j^b = u_1^a u_{1j} + u_2^a u_{2j} e^{i\theta_j}. \tag{4.16}$$

Hence

$$p_C^b(\beta_j) = |v_j^b|^2 = |u_1^a u_{1j} + u_2^a u_{2j} e^{i\theta_j}|^2. \tag{4.17}$$

This is the *interference representation of probabilities* that is used, e.g., in quantum formalism.[4]

[4] By starting with the general representation (4.10) we obtain $v_j^b = u_1^a u_{1j} e^{i\xi_{1j}} + u_2^a u_{2j} e^{i\xi_{2j}}$ and the interference representation $p_C^b(\beta_j) = |v_j^b|^2 = |u_1^a u_{1j} e^{i\xi_{1j}} + u_2^a u_{2j} e^{i\xi_{2j}}|^2$.

4.4 Hilbert Space Representation of a-Observable

We would like to have Born's rule and the Hilbert space representation not only for the b-observable, but also for the a-observable. Therefore we should introduce in a natural way a basis $e^a = \{e^a_\alpha\}_{\alpha \in X_a}$ corresponding to the a-observable in the Hilbert space $\mathcal{H}_b^{b|a}$. The formula of interference of probabilities written in the form of superposition (4.16) will play a crucial role in introducing the a-basis in $\mathcal{H}_b^{b|a}$.

4.4.1 Conventional Quantum and Quantum-Like Representations

As we shall see, we cannot be lucky in the general case. In fact, by starting from two arbitrary (probabilistically conjugate) observables a and b we constructed the complex Hilbert space representation for the b-observable which was more general than the standard quantum representation. In our (more general) representation the conjugate observable a need not be mapped into a self-adjoint operator.

We recall that in quantum mechanics both reference observables (the position observable and the momentum observable) are represented by self-adjoint operators. We should find conditions which would guarantee the possibility to represent both reference observables by self-adjoint operators and to have Born's rule for both of them.

QM implies that for any pair of observables a and b which are represented by self-adjoint operators $\hat{a}$ and $\hat{b}$, the corresponding transition probabilities are symmetrically conditioned, see Chap. 3, Sect. 3.6: $p^{b|a}(\beta|\alpha) = p^{a|b}(\alpha|\beta)$. In the Växjö model we do not assume that in general probabilities are symmetrically conditioned. It can be that for some probabilistic data which was collected, e.g., in sociology or psychology, we shall have

$$p^{b|a}(\beta|\alpha) \equiv \mathbf{P}(b = \beta|C_\alpha) \neq p^{a|b}(\alpha|\beta) \equiv \mathbf{P}(a = \alpha|C_\beta).$$

Here C_α and C_β are contexts corresponding to selections with respect to fixed values $a = \alpha$ and $b = \beta$. Really there are no reasons to expect symmetrical conditioning for any pair of reference observables. In QM the reference observables were chosen in a very special way to guarantee symmetrical conditioning.

Nevertheless, starting with the Växjö model we can apply QLRA for any pair of probabilistically conjugate observables and represent the set of contexts $\mathscr{C}^{\mathrm{tr}}_{b|a,a}$ in the complex Hilbert space $\mathscr{H}_b^{b|a}$ where the b-observable is represented by a self-adjoint operator and Born's rule holds.

We now would like to find conditions for the matrix $\mathbf{P}^{b|a}$ which would provide a possibility to represent the a-observable by a self-adjoint operator in such a way that Born's rule would hold. The easiest way is to borrow the QM-condition, namely, symmetric conditioning. We shall see that this condition is really sufficient to construct representation for which we are looking, see Sect. 4.6. However, we do not borrow directly the QM-condition. We would like to derive a natural necessary and sufficient condition internally from the Växjö model. We shall see that this condition is essentially weaker than symmetric conditioning.

4.4.2 *a*-Basis from Interference

For any (trigonometric) context C_0, we can represent (by using the expansion (4.16)) the corresponding wave function $\psi = \psi_{C_0}$ in the form

$$\psi = u_1^a e_{\alpha_1}^a + u_2^a e_{\alpha_2}^a, \tag{4.18}$$

where

$$e_{\alpha_1}^a = \begin{pmatrix} u_{11} \\ u_{12} \end{pmatrix}, \qquad e_{\alpha_2}^a = \begin{pmatrix} e^{i\theta_1} u_{21} \\ e^{i\theta_2} u_{22} \end{pmatrix}. \tag{4.19}$$

The system of vectors $\{e^a_{\alpha_i}\}$ will be used to represent the a-observable in the Hilbert space $\mathcal{H} \equiv \mathcal{H}_b^{b|a}$.[5]

We suppose that vectors $\{e^a_{\alpha_i}\}$ are linearly independent, so $\{e^a_{\alpha_i}\}$ is a basis in $\mathcal{H}$. We have

$$e^a_{\alpha_1} = v_{11} e^b_{\beta_1} + v_{12} e^b_{\beta_2}, \qquad e^a_{\alpha_2} = v_{21} e^b_{\beta_1} + v_{22} e^b_{\beta_2}.$$

Here

$$V \equiv V^{b|a}_{b \to a} = (v_{ij}) = \begin{pmatrix} u_{11} & u_{12} \\ e^{i\theta_1} u_{21} & e^{i\theta_2} u_{22} \end{pmatrix}$$

is the matrix corresponding to the transformation from the b-basis to the a-basis

$$\begin{pmatrix} e^a_{\alpha_1} \\ e^a_{\alpha_2} \end{pmatrix} = V \begin{pmatrix} e^b_{\beta_1} \\ e^b_{\beta_2} \end{pmatrix}.$$

It is interesting to notice the expressions of the scalar products for basis vectors

$$V = (v_{ij}) = \begin{pmatrix} \langle e^a_{\alpha_1}, e^b_{\beta_1} \rangle & \langle e^a_{\alpha_1}, e^b_{\beta_2} \rangle \\ \langle e^a_{\alpha_2}, e^b_{\beta_1} \rangle & \langle e^a_{\alpha_2}, e^b_{\beta_2} \rangle \end{pmatrix}$$

$$= \begin{pmatrix} \sqrt{p^{b|a}(\beta_1|\alpha_1)} & \sqrt{p^{b|a}(\beta_2|\alpha_1)} \\ e^{i\theta_1}\sqrt{p^{b|a}(\beta_1|\alpha_2)} & e^{i\theta_2}\sqrt{p^{b|a}(\beta_2|\alpha_2)} \end{pmatrix}. \tag{4.20}$$

The crucial point is that the matrix $V^{b|a}_{b \to a}$ can be constructed by using only probabilistic data; even phases are purely probabilistic phases.

4.4.3 Necessary and Sufficient Conditions for Born's Rule

We would like to find a class of matrices $V^{b|a}_{b \to a}$ such that Born's rule also holds for the a-basis

$$p^a_C(\alpha) = |\langle \psi_C, e^a_\alpha \rangle|^2. \tag{4.21}$$

[5] We recall that our construction is not symmetric with respect to the reference observables a and b. The observable b is the "fundamental observable" and a is its dual, cf. views of De Broglie and Bohm.

We have Born's rule (4.21) iff $\{e^a_\alpha\}$ was an *orthonormal basis*, i.e., the $V^{b|a}_{b\to a}$ was a *unitary* matrix.

A matrix $P = (p_{ij})$ is *double stochastic* if it is stochastic, i.e., $p^{b|a}_{j1} + p^{b|a}_{j2} = 1$, and, moreover,

$$p^{b|a}_{1j} + p^{b|a}_{2j} = 1, \quad j = 1, 2. \tag{4.22}$$

We recall, see Chap. 3, Sect. 3.6, that any symmetrically conditioned matrix $\mathbf{P}^{b|a}$ is double stochastic. But in general double stochasticity does not imply symmetrical conditioning. The following proposition can be easily proved by direct calculations.

Proposition 4.1. *In the two-dimensional case (i.e., for dichotomous observables), the matrix* $V^{b|a}_{b\to a}$ *is unitary iff the matrix* $\mathbf{P}^{b|a}$ *is double stochastic and additionally*

$$e^{i\theta_1} = -e^{i\theta_2}$$

or

$$\theta_{C_0}(\beta_1) - \theta_{C_0}(\beta_2) = \pi \bmod 2\pi. \tag{4.23}$$

We remark that the constraints (4.23) on phases and the double stochasticity constraint are not independent.

Lemma 4.1. *Let the matrix of transition probabilities* $\mathbf{P}^{b|a}$ *be double stochastic. Then*

$$\cos\theta_C(\beta_2) = -\cos\theta_C(\beta_1) \tag{4.24}$$

for any context $C \in \mathscr{C}^{\mathrm{tr}}$.

Proof. By Lemma 3.1, Chap. 3, we have

$$\sum_{\beta\in X_b} \cos\theta_C(\beta)\sqrt{\Pi_{\alpha\in X_a} p^a_C(\alpha) p^{b|a}(\beta|\alpha)} = 0.$$

But, for a double stochastic matrix $\mathbf{P}^{b|a} = (p^{b|a}(\alpha|\beta))$, we have

$$\Pi_{\alpha\in X_a} p^a_C(\alpha_1) p^{b|a}(\beta_1|\alpha) = \Pi_{\alpha\in X_a} p^a_C(\alpha_2) p^{b|a}(\beta_2|\alpha).$$

Since we work with probabilistically conjugate reference observables a, b and the context C is a-nondegenerate, all probabilities are strictly positive. Therefore we obtain (4.24).

4.4.4 Choice of Probabilistic Phases

By Lemma 4.1 we have two different possibilities to choose phases

$$\theta_{C_0}(\beta_1) + \theta_{C_0}(\beta_2) = \pi \quad \text{or} \quad \theta_{C_0}(\beta_1) - \theta_{C_0}(\beta_2) = \pi \bmod 2\pi.$$

By (4.23) to obtain Born's rule for the a-observable we should choose phases $\theta_{C_0}(\beta_i), i = 1, 2$, in such a way that

$$\theta_{C_0}(\beta_2) = \theta_{C_0}(\beta_1) + \pi. \tag{4.25}$$

If $\theta_{C_0}(\beta_1) \in [0, \pi]$ then $\theta_{C_0}(\beta_2) \in [\pi, 2\pi]$ and vice versa.

Lemma 4.1 is very important. If the matrix $\mathbf{P}^{b|a}$ is double stochastic we can always choose $\theta_{C_0}(\beta_j), j = 1, 2$, satisfying (4.25). Hence we can always assume that QLRA produces complex amplitudes of the form

$$\psi(\beta_1) = \sqrt{p_C^a(\alpha_1) p^{b|a}(\beta_1|\alpha_1)} + e^{i\theta_C(\beta_1)} \sqrt{p_C^a(\alpha_2) p^{b|a}(\beta_1|\alpha_2)}, \tag{4.26}$$

$$\psi(\beta) = \sqrt{p_C^a(\alpha_1) p^{b|a}(\beta_2|\alpha_1)} - e^{i\theta_C(\beta_1)} \sqrt{p_C^a(\alpha_2) p^{b|a}(\beta_2|\alpha_2)}. \tag{4.27}$$

We now come back to Remark 4.1. Suppose that, for a context C_0, we have chosen the phases $\theta_{C_0}(\beta_1)$ and $\theta_{C_0}(\beta_2)$ satisfying (4.25). We denote this representation of C_0 by a complex probability amplitude by Choice 1. In Remark 4.1 we presented three other choices created by varying the signs of phases. We see that only phases given by the Choice 4 also satisfy the condition (4.25). Thus if we want to have a natural representation (with Born's rule for both reference observables) we should take away Choices 2, 3. So the arbitrariness in choosing a complex amplitude for a fixed context is essentially reduced.

4.4.5 Contextual Dependence of a-Basis

The delicate feature of the presented construction of the a-representation is that the basis $\{e^a_\alpha\}$ depends on the context C_0:

$$e^a_\alpha = e^a_\alpha(C_0).$$

And Born's rule, in fact, has the form

$$p^a_{C_0}(\alpha) = |\langle \psi_{C_0}, e^a_\alpha(C_0)\rangle|^2, \quad \alpha \in X_a.$$

We would like to use (as in the conventional quantum formalism) one fixed a-basis for all contexts $C \in \mathscr{C}^{\rm tr}$. We may try to use for all contexts $C \in \mathscr{C}^{\rm tr}$ the basis $e^a_\alpha \equiv e^a_\alpha(C_0)$ corresponding to one fixed context C_0. We shall see that this is really the most fruitful strategy.

Lemma 4.2. *Let the matrix of transition probabilities* $\mathbf{P}^{b|a}$ *be double stochastic and let for any context* $C \in \mathscr{C}^{\rm tr}$ *phases* $\theta_C(\beta)$, $\beta \in X_b$, *be chosen as*

$$\theta_C(\beta_2) = \theta_C(\beta_1) + \pi \bmod 2\pi. \tag{4.28}$$

Then for any context $C \in \mathscr{C}^{\rm tr}$ *we have Born's rule for the basis* $e^a_\alpha \equiv e^a_\alpha(C_0)$ *constructed for a fixed context* C_0

$$p^a_C(\alpha) = |\langle \psi_C, e^a_\alpha\rangle|^2, \quad \alpha \in X_a. \tag{4.29}$$

Proof. Let C_0 be some fixed context. We take the basic $\{e^a_{\alpha_j}(C_0)\}$ (and the matrix $V(C_0)$) corresponding to this context. For any $C \in \mathscr{C}^{\rm tr}$, we would like to represent the wave function ψ_C as

$$\psi_C = v_1^a(C)e_{\alpha_1}^a(C_0) + v_2^a(C)e_{\alpha_2}^a(C_0), \quad \text{where } |v_j^a(C)|^2 = p_C^a(\alpha_j). \tag{4.30}$$

It is clear that, for any $C \in \mathscr{C}^{\mathrm{tr}}$, we can represent the wave function as

$$\psi_C(\beta_1) = u_1^a(C)v_{11}(C_0) + e^{i[\theta_C(\beta_1)-\theta_{C_0}(\beta_1)]}u_2^a(C)v_{12}(C_0),$$

$$\psi_C(\beta_2) = u_1^a(C)v_{21}(C_0) + e^{i[\theta_C(\beta_2)-\theta_{C_0}(\beta_2)]}u_2^a(C)v_{22}(C_0).$$

Thus to obtain (4.30) we should have

$$\theta_C(\beta_1) - \theta_{C_0}(\beta_1) = \theta_C(\beta_2) - \theta_{C_0}(\beta_2) \bmod 2\pi \tag{4.31}$$

for any pair of contexts C_0 and C_1. By using the relations (4.28) between phases $\theta_C(\beta_1), \theta_C(\beta_2)$ and $\theta_{C_0}(\beta_1), \theta_{C_0}(\beta_2)$ we obtain

$$\theta_C(\beta_2) - \theta_{C_0}(\beta_2) = (\theta_C(\beta_1) + \pi - \theta_{C_0}(\beta_1) - \pi) = \theta_C(\beta_1) - \theta_{C_0}(\beta_1) \bmod 2\pi.$$

The constraint (4.28) essentially restricted the class of complex amplitudes which can be used to represent a context $C \in \mathscr{C}^{\mathrm{tr}}$. Any C can be represented by only two amplitudes $\psi(x)$ and $\bar{\psi}(x)$ corresponding to the two possible choices of $\theta_C(\beta_1)$: in $[0, \pi]$ or $(\pi, 2\pi)$.

4.4.6 *Existence of Quantum-Like Representation with Born's Rule for Both Reference Observables*

By Lemma 4.2 we obtain the following result playing the fundamental role in our approach.

Theorem 4.1. *We can construct the QL (complex Hilbert space) representation of the set of trigonometric contexts $\mathscr{C}^{\mathrm{tr}}$ such that Born's rule holds true for both reference observables a, b (which are assumed to be probabilistically conjugate) iff the matrix $\mathbf{P}^{b|a}$ is double stochastic.*

If the matrix $\mathbf{P}^{b|a}$ is double stochastic, then we have the QL representation not only for the contextual expectation of the observable b, see (4.14), but also for the observable a

$$E[a|C] = \sum_{\alpha \in X_a} \alpha p_C^a(\alpha) = \sum_{\alpha \in X_a} \alpha |\langle \psi_C, e_\alpha^a \rangle|^2 = \langle \hat{a} \psi_C, \psi_C \rangle, \tag{4.32}$$

where the self-adjoint operator (symmetric matrix) $\hat{a} : E_b \to E_b$ is determined by its eigenvectors: $\hat{a} e_\alpha^a = \alpha e_\alpha^a$. By (4.32) it is natural to represent the observable a by the operator $\hat{a}$.

We also remark that in the case of double stochastic $\mathbf{P}^{b|a}$ the scalar products for basis vectors have the form

$$\begin{aligned} v_{11} &= \langle e_{\alpha_1}^a, e_{\beta_1}^b \rangle = \sqrt{p^{b|a}(\beta_1|\alpha_1)}, \\ v_{12} &= \langle e_{\alpha_1}^a, e_{\beta_2}^b \rangle = \sqrt{p^{b|a}(\beta_2|\alpha_1)}, \end{aligned} \tag{4.33}$$

$$\begin{aligned} v_{21} &= \langle e_{\alpha_2}^a, e_{\beta_1}^b \rangle = e^{i\theta} \sqrt{p^{b|a}(\beta_1|\alpha_2)}, \\ v_{22} &= \langle e_{\alpha_2}^a, e_{\beta_2}^b \rangle = -e^{i\theta} \sqrt{p^{b|a}(\beta_2|\alpha_2)}. \end{aligned} \tag{4.34}$$

As always, we denote the unit sphere in the Hilbert space $\mathscr{H}$ by the symbol S. In general, i.e., for an arbitrary contextual probabilistic model of reality, there are no reasons to expect that the representation map $J^{b|a} : \mathscr{C}^{\text{tr}} \to S$ should to be one-to-one, i.e., surjection and injection. We shall study the question about injectivity of the map $J^{b|a}$ in Sect. 4.5.2.

Regarding surjectivity we can say that in principle in some physical (or mental, or economic) model the set of context $\mathscr{C}$ may not be large enough to cover the whole unit sphere S of the complex Hilbert space. However, in the conventional quantum model it is claimed that each quantum state can be prepared on the basis of some complex of physical conditions.

4.4.7 "Pathologies"

We remark that, although Theorem 4.1 guarantees existence of the QL representation with Born's rule for both reference observables, this representation may have features which differ essentially from features of the conventional quantum representation.

In particular, contexts C_β need not belong to the family of trigonometric contexts. In such a case, although the scalar product $\langle e^b_\beta, e^a_\alpha\rangle$ is well defined and, moreover, $|\langle e^b_\beta, e^a_\alpha\rangle|^2 = p^{b|a}(\beta|\alpha)$, we cannot write Born's rule in our contextual form: $|\langle \psi_{C_\beta}, e^a_\alpha\rangle|^2 = p^{b|a}(\beta|\alpha)$ (because QLRA cannot be applied to the context C_β). In principle, one might just formally extend the domain of application of QLRA by setting

$$J^{b|a}(C_\beta) = e^b_\beta. \tag{4.35}$$

Situation is even worse for contexts $C_\alpha, \alpha \in X_a$. They are not a-nondegenerate: $\mathbf{P}(a = \alpha_i|C_{\alpha_j}) = 0,\ i \neq j$. Therefore QLRA cannot be applied to C_α. Thus the image $J^{b|a}(C_\alpha)$ cannot be defined by (4.8). In principle, one might just formally extend the domain of application of QLRA by setting

$$J^{b|a}(C_\alpha) = e^a_\alpha. \tag{4.36}$$

However, if the model is not symmetrically conditioned, then such a definition would imply the following pathology

$$|\langle J^{b|a}(C_\alpha), e^b_\beta\rangle|^2 = |\langle e^a_\alpha, e^b_\beta\rangle|^2 = |\langle e^b_\beta, e^a_\alpha\rangle|^2 = p^{b|a}(\beta|\alpha).$$

Hence

$$|\langle J^{b|a}(C_\alpha), e^b_\beta\rangle|^2 \neq p^{a|b}(\alpha|\beta). \tag{4.37}$$

It is clear that the latter problem would disappear if one considers only Växjö models with symmetrically conditioned reference observables. It is surprising that such a restriction would also eliminate the problem with non-trigonometrical behavior of contexts C_β, see Sect. 4.6.

4.5 Properties of Mapping of Trigonometric Contexts into Complex Amplitudes

4.5.1 Classical-Like Contexts

Suppose that, for some context $C \in \mathscr{C}^{\mathrm{tr}}$, the reference observables are not $b|a$-supplementary with respect to C (see Chap. 3, Definition 3.7). Thus

$$\delta(\beta|a, C) = 0, \quad \beta \in X_b.$$

Thus even

$$\lambda(\beta|a, C) = 0, \quad \beta \in X_b.$$

Hence $\theta_C(\beta_1) = \frac{\pi}{2}$ or $\theta_C(\beta_1) = \frac{3}{2}\pi$. In the first case we have

$$\begin{aligned} \psi_C(\beta_1) &= \sqrt{p_C^a(\alpha_1)p(\beta_1|\alpha_1)} + i\sqrt{p_C^a(\alpha_2)p(\beta_1|\alpha_2)}, \\ \psi_C(\beta_2) &= \sqrt{p_C^a(\alpha_1)p(\beta_2|\alpha_1)} - i\sqrt{p_C^a(\alpha_2)p(\beta_2|\alpha_2)}. \end{aligned} \tag{4.38}$$

The second choice of phases gives the representation of C by the complex amplitude ϕ_C which is conjugate to (4.38): $\phi_C = \overline{\psi_C}$. We set

$$\mathscr{C}_{\mathrm{CL}}^{\mathrm{tr}} = \{C \in \mathscr{C}^{\mathrm{tr}} : \delta(\beta|a, C) = 0\}.$$

These are trigonometric contexts for which the reference observables are not $b|a$-supplementary. We call them CL-contexts.

4.5.2 Non-Injectivity of Representation Map

Let $C_1, C_2 \in \mathscr{C}^{\rm tr}$ be contexts such that the probability distributions of the reference observables a and b under C_1 and C_2 coincide

$$p^a_{C_1}(\alpha) = p^a_{C_2}(\alpha), \quad \alpha \in X_a, \qquad p^b_{C_1}(\beta) = p^b_{C_2}(\beta), \quad \beta \in X_b.$$

In such a case $\lambda(\beta|a, C_1) = \lambda(\beta|a, C_2)$ and $\theta(\beta|a, C_1) = \pm\theta(\beta|a, C_2)$. If the probability distributions coincide only for a pair of contexts (C_1, C_2), then we can represent C_1 and C_2 by two different complex amplitudes, ψ_{C_1} and $\psi_{C_2} = \bar{\psi}_{C_1}$. But if the probability distributions coincide for a triple of contexts (C_1, C_2, C_3), then it is impossible to represent them by different complex amplitudes. We should choose $\psi_{C_3} = \psi_{C_1}$ or $\psi_{C_3} = \psi_{C_2}$; so $J^{b|a}(C_3) = J^{b|a}(C_1)$ or $J^{b|a}(C_3) = J^{b|a}(C_2)$. Thus in general the map $J^{b|a}$ is not injective.

4.6 Non-Double Stochastic Matrix: Quantum-Like Representations

Of course, for arbitrary (probabilistically conjugate) observables a and b the matrix $\mathbf{P}^{b|a}$ need not be double stochastic. Therefore the matrix $V^{b|a}_{b\to a}$ for transition from the b-observable (which could be interpreted, cf. De Broglie and Bohm—Remark 1.1, Part I: Chap. 1, as the fundamental observable) to the "dual observable" a can be nonunitary. In this case we could not obtain Born's rule in the Hilbert space $\mathscr{H}^{b|a}_b$ both for the b and a observables.

We now assume that the $\mathbf{P}^{b|a}$ is not double stochastic. For each reference observable we should introduce its own scalar product and corresponding Hilbert space in that Born's rule holds true

$$\mathcal{H}_b^{b|a} = (E_b, \langle\cdot,\cdot\rangle_b), \qquad \mathcal{H}_a^{b|a} = (E_b, \langle\cdot,\cdot\rangle_a), \tag{4.39}$$

where scalar products on the complex linear spaces E_b are given by

$$\langle\psi,\psi\rangle_b = \sum_j v_j^b \bar{w}_j^b \quad \text{for } \psi = \sum_j v_j^b e_{\beta_j}^b, \psi = \sum_j w_j e_{\beta_j}^b,$$

and

$$\langle\psi,\psi\rangle_a = \sum_j v_j^a \bar{w}_j^a \quad \text{for } \psi = \sum_j v_j^a e_{\alpha_j}^a, \psi = \sum_j w_j^a e_{\alpha_j}^a.$$

We have Born's rules with respect to these scalar products

$$p_C^b(\beta) = |\langle\psi_C, e_\beta^b\rangle_b|^2, \qquad p_C^a(\alpha) = |\langle\psi_C, e_\alpha^a\rangle_a|^2.$$

The reference observables b and a are represented by symmetric matrices in the Hilbert spaces $\mathcal{H}_b^{b|a}$ and $\mathcal{H}_a^{b|a}$, respectively. Thus we do not have even mathematical equivalence (in the sense of unitary equivalence) of representations of a and b, cf. the discussion on physical nonequivalence for the position and momentum representations in QM, cf. Sect. 4.3.2. But the appearance of different Hilbert spaces (4.39) is not the end of mathematical difficulties in the case in that the $\mathbf{P}^{b|a}$ is not double stochastic.

As we have already discussed, the crucial difficulty is that $e_\alpha^a = e_\alpha^a(C_0)$. In fact, for any context $C_0 \in \mathcal{C}^{\text{tr}}$ we constructed its own Hilbert space representation for the a-observable: $\mathcal{H}_a^{b|a} = \mathcal{H}_a^{b|a}(C_0)$. In the same way as in Sect. 4.3 we obtain that we would be able to use the same representation for contexts C and C_0 if the condition (4.31) holds true. Thus we should have

$$\theta_C(\beta_2) = \theta_C(\beta_1) + \alpha \quad \text{and} \quad \theta_{C_0}(\beta_2) = \theta_{C_0}(\beta_1) + \gamma \bmod 2\pi,$$

where γ is some phase (if $\mathbf{P}^{b|a}$ is double stochastic then $\gamma = \pi$).

Theorem 4.2. *Suppose that $\mathbf{P}^{b|a}$ is not double stochastic and $\mathcal{C}^{\text{tr}} \neq \mathcal{C}_0^{\text{tr}}$. Then there is no such γ that*

$$\theta_C(\beta_2) = \theta_C(\beta_1) + \gamma \tag{4.40}$$

for all contexts $C \in \mathscr{C}^{\text{tr}}$.

To prove this theorem we need the following generalization of Lemma 4.1 for the case in which the $\mathbf{P}^{b|a}$ is not double stochastic.

Lemma 4.3. *For any context* $C \in \mathscr{C}^{\text{tr}}$, *the following equality holds true*

$$\cos\theta_C(\beta_2) = -k\cos\theta_C(\beta_1) \tag{4.41}$$

where

$$k \equiv k^{b|a} = \sqrt{\frac{p_{11}^{b|a} p_{21}^{b|a}}{p_{12}^{b|a} p_{22}^{b|a}}}.$$

It is also easy to obtain:

Proposition 4.2. *The coefficient* $k^{b|a} = 1$ *iff* $\mathbf{P}^{b|a}$ *is double stochastic.*

Proof of Theorem. By Lemma 4.3 we have $-k\cos\theta_C(\beta_1) = \cos(\theta_C(\beta_1) + \gamma)$. We take $C = \Omega$ and obtain $\cos(\theta_\Omega(\beta_1) + \gamma) = 0$. But $\theta_\Omega(\beta_1) = \pm\frac{\pi}{2}$. Thus $\theta_\Omega(\beta_1) + \gamma = \pm\frac{\pi}{2}$ and $\gamma = 0, \pi \mod 2\pi$.

Since $\mathscr{C}^{\text{tr}} \neq \mathscr{C}_0^{\text{tr}}$ there exists a context C such that $\cos\theta_C(\beta_1) \neq 0$. If $\gamma = 0$ then $\cos\theta_C(\beta_1)(k+1) = 0$. This contradicts the positivity of k. Let $\gamma = \pi$. Then $\cos\theta_C(\beta_1)(k-1) = 0$. Thus $k = 1$. But this implies that $\mathbf{P}^{b|a}$ is double stochastic.

Despite Theorem 4.2, we can still hope that there can be found some extended family of contexts such that (4.40) would hold true for contexts from that family. But it is impossible.

Proposition 4.3. *Let condition* (4.40) *hold true for two contexts* C_1, C_2 *such that*

$$|\lambda(\beta_1|a, C_1)| \neq |\lambda(\beta_1|a, C_2)|. \tag{4.42}$$

Then $\mathbf{P}^{b|a}$ *is double stochastic.*

Proof. We set $\theta = \theta_{C_1}(\beta_1)$ and $\theta' = \theta_{C_2}(\beta_1)$. We have

$$-k\cos\theta = \cos(\theta + \gamma), \qquad -k\cos\theta' = \cos(\theta' + \gamma).$$

Thus

$$-k\cos\frac{\theta + \theta'}{2}\cos\frac{\theta - \theta'}{2} = \cos\left(\frac{\theta + \theta'}{2} + \gamma\right)\cos\frac{\theta - \theta'}{2}.$$

By (4.42) we have that $\cos\frac{\theta-\theta'}{2} \neq 0$ and hence $-k\cos\frac{\theta+\theta'}{2} = \cos(\frac{\theta+\theta'}{2} + \gamma)$. We also have

$$k\sin\frac{\theta + \theta'}{2}\sin\frac{\theta - \theta'}{2} = -\sin\left(\frac{\theta + \theta'}{2} + \alpha\right)\sin\frac{\theta - \theta'}{2}.$$

By (4.42) we have that $\sin\frac{\theta-\theta'}{2} \neq 0$ and hence $-k\sin\frac{\theta+\theta'}{2} = \sin(\frac{\theta+\theta'}{2} + \gamma)$. Thus $k^2 = 1$ and hence $k = 1$. Hence the matrix $\mathbf{P}^{b|a}$ is double stochastic.

Thus if $\mathbf{P}^{b|a}$ is not double stochastic, then every surface

$$M_t = \{C \in \mathscr{C}^{\text{tr}} : |\lambda(\beta_1|\alpha, C)| = t\}, \quad 0 \le t \le 1,$$

in the space of contexts is represented in its own Hilbert space $\mathscr{H}_a(t)$.

4.7 Noncommutativity of Operators Representing Observables

Let the matrix of probabilities $\mathbf{P}^{b|a}$ be double stochastic. We consider in this section the case of real-valued observables. Here the ranges of observables b and a are subsets of $\mathbf{R}$. We set

$$q_1 = \sqrt{p_{11}^{b|a}} = \sqrt{p_{22}^{b|a}}$$

and

$$q_2 = \sqrt{p_{12}^{b|a}} = \sqrt{p_{21}^{b|a}}.$$

Thus the vectors of the a-basis, see (4.19), have the following form

$$e^a_{\alpha_1} = \begin{pmatrix} q_1 \\ q_2 \end{pmatrix}, \qquad e^a_{\alpha_2} = \begin{pmatrix} e^{i\theta_1} q_2 \\ e^{i\theta_2} q_1 \end{pmatrix}.$$

Since $\theta_2 = \theta_1 + \pi$, we get

$$e^a_{\alpha_2} = e^{i\theta_2} \begin{pmatrix} -q_2 \\ q_1 \end{pmatrix}.$$

We now find matrices of operators $\hat{a}$ and $\hat{b}$ in the b-representation. The latter one is diagonal. For $\hat{a}$ we have

$$\hat{a} = V^\star \operatorname{diag}(\alpha_1, \alpha_2) V,$$

where V is the matrix of transition from the b-basis to the a-basis. Thus

$$\hat{a} = \begin{pmatrix} \alpha_1 q_1^2 + \alpha_2 q_2^2 & (\alpha_1 - \alpha_2) q_1 q_2 \\ (\alpha_1 - \alpha_2) q_1 q_2 & \alpha_1 q_2^2 + \alpha_2 q_1^2 \end{pmatrix}.$$

We remark that by varying the matrix $\mathbf{P}^{b|a}$ we can obtain any symmetric matrix with real coefficients. We do not obtain matrices with complex coefficients (as a consequence of the special choice of a and b-bases).

Hence

$$[\hat{b}, \hat{a}] = \hat{m},$$

where

$$\hat{m} = \begin{pmatrix} 0 & (\alpha_1 - \alpha_2)(\beta_2 - \beta_1) q_1 q_2 \\ (\alpha_1 - \alpha_2)(\beta_2 - \beta_1) q_1 q_2 & 0 \end{pmatrix}.$$

Since $\alpha_1 \neq \alpha_2, \beta_1 \neq \beta_2$ and $q_j \neq 0$, we have $\hat{m} \neq 0$.

4.8 Symmetrically Conditioned Observables

We recall that in quantum mechanics, matrices $\mathbf{P}^{b|a}$ and $\mathbf{P}^{a|b}$ always satisfy the following condition of the interchange symmetry

$$p^{b|a}(\beta|\alpha) = p^{a|b}(\alpha|\beta). \tag{4.43}$$

This is a consequence of symmetry of the scalar product. We recall that in Chap. 3 we called arbitrary reference observables a and b (i.e., having no direct relation to QM) satisfying this condition *symmetrically conditioned.* We recall that by Lemma 3.3, Chap. 3, symmetrically conditioned reference observables are always *probabilistically balanced,* i.e., both matrices $\mathbf{P}^{b|a}$ and $\mathbf{P}^{a|b}$ are double stochastic. In this section we would like to study some special features of our representation of trigonometric contexts in this case.

4.8.1 b-Selections Are Trigonometric Contexts

Theorem 4.3. *Let the matrix* $\mathbf{P}^{b|a}$ *be double stochastic. The contexts* $C_\beta, \beta \in X_b$, *belong to* $\mathcal{C}^{\rm tr}$ *iff the reference observables a and b are symmetrically conditioned.*

Proof. (A) We have

$$\lambda(\beta_2|a, C_{\beta_1}) = -\frac{\mu_1^2 + \mu_2^2}{2\mu_1\mu_2},$$

where $\mu_j = \sqrt{p^a_{C_{\beta_1}}(\alpha_j) p^{b|a}(\beta_2|\alpha_j)}$. So $\lambda(\beta_2|a, C_{\beta_1}) \geq 1$ and we have the trigonometric behavior only in the case $\mu_1 = \mu_2$. Thus

$$p^a_{C_{\beta_1}}(\alpha_1) p^{b|a}(\beta_2|\alpha_1) = p^a_{C_{\beta_1}}(\alpha_2) p^{b|a}(\beta_2|\alpha_2).$$

In this case $\lambda(\beta_2|a, C_{\beta_1}) = -1$ and hence $\theta(\beta_2|a, C_{\beta_1}) = \pi$, and consequently $\theta(\beta_1|a, C_{\beta_1}) = 0$. We point out that $p^a_{C_\beta}(\alpha) = p^{a|b}(\alpha|\beta)$. Thus we have

$$p^{a|b}(\alpha_1|\beta_1) p^{b|a}(\beta_2|\alpha_1) = p^{a|b}(\alpha_2|\beta_1) p^{b|a}(\beta_2|\alpha_2). \tag{4.44}$$

In the same way by using conditioning with respect to C_{β_2} we obtain

$$p^{a|b}(\alpha_1|\beta_2) p^{b|a}(\beta_1|\alpha_1) = p^{a|b}(\alpha_2|\beta_2) p^{b|a}(\beta_1|\alpha_2).$$

By using double stochasticity of $\mathbf{P}^{b|a}$ we can rewrite the last equality as

$$p^{a|b}(\alpha_1|\beta_2) p^{b|a}(\beta_2|\alpha_2) = p^{a|b}(\alpha_2|\beta_2) p^{b|a}(\beta_2|\alpha_1). \tag{4.45}$$

Thus by (4.44) and (4.45) we have

$$\frac{p^{a|b}(\alpha_1|\beta_2)}{p^{a|b}(\alpha_2|\beta_1)} = \frac{p^{a|b}(\alpha_2|\beta_2)}{p^{a|b}(\alpha_1|\beta_1)}.$$

Hence $p^{a|b}(\alpha_1|\beta_2) = tp^{a|b}(\alpha_2|\beta_1)$ and $p^{a|b}(\alpha_2|\beta_2) = tp^{a|b}(\alpha_1|\beta_1), t > 0$. But $1 = p^{a|b}(\alpha_1|\beta_2) + p^{a|b}(\alpha_2|\beta_2) = t[p^{a|b}(\alpha_2|\beta_1) + p^{a|b}(\alpha_1|\beta_1)] = t$. We proved that the matrix $P^{a|b}$ is also double stochastic. Thus the reference observables a and b are probabilistically balanced. We now prove even more: they are symmetrically conditioned. By the equality (4.44) there exists $k > 0$ such that

$$\frac{p^{a|b}(\alpha_1|\beta_1)}{p^{a|b}(\alpha_2|\beta_1)} = \frac{p^{b|a}(\beta_2|\alpha_2)}{p^{b|a}(\beta_2|\alpha_1)} = k. \tag{4.46}$$

Thus

$$p^{a|b}(\alpha_1|\beta_1) = kp(\alpha_2|\beta_1), \qquad p^{b|a}(\beta_2|\alpha_2) = kp^{b|a}(\beta_2|\alpha_1).$$

But $1 = p^{a|b}(\alpha_1|\beta_1) + p^{a|b}(\alpha_2|\beta_1) = (k+1)p^{a|b}(\alpha_2|\beta_1)$ (because $P^{a|b}$ is a stochastic matrix) and $1 = p^{b|a}(\beta_2|\alpha_2) + p^{b|a}(\beta_2|\alpha_1) = (k+1)p^{b|a}(\beta_2|\alpha_1)$ (because $P^{b|a}$ is a double stochastic matrix). Thus

$$p^{a|b}(\alpha_2|\beta_1) = p^{b|a}(\beta_2|\alpha_1) = p^{b|a}(\beta_1|\alpha_2), \tag{4.47}$$

$$p^{a|b}(\alpha_1|\beta_1) = p^{b|a}(\beta_2|\alpha_2) = p^{b|a}(\beta_1|\alpha_1) \tag{4.48}$$

(we have used again that $P^{b|a}$ is double stochastic). Finally, by using double stochasticity of $P^{a|b}$ we obtain

$$p^{a|b}(\alpha_1|\beta_2) = p^{a|b}(\alpha_2|\beta_1) = p^{b|a}(\beta_2|\alpha_1), \tag{4.49}$$

$$p^{a|b}(\alpha_2|\beta_2) = p^{a|b}(\alpha_1|\beta_1) = p^{b|a}(\beta_2|\alpha_2). \tag{4.50}$$

(B) Now let the reference observables a and b be symmetrically conditioned. Lemma 3.3, Chap. 3 of this part, implies that they are probabilistically balanced. Therefore

$$p^{a|b}(\alpha_1|\beta_1)p^{b|a}(\beta_2|\alpha_1) = p^{a|b}(\alpha_2|\beta_2)p^{b|a}(\beta_1|\alpha_2) = p^{a|b}(\alpha_2|\beta_1)p^{b|a}(\beta_2|\alpha_2). \tag{4.51}$$

Thus we obtained the equality (4.44). It implies that C_{β_1} belongs to $\mathscr{C}^{\rm tr}$. In the same way we prove that C_{β_2} belongs to $\mathscr{C}^{\rm tr}$.

Lemma 4.4. *Let the reference observables a and b be symmetrically conditioned. Then*

$$\lambda(\beta|a, C_\beta) = 1, \quad \beta \in X_b. \tag{4.52}$$

Proof. Here $\delta(\beta|a, C_\beta) = 1 - p^{b|a}(\beta|\alpha_1)p^{a|b}(\alpha_1|\beta) - p^{b|a}(\beta|\alpha_2)p^{a|b}(\alpha_2|\beta) = 1 - p^{a|b}(\alpha_1|\beta)^2 - p^{a|b}(\alpha_2|\beta)^2 = 2p^{a|b}(\alpha_1|\beta)p^{a|b}(\alpha_2|\beta)$. Hence

$$\lambda(\beta|a, C_\beta) = \frac{\delta(\beta|a, C_\beta)}{2\sqrt{p_{C_\beta}(\alpha_1)p_{C_\beta}(\alpha_2)p^{b|a}(\beta|\alpha_1)p^{b|a}(\beta|\alpha_2)}}.$$

We now remark that a and b are symmetrically conditioned. Thus $\lambda(\beta|a, C_\beta) = 1$.

By (4.52) we have

$$\lambda(\beta_i|a, C_{\beta_j}) = -1, \quad i \neq j.$$

Thus

$$\theta(\beta_i|a, C_{\beta_i}) = 0 \quad \text{and} \quad \theta(\beta_i|a, C_{\beta_j}) = \pi, \quad i \neq j.$$

Proposition 4.4. *Let the reference observables a and b be symmetrically conditioned. Then*

$$J^{b|a}(C_{\beta_j})(\beta) = \delta(\beta_j - \beta), \quad \beta \in X_b, \quad \textit{and}$$

$$J^{a|b}(C_{\alpha_j})(\alpha) = \delta(\alpha_j - \alpha), \quad \alpha \in X_a.$$

Proof. Since $\theta(\beta_1|a, C_{\beta_1}) = 0$ we have

$$\begin{aligned} J^{b|a}(C_{\beta_1})(\beta_1) &= \sqrt{p^{a|b}(\alpha_1|\beta_1)p^{b|a}(\beta_1|\alpha_1)} + e^{i0}\sqrt{p^{a|b}(\alpha_2|\beta_1)p^{b|a}(\beta_1|\alpha_2)} \\ &= p^{a|b}(\alpha_1|\beta_1) + p^{a|b}(\alpha_2|\beta_1) = 1. \end{aligned}$$

Since $\theta(\beta_2|a, C_{\beta_1}) = \pi$ we have

$$J^{b|a}(C_{\beta_1})(\beta_2) = \sqrt{p^{a|b}(\alpha_1|\beta_1)p^{b|a}(\beta_2|\alpha_1)} + e^{i\pi}\sqrt{p^{a|b}(\alpha_2|\beta_1)p^{b|a}(\beta_2|\alpha_2)}$$
$$= \sqrt{p^{a|b}(\alpha_1|\beta_1)}(\sqrt{p^{b|a}(\beta_2|\alpha_1} - \sqrt{p^{a|b}(\alpha_2|\beta_1)}) = 0.$$

Thus in the case of symmetrically conditioned reference observables a and b we have

$$J^{b|a}(C_\beta) = e^b_\beta, \beta \in X_b$$

and Born's rule has the form

$$p^b_C(\beta) = |\langle \psi_C, \psi_{C_\beta}\rangle|^2. \tag{4.53}$$

4.8.2 Extension of Representation Map

We can formally extend the map $J^{b|a}$ to contexts C_α by (4.36). We set

$$\overline{\mathscr{C}^{\text{tr}}} = \mathscr{C}^{\text{tr}} \bigcup_{\alpha \in X_a} C_\alpha.$$

Thus we have constructed the Hilbert space representation $J^{b|a} : \overline{\mathscr{C}}^{\text{tr}} \to S$.

The domain of definition of QLRA contains now the selection contexts for both reference observables.

4.9 Formalization of the Notion of Quantum-Like Representation

We have constructed the representation of trigonometric contexts by normalized vectors of complex Hilbert space and the corresponding representation of reference observables by self-adjoint operators. It would be interesting to extend this representation to larger sets of observables $\mathscr{O}$. We start with formalization of the notion of the QL-representation.

For the unit sphere S of complex Hilbert space $\mathscr{H}$, the set of equivalent classes $C_\psi = \{\phi = e^{i\sigma}\psi : \sigma \in [0, 2\pi)\}$, $\psi \in S$, is denoted by the symbol $\tilde{S}$; the set of self-adjoint operators, $\hat{u} : \mathscr{H} \to \mathscr{H}$, is denoted by the symbol $L_s(\mathscr{H})$. The symbol P^u_γ denotes the spectral projector onto the eigenspace of the operator $\hat{u}$ corresponding to its eigenvalue γ.

Definition 4.4. *Consider an arbitrary Växjö model. Let two observables $a, b \in \mathscr{O}$ be fixed—reference observables. A QL representation of this model (corresponding to these reference observables) is defined by a pair of maps with the domains of definition $\mathscr{C}_{J_1} \subset \mathscr{C}$ and $\mathscr{O}_{J_2} \subset \mathscr{O}$, respectively: $J_1 : \mathscr{C}_{J_1} \to \tilde{S}$, and $J_2 : \mathscr{O}_{J_2} \to L_s(\mathscr{H})$. These maps have the following properties:*

(AV) *For any observable $d \in \mathscr{O}_{J_2}$ and any context $C \in \mathscr{C}_{J_1}$ the contextual and quantum averages coincide*

$$E[d|C] = \langle J_2(d) J_1(C), J_1(C)\rangle \tag{4.54}$$

(if $J_1(C)$ belongs to the domain of definition of the operator $J_2(d)$).

(RO) *Both reference observables $u = a, b$ belong to $\mathscr{O}_{J_2}$, the corresponding selection contexts C^u_γ, $\gamma \in X_u$, belong to $\mathscr{C}_{J_1}$. Moreover,*

   (a) *the range of values X_u of the observable u coincides with the spectrum of the corresponding operator $\hat{u}$;*

   (b) *the contextual probability distribution coincides with the corresponding quantum probability distribution given by the Born rule*

$$\mathbf{P}(u = \gamma | C) = \| P^u_\gamma J_1(C) \|^2, \quad C \in \mathscr{C}_{J_1}. \tag{4.55}$$

If the operator $\hat{u} = J_2(u)$ has nondegenerate (purely discrete) spectrum, then

$$P^u_\gamma = J_1(C^u_\gamma) \otimes J_1(C^u_\gamma), \quad \gamma \in X_u. \tag{4.56}$$

If (a) and (b) hold for any observable $d \in \mathcal{O}_{J_2}$ (and not only for the reference observables $u = a, b$), then the QL-representation is called *strong*. From consideration in Sect. 4.6, we obtain:

Theorem 4.4. *Let the reference observables a and b be symmetrically conditioned (as well as probabilistically conjugate and dichotomous). Then the pair of maps* $J_1 \equiv J^{a|b} : \overline{\mathcal{C}^{\rm tr}} \to \tilde{S}$ *and* $J_2 : \{a, b\} \to L_s(\mathcal{H}_b^{b|a})$, $J_2(a) = \hat{a}$, $J_2(b) = \hat{b}$, *where the operators* $\hat{a}, \hat{b}$ *were defined in Sect. 4.7, give the strong QL representation.*

Of course, we would be more happy to construct strong representations. However, simple examples with $\mathcal{O}_{J_2}$ larger than just the set of reference observables $\{a, b\}$ will show, see Chap. 4, that, for an arbitrary observable $d \in \mathcal{O}_{J_2}$, one cannot expect more than coincidence of the contextual ("classical") and QL averages. Only for the reference observables a and b are the probability distributions also preserved.

We now proceed with *probabilistically conjugate and symmetrically conditioned dichotomous* reference observables. We also assume that all observables belonging to $\mathcal{O}$ take values in $\mathbf{R}$. We would like to extend the QL-representation given by Theorem 4.4.

Proposition 4.5. *For any map* $f : \mathbf{R} \to \mathbf{R}$, *we have*:

$$E[f(a)|C] = \langle f(\hat{a}) \rangle_{\psi_C} \equiv \langle f(\hat{a}) J^{a|b}(C), J^{a|b}(C) \rangle, \tag{4.57}$$

$$E[f(b)|C] = \langle f(\hat{b}) \rangle_{\psi_C} \equiv \langle f(\hat{b}) J^{a|b}(C), J^{a|b}(C) \rangle \tag{4.58}$$

for any context $C \in \overline{\mathcal{C}^{\rm tr}}$.

Proof. Since in the b-representation the Born rule holds, we obtain

$$E[f(b)|C] = \sum_{\beta \subset X_b} f(\beta) p_c^b(\beta) = \sum_{\beta \subset X_b} f(\beta) |\langle \psi_C, e_\beta^b \rangle|^2 = \langle f(\hat{b}) \rangle_{\psi_C},$$

•

where $\psi_C = J^{a|b}(C)$. We have the same result as for the $f(\hat{a})$, since we have the Born probability rule both for b and a (because the matrix $\mathbf{P}^{b|a}$ is double stochastic).

Proposition 4.6. *Let $f, g : \mathbf{R} \to \mathbf{R}$ be two arbitrary functions. Then*

$$E[f(a) + g(b)|C] = \langle f(\hat{a}) + g(\hat{b})\rangle_{\psi_C} \tag{4.59}$$

for any context $C \in \overline{\mathscr{C}^{\mathrm{tr}}}$.

Proof. By using linearity of the mathematical expectation, and linearity of the Hilbert space scalar product we obtain

$$\begin{aligned} E[f(a) + g(b)|C] &= E[f(a|C] + E[g(b)|C] \\ &= \langle f(\hat{a})\rangle_{\psi_C} + \langle g(\hat{b})\rangle_{\psi_C} = \langle f(\hat{a}) + g(\hat{b})\rangle_{\psi_C}. \end{aligned}$$

Denote the linear space of all observables of the form $d = f(a) + g(b)$ by the symbol $\mathscr{O}_+(a, b)$. We assume that

$$\mathscr{O}_+(a, b) \subset \mathscr{O}. \tag{4.60}$$

Proposition 4.7. *The map*

$$J_2^{a|b} : \mathscr{O}_+(a, b) \to L_s(\mathscr{H}_b^{b|a}), \qquad d = f(a)+g(b) \to \hat{d} = f(\hat{a})+g(\hat{b}), \tag{4.61}$$

preserves the conditional expectation:

$$E[d|C] = \langle J_2^{a|b}(d)\rangle_{\psi_C} \equiv \langle J^{a|b}(d)J^{a|b}(C), J^{a|b}(C)\rangle \tag{4.62}$$

for any context $C \in \overline{\mathscr{C}^{\mathrm{tr}}}$.

As a consequence of this proposition, we have

Theorem 4.5. *Let the reference observables a and b be symmetrically conditioned (as well as probabilistically conjugate and dichotomous) and let* (4.60) *take place.*

Then maps $J_1 \equiv J^{a|b} : \overline{\mathscr{C}^{\rm tr}} \to \tilde{S}$ *and* $J_2 \equiv J_2^{a|b} : \mathscr{O}_+(a,b) \to L_s(\mathscr{H}_b^{b|a})$ *give the QL representation.*

The transformation $J_2^{a|b}$ preserves the conditional expectation for observables $d \in \mathscr{O}_+(a,b)$. In general we cannot expect anything more, since in general $J_2^{a|b}$ does not preserve probability distributions. It preserves them only for the reference observables, see Chap. 6 for a counterexample.

The important problem is to extend the map $J_2^{a|b}$ to even a larger class of observables with preserving (at least) the averages. It might be natural to define (as we always do in the conventional quantum formalism):

$$J_2^{a|b}(f)(\hat{a},\hat{b}) = f(\hat{a},\hat{b})$$

where $f(\hat{a},\hat{b})$ is the pseudo differential operator with the Weyl symbol $f(a,b)$. We shall see in Chap. 6 that already for the function

$$f(a,b) = ab \to f(\hat{a},\hat{b}) = (\hat{a}\hat{b} + \hat{b}\hat{a})/2$$

even the equality (4.62) is violated.

4.10 Domain of Application of Quantum-Like Representation Algorithm

In this section we collect conditions providing the possibility to apply QLRA and obtain a natural QL representation of probabilities.

(R1) The reference observables a and b are symmetrically conditioned:[6]

$$p^{b|a}(\beta|\alpha) - p^{a|b}(\alpha|\beta).$$

[6] This condition induces symmetry of the scalar product and equivalence of the $b|a$ and $a|b$ representations.

(R2) The reference observables a and b are probabilistically conjugate (mutually nondegenerate)[7]

$$p^{a|b}(\alpha|\beta) > 0, \qquad p^{b|a}(\beta|\alpha) > 0.$$

(R2a) Context C is nondegenerate with respect to both reference observables a and b:

$$p_C^b(\beta) > 0, \qquad p_C^a(\alpha) > 0.$$

Suppose that also the following conditions hold:

(R3) Coefficients of supplementarity are bounded by 1[8]

$$\left|\frac{p_C^b(\beta) - \sum_\alpha p_C^a(\alpha)p^{b|a}(\beta|\alpha)}{2\sqrt{\prod_\alpha p_C^a(\alpha)p^{b|a}(\beta|\alpha)}}\right| \leq 1,$$
$$\left|\frac{p_C^a(\alpha) - \sum_\beta p_C^b(\beta)p^{a|b}(\alpha|\beta)}{2\sqrt{\prod_\alpha p_C^b(\beta)p^{a|b}(\alpha|\beta)}}\right| \leq 1.$$

Under these conditions we can apply QLRA (to probabilistic data). The QL representation $\mathcal{H}_b^{b|a}$, see (4.13), is unitary equivalent to the representation $\mathcal{H}_a^{b|a}$, see (4.39). Thus we can identify these two representations. In the same way we can identify the representations $\mathcal{H}_a^{a|b}$ and $\mathcal{H}_b^{a|b}$.

---

[7] This condition induces noncommutativity of operators $\hat{a}$ and $\hat{b}$ representing these observables.

[8] This condition induces the QL-representation of the context C in the complex Hilbert space. Thus complex numbers appear due to this condition.

Chapter 5
Ensemble Representation of Contextual Statistical Model

In this chapter we consider the representation of contexts (e.g., physical) by ensembles of systems. The basic formulas of this chapter can be obtained as particular cases of the formulas which have been derived in the general contextual framework in Chap. 3. Nevertheless, we present the *direct frequency derivations which are based on the ensemble representation of contexts.* These derivations are essentially longer and more technical than the abstract contextual derivations. However, by operating with systems and their *objective properties* one gets a better feeling of the classical (frequency-)probabilistic basis of the interference phenomena.

5.1 Systems and Contexts

We start with a contextual probabilistic model (Växjö model) based on the contextual space

$$\mathcal{P}_{\text{cont}} = (\mathcal{C}, \mathcal{O}, \pi),$$

where $\mathcal{C}$ and $\mathcal{O}$ are the collections of contexts and observables, π is the collection of corresponding probability distributions. We now consider a class of Växjö models such that *contexts $C \in \mathcal{C}$ can be represented by ensembles of systems.*

Let S be a set where elements are considered as systems (e.g., physical or social). It is assumed that any context $C \in \mathscr{C}$ is represented by an ensemble of systems—elements of S.

It is important to point out that observables belonging to $\mathcal{O}$ could be considered as *objective properties of systems.* Thus the interference of probabilities can appear even in the model based on the objective interpretation of observers.

This chapter is also very important for applications outside the domain of quantum mechanics. In such applications one operates with systems (typically macroscopic) having objective properties.

To describe completely all probabilistic features of a context C, the corresponding ensemble should be infinite. To escape operating with infinite ensembles, we represent any context $C \in \mathscr{C}$ by a family of finite ensembles $\mathbf{S}_C = \{S_C\}$ such that

$$\overline{\lim}_{S_C \in \mathbf{S}_C} |S_C| = \infty,$$

where $|S_C|$ denotes the number of elements in S_C.

It is assumed that, for any observable (property) $d \in \mathcal{O}$, contextual probabilities of the type $\mathbf{P}(d = \gamma|C)$, where $\gamma \in X_d$, can be represented as the frequency probabilities. Thus, for any value $\gamma \in X_d$, the limit

$$\mathbf{P}(d = \gamma|C) = \lim_{S_C \in \mathbf{S}_C : |S_C| \to \infty} \mathbf{P}_{S_C}(d = \gamma) \tag{5.1}$$

exists. Here

$$\mathbf{P}_{S_C}(d = \gamma) \equiv \nu(d = \gamma; S_C) = \frac{|\{s \in S_C : d(s) = \gamma\}|}{|S_C|}$$

is the proportion of the number of elements in the ensemble S_C having the property $d = \gamma$ to the total number of elements in this ensemble (frequency).

Definition 5.1. *The ensemble representation of a contextual probabilistic model is given by a pair*

$$E(M) = (S, \mathbf{S}(\mathscr{C})), \tag{5.2}$$

where S is a set of systems, and

$$\mathbf{S}(\mathscr{C}) = \{\mathbf{S}_C : C \in \mathscr{C}\}$$

is a set of families $\mathbf{S}_C = \{S_C\}$ of ensembles representing contexts belonging to the collection $\mathscr{C}$.

5.2 Interference of Probabilities: Ensemble Derivation

We again restrict our consideration to two (dichotomous) reference observables, $b = \beta_1, \beta_2$ and $a = \alpha_1, \alpha_2$. We assume that they are probabilistically conjugate, see Chap. 3 of this part, Definition 3.6.

Let us consider a context C. It is assumed that it is a-nondegenerate. The context C is represented by the family $\mathbf{S}_C = \{S_C\}$ of finite ensembles of systems. We fix one of such ensembles, say S_C. The number of elements $|S_C|$ in the ensemble S_C is denoted by

$$N \equiv N_C.$$

Let C_{α_1} and C_{α_2} be the contexts corresponding to α_1 and α_2-selections. It is assumed that elements of the ensemble S_C can interact with each of these contexts. These interactions are nothing else than measurements of the a-observable and selections of systems giving the results α_1 and α_2, respectively. Interactions (in fact, selections) of systems with contexts C_{α_i} produce a new statistical ensemble

$$S_{\alpha_i} \equiv S_{C_{\alpha_i}}.$$

As was mentioned, contexts C_{α_i} can be considered as filters with respect to the property a: the context C_{α_i} selects elements of the ensemble S_C such that $a(s) = \alpha_i$, $s \in S_C$ $(i = 1, 2)$. Such a procedure justifies the assumption that the number of elements in the ensemble S_{α_i} can be chosen equal to the number of elements, N_i, in the original ensemble S_C having the property $a(s) = \alpha_i (i = 1, 2)$. Here we used the assumption that the a-observations are about the objective property, say a, of systems. Thus everywhere below

$$|S_{\alpha_i}| = N_i, \quad i = 1, 2.$$

The crucial point of our considerations is that *in general we could not select, for example, elements with the property $a = \alpha_1$ without disturbing the property b*. In general the sub-ensemble

$$S_{\alpha_i \beta_j} = \{s \in S_{\alpha_i} : b(s) = \beta_j\}$$

of the ensemble S_{α_i} does not coincide with the sub-ensemble

$$S^{(0)}_{\alpha_i \beta_j} = \{s \in S_C : a(s) = \alpha_i, b(s) = \beta_j\}$$

of the original ensemble S_C.

We remark that, e.g., in *quantum physics* the ensembles $S_{\alpha_i \beta_j}$ can always be prepared in the experimental framework, but ensembles $S^{(0)}_{\alpha_i \beta_j}$ could not be prepared experimentally for some pairs of observable, so-called incompatible observables (which are represented in the quantum formalism by noncommutative operators).

We set

$$n_{ij} = |S^{(0)}_{\alpha_i \beta_j}|$$

and

$$m_{ij} = |S_{\alpha_i \beta_j}|.$$

These are numbers of elements in the sub-ensembles $S^{(0)}_{\alpha_i \beta_j}$ and $S_{\alpha_i \beta_j}$, respectively. We also set

$$N_i = |\{s \in S_C : a(s) = \alpha_i\}|,$$

$$n_j = |\{s \in S_C : b(s) = \beta_j\}|.$$

These are numbers of elements in the original ensemble S_C having, respectively, properties $a(s) = \alpha_i, i = 1, 2$ and $b(s) = \beta_j, j = 1, 2$. We note that everywhere below the first number, i, in the index pair ij is related to the property a and the second one, j, to the property b.

We consider relative frequencies for $b = \beta_j$ and $a = \alpha_i$ in the ensemble S_C

$$\nu_N(\beta_j; S_C) = \frac{n_j}{N}, \qquad \nu_N(\alpha_j; S_C) = \frac{N_i}{N}$$

and for $b = \beta_j$ in the ensembles S_{α_i}

$$\nu_{N_i}(\beta_j; S_{\alpha_i}) = \frac{m_{ij}}{N_i}$$

and the corresponding probabilities

$$p_j^b \equiv \mathbf{P}(b = \beta_j | C) = \lim_{N \to \infty} \nu_N(\beta_j; S_C);$$

$$p_i^a \equiv \mathbf{P}(a = \alpha_i | C) = \lim_{N \to \infty} \nu_N(\alpha_j; S_C);$$

$$p_{ij}^{b|a} \equiv \mathbf{P}(b = \beta_j | C_{\alpha_i}) = \lim_{N_i \to \infty} \nu_{N_i}(\beta_j; S_{\alpha_i}).$$

Since in general n_{ij} do not equal m_{ij} (even asymptotically when $N \to \infty$), we do not have the conventional formula of total probability. In general

$$p_j^b \neq p_1^a p_{11}^{b|a} + p_2^a p_{21}^{b|a}.$$

We want to investigate the various forms probabilities p_j^b can take, depending on perturbations induced by context transitions $S_C \to S_{\alpha_1}, S_{\alpha_2}$. In the general case we have

$$\begin{aligned}
\nu_N(\beta_j; S_C) &= \frac{n_j}{N} = \frac{n_{1j}}{N} + \frac{n_{2j}}{N} = \frac{m_{1j}}{N} + \frac{m_{2j}}{N} + \delta_j^{(N)} \\
&= \frac{N_1}{N} \cdot \frac{m_{1j}}{N_1} + \frac{N_2}{N} \cdot \frac{m_{2j}}{N_2} + \delta_j^{(N)} \\
&= \nu_N(\alpha_1; S_C)\nu_{N_1}(\beta_j; S_{\alpha_1}) + \nu_N(\alpha_2; S_C)\nu_{N_2}(\beta_j; S_{\alpha_2}) + \delta_j^{(N)},
\end{aligned}$$

where the perturbation term (which appears due to the transition from the ensemble S_C to ensembles S_{α_1} and S_{α_2}) has the form

$$\delta_j^{(N)} \equiv \delta_j^{(N)}(\beta_j|a, C) = \frac{1}{N}[(n_{1j} - m_{1j}) + (n_{2j} - m_{2j})].$$

We remark that there exists the limit

$$\delta_j = \lim_{N\to\infty} \delta_j^{(N)} = p_j^b - (p_1^a p_{1j}^{b|a} + p_2^a p_{2j}^{b|a}),$$

since

$$\delta_j^{(N)} = \nu_N(\beta_j; S_C) - \nu_N(\alpha_1; S_C)\nu_{N_1}(\beta_j; S_{\alpha_1}) + \nu_N(\alpha_2; S_C)\nu_{N_2}(\beta_j; S_{\alpha_2}).$$

The limits of all frequencies in the right-hand side of the last equality exist as a consequence of the principle of statistical stabilization, see Part I: Chap. 2. Thus in general we have

$$p_j^b = p_1^a p_{1j}^{b|a} + p_2^a p_{2j}^{b|a} + \delta_j,$$

where

$$\delta_j = \lim_{N\to\infty} \frac{1}{N}[(n_{1j} - m_{1j}) + (n_{2j} - m_{2j})].$$

It is useful, cf. Part I: Chap. 2, to make normalization by setting

$$\delta_j = 2\sqrt{p_1^a p_{1j} p_2^a p_{2j}}\lambda_j, \quad j = 1, 2.$$

The coefficients λ_j are nothing else than coefficients of supplementarity. By repeating arguments of Chap. 3 we see that there are three possibilities.

(1) $|\lambda_j| \leq 1$; (2) $|\lambda_j| > 1$;

(3) $|\lambda_1| \leq 1$ and $|\lambda_2| > 1$ or $|\lambda_1| > 1$ and $|\lambda_2| \leq 1$.

In the first case we can always represent the coefficient as $\lambda_j = \cos\theta_j$, $j = 1, 2$; in the second case—as $\lambda_j = \pm\cosh\theta_j$, $j = 1, 2$; in the third case—as $\lambda_1 = \cos\theta_1$ and $\lambda_2 = \pm\cosh\theta_2$ or vice versa. Here the probabilistic phases $\theta_j \equiv$

$\theta_j(\beta_j|a, C)$. Probabilistic behaviors of these types are called *trigonometric, hyperbolic and hyper-trigonometric* behaviors, respectively.

We have studied the general case. There are three contexts C, C_{α_1} and C_{α_2} such that C_{α_1} and C_{α_2} are selections with respect to values $a = \alpha_1$ and $a = \alpha_2$. The general probabilistic transformation induced by transitions $C \to C_{\alpha_i}$, $i = 1, 2$, has the form

$$p_j^b = p_1^a p_{1j}^{b|a} + p_2^b p_{2j}^{b|a} \pm 2\sqrt{p_1^a p_{1j}^{b|a} p_2^c p_{2j}^{b|a}}\, \lambda_j \tag{5.3}$$

where $\lambda_j = \cos\theta_j$ or $\lambda_j = \cosh\theta_j$, or $\lambda_1 = \cos\theta_1$ and $\lambda_2 = \pm\cosh\theta_2$ or vice versa. Here

$$\lambda_j = \lim_{N\to\infty} \lambda_j^{(N)}, \quad \text{where } \lambda_j^{(N)} = \frac{1}{2\sqrt{m_{1j}m_{2j}}}[(n_{1j} - m_{1j}) + (n_{2j} - m_{2j})]. \tag{5.4}$$

We have seen in Chap. 3, Lemma 3.1, that the coefficients λ_1 and λ_2 are connected by the "condition of orthogonality" (in the quantum formalism this is the real condition of orthogonality in the complex Hilbert space)

$$\sqrt{p_1^a p_{11}^{b|a} p_2^a p_{22}^{b|a}}\, \lambda_1 + \sqrt{p_1^a p_{12}^{b|a} p_2^a p_{22}^{b|a}}\, \lambda_2 = 0.$$

Since the original context C is a-nondegenerate, probabilities p_i^a, $i = 1, 2$, are strictly positive, this implies

$$\sqrt{p_{11}^{b|a} p_{21}^{b|a}}\, \lambda_1 + \sqrt{p_{12}^{b|a} p_{22}^{b|a}}\, \lambda_2 = 0. \tag{5.5}$$

We set

$$K = \sqrt{\frac{p_{12}^{b|a} p_{22}^{b|a}}{p_{11}^{b|a} p_{21}^{b|a}}}. \tag{5.6}$$

We get

$$\lambda_1 = -K\lambda_2.$$

We observe that probabilities p_j^a are not involved in the condition of orthogonality (5.5). In particular, in the trigonometric case we always have

$$\cos\theta_1 = -K\cos\theta_2 \tag{5.7}$$

in the hyperbolic case we have

$$\cosh\theta_1 = K\cosh\theta_2 \tag{5.8}$$

(here $\lambda_1 = \pm\cosh\theta_1$ and $\lambda_2 = \mp\cosh\theta_2$). In the hyper-trigonometric case we have

$$\cos\theta_1 = \pm K\cosh\theta_2 \quad \text{or} \quad \cosh\theta_1 = \pm K\cos\theta_2. \tag{5.9}$$

5.3 Classical and Nonclassical Probabilistic Behaviors

In this section we consider contexts that produce relatively small coefficients of supplementarity, so-called trigonometric contexts

$$|\lambda_j| \leq 1, \quad j = 1, 2.$$

In the language of ensemble frequencies this means that for $j = 1, 2$

$$\left|\frac{(n_{1j} - m_{1j}) + (n_{2j} - m_{2j})}{2\sqrt{m_{1j}m_{2j}}}\right| \leq 1, \quad N \to \infty. \tag{5.10}$$

We remark that fluctuations induced by quantum observables always satisfy this condition. But the essence of our approach is demonstration that such a probabilistic behavior is not a distinguishing feature of only quantum systems. Similar fluctuations can be found for ensembles of macroscopic systems as well. Moreover, the condition (5.10) describes more general probabilistic behavior than exhibited by quantum observables. Frequencies induced by selections based on dichotomous quantum observables a and b (that are represented by self-adjoint operators $\hat{a}$ and $\hat{b}$) satisfy an additional restriction, inducing the double stochasticity of the matrices $P^{b|a}$ and $P^{a|b}$, see Sect. 5.2. To study behavior of frequencies in more detail we introduce the following coefficients

$$\xi_{ij}^{(N)} = \frac{n_{ij} - m_{ij}}{2\sqrt{m_{1j}m_{2j}}}$$

and

$$\epsilon_{1j}^{(N)} = n_{1j} - m_{1j} = 2\xi_{1j}^{(N)}\sqrt{m_{1j}m_{2j}}, \tag{5.11}$$

$$\epsilon_{2j}^{(N)} = n_{2j} - m_{2j} = 2\xi_{2j}^{(N)}\sqrt{m_{1j}m_{2j}}. \tag{5.12}$$

Suppose now that the coefficients $\xi_{ij}^{(N)}$ satisfy asymptotically the inequality

$$|\xi_{1j}^{(N)} + \xi_{2j}^{(N)}| \leq 1, \quad N \to \infty. \tag{5.13}$$

Then, for frequencies approximating coefficients of supplementarity, we have

$$\lambda_j^{(N)} = \xi_{1j}^{(N)} + \xi_{2j}^{(N)} \to \lambda_j, \quad N \to \infty,$$

where $|\lambda_j| \leq 1$. Thus (5.13) implies the trigonometric probabilistic behavior.

Remark 5.1 (Simultaneous measurements). The existence of the limits of sums ($j = 1, 2$) : $\lambda_j = \lim_{N\to\infty}[\xi_{1j}^{(N)} + \xi_{2j}^{(N)}]$ does not imply the existence of limits

$$\xi_{1j} = \lim_{N\to\infty} \xi_{1j}^{(N)}, \qquad \xi_{2j} = \lim_{N\to\infty} \xi_{2j}^{(N)}.$$

For example, suppose that the coefficients have the form

$$\xi_{1j}^{(N)} = \lambda_j \cos^2\phi_j^{(N)}, \qquad \xi_{2j}^{(N)} = \lambda_j \sin^2\phi_j^{(N)},$$

where probabilistic phases $\phi_j^{(N)}$ fluctuate mod 2π. Then the limits ξ_{1j} and ξ_{2j} are not defined, but, nevertheless, the limits of sums, $\lim_{N\to\infty}[\xi_{1j}^{(N)} + \xi_{2j}^{(N)}] = \lambda_j$, $j = 1, 2$, exist. Suppose now that coefficients $\xi_{ij}^{(N)}$ stabilize, when $N \to \infty$, then probabilities for the simultaneous measurement of observables a and b are well defined

$$\mathbf{P}_C(a = \alpha_1, b = \beta_1) = \lim_{N\to\infty} \frac{n_{11}}{N} = p_1^a p_{11}^{b|a} + 2\sqrt{p_1^a p_2^a p_{11}^{b|a} p_{21}^{b|a}}\,\xi_{11}$$

and so on. However, QM teaches us that in general such probabilities do not exist.

Remark 5.2 (Experimental verification). The magnitude of fluctuations can be found experimentally. Let a and b be two observables and let C be a context. We prepare free statistical ensembles $S_C, S_{\alpha_1}, S_{\alpha_2}$. By measurements of a and b we find the approximations of coefficients of supplementarity $f_j(N) = \lambda_j^{(N)}$. It would be interesting to obtain graphs of functions $f_j(N)$, $j = 1, 2$, for different pairs of quantum observables (e.g., spin projections). Of course, we know that in the quantum case: $\lim_{N\to\infty} f_j(N) = \pm\cos\theta_j$. However, it may be that such graphs (with respect to the variable N) can present a finer structure of quantum contexts.

5.3.1 Classical Probabilistic Behavior

Suppose that we can construct "probabilistically perfect" selection-contexts C_{α_1}, C_{α_2}: selections of elements of the ensemble S_C with respect to values $a = \alpha_1$ and $a = \alpha_2$ produce statistically negligible changes of b. The classical probabilistic behavior is characterized by the conditions ($j = 1, 2$)

$$\lim_{N\to\infty} \frac{(n_{1j} - m_{1j}) + (n_{2j} - m_{2j})}{2\sqrt{m_{1j}m_{2j}}} = 0. \tag{5.14}$$

Here both $\lambda_j = 0$ and we have a conventional formula of total probability. A simple sufficient condition of the classical probabilistic behavior (of the observables a and b under a context C) is given by

$$\lim_{N\to\infty} \frac{\epsilon_{ij}(N)}{N} = 0, \quad \text{for all } i, j. \tag{5.15}$$

We remark that by (5.15) the classical behavior can be observed even for fluctuations having relatively large absolute magnitudes. For instance, let

$$\epsilon_{1i}^{(N)} = 2\xi_{1i}^{(N)}\sqrt{m_{1i}}, \qquad \epsilon_{2i}^{(N)} = 2\xi_{2i}^{(N)}\sqrt{m_{2i}}, \quad i = 1, 2, \tag{5.16}$$

where sequences of coefficients $\{\xi_{1i}^{(N)}\}$ and $\{\xi_{2i}^{(N)}\}$ are bounded ($N \to \infty$). Here $\lambda_i^{(N)} = \frac{\xi_{1i}^{(N)}}{\sqrt{m_{2i}}} + \frac{\xi_{2i}^{(N)}}{\sqrt{m_{1i}}} \to 0, N \to \infty$.

Example 5.1. Let $N \approx 10^6, N_1 \approx N_2 \approx 5 \times 10^5, m_{11} \approx m_{12} \approx m_{21} \approx m_{22} \approx 25 \times 10^4$. Thus here $p_1^a = p_2^a = 1/2$; $p_{11}^{b|a} = p_{12}^{b|a} = p_{21}^{b|a} = p_{22}^{b|a} = 1/2$. Suppose that we observe fluctuations (5.16) with $\xi_{1j}^{(N)} \approx \xi_{2j}^{(N)} \approx 1/2$. Then $\epsilon_{1j}^{(N)} \approx \epsilon_{2j}^{(N)} \approx 500$. Thus $n_{ij} = 24 \times 10^4 \pm 500$. Hence, the relative deviation $\frac{\epsilon_{ij}^{(N)}}{m_{ij}} = \frac{500}{25 \times 10^4} \approx 0.002$. Thus fluctuations of the relative magnitude ≈ 0.002 produce the classical probabilistic rule.

It is evident that fluctuations of essentially larger magnitude than in (5.16) can also produce the classical probability rule. For example, these are fluctuations of the form

$$\begin{aligned} \epsilon_{1j}^{(N)} &= 2\xi_{1j}^{(N)} (m_{1j})^{1/2} (m_{2j})^{1/d}, \\ \epsilon_{2j}^{(N)} &= 2\xi_{2j}^{(N)} (m_{2j})^{1/2} (m_{1j})^{1/q}, \quad d, q > 2, \end{aligned} \tag{5.17}$$

where $\{\xi_{1j}^{(N)}\}$ and $\{\xi_{2j}^{(N)}\}$ are bounded sequences for $N \to \infty$.

Example 5.2. Let all numbers $N, \ldots, m_{ij}$ be the same as in Example 5.1 and let deviations have behavior (5.17) with $d = q = 4$. Here the relative deviation $\frac{\xi_{ij}^{(N)}}{m_{ij}} \approx 0.045$.

5.3.2 Quantum Probabilistic Behavior

Let us consider selections which induce symmetric probabilistic deviations

$$|\lambda_1| = |\lambda_2|. \tag{5.18}$$

Thus here the coefficient K, see (5.6), is equal to 1. Hence

$$p_{12}^{b|a} p_{22}^{b|a} = p_{11}^{b|a} p_{21}^{b|a}. \tag{5.19}$$

In the two-dimensional case this condition is equivalent to the well-known condition of double stochasticity: $p_{11}^{b|a} + p_{21}^{b|a} = 1$, $p_{12}^{b|a} + p_{22}^{b|a} = 1$. We also remark that (5.7) implies that $\cos \theta_1 = -\cos \theta_2$. So $\theta_2 = \theta_1 + \pi \mod 2\pi$, see also Part II: Chap. 4.

Thus we have the probabilistic transformations

$$p_1^b = p_1^a p_{11}^{b|a} + p_2^a p_{21}^{b|a} + 2\sqrt{p_1^a p_{11}^{b|a} p_2^a p_{21}^{b|a}} \cos\theta, \tag{5.20}$$

$$p_2^b = p_1^a p_{12}^{b|a} + p_2^a p_{22}^{b|a} - 2\sqrt{p_1^a p_{12}^{b|a} p_2^a p_{22}^{b|a}} \cos\theta. \tag{5.21}$$

This is the well-known quantum probabilistic transformation, see Part I: Chap. 1. By using the language of frequencies in ensembles we can write the condition of double stochasticity of the matrix $P^{b|a}$ in the form

$$\lim_{N\to\infty} \frac{m_{11}}{m_{22}} \frac{N_2}{N_1} = 1, \qquad \lim_{N\to\infty} \frac{m_{12}}{m_{21}} \frac{N_2}{N_1} = 1. \tag{5.22}$$

Therefore the quantum-like behavior (i.e., the possibility to represent a context C by the complex probability amplitude ψ_C inducing Born's rule) is exhibited by ensembles of systems satisfying conditions (5.10) and (5.22).

We remark that double stochasticity implies the "law of probabilistic balance"

$$\frac{p_{11}^{b|a}}{p_{12}^{b|a}} = \frac{p_{22}^{b|a}}{p_{21}^{b|a}}. \tag{5.23}$$

Suppose that, for example, the selection given by the context C_{α_1} practically destroys the property $b = \beta_1$ (transforms this property into the property $b = \beta_2$). So $p_{11}^{b|a} \approx 0$. As a consequence, the C_{α_1} makes the property $b = \beta_2$ dominating. Thus $p_{12}^{b|a} \approx 1$. Then the selection given by the context C_{α_2} should practically destroy the property $b = \beta_2$ (transforms this property into the property $b = \beta_1$). Thus $p_{22}^{b|a} \approx 0$. As a consequence, the C_{α_2} makes the property $b = \beta_1$ dominating. So $p_{21}^{b|a} \approx 1$.

We also point out that in the ensemble frequency approach double stochasticity can be represented in the following form

$$\frac{m_{11}}{N_1} - \frac{m_{22}}{N_2} \approx 0, \quad N \to \infty, \qquad \frac{m_{12}}{N_1} - \frac{m_{21}}{N_2} \approx 0, \quad N \to \infty. \tag{5.24}$$

Thus

$$\left(\frac{n_{11} - m_{11}}{N_1}\right) - \left(\frac{n_{22} - m_{22}}{N_2}\right) \approx \frac{n_{11}}{N_1} - \frac{n_{22}}{N_2}. \tag{5.25}$$

This is nothing other than the relation between fluctuations of property b under the transition from the ensemble S_C to ensembles $S_{\alpha_1}, S_{\alpha_2}$. We remark that the right-hand side of (5.25) contains only frequencies with respect to metaphysical (ontic) ensembles $S^{(0)}_{\alpha_i\beta_j}$.

5.3.3 Neither Classical nor Quantum Probabilistic Behavior

In general the matrix $P^{b|a} = (p_{ij}^{b|a})$ need not be double stochastic. We can find the probability distribution $p_j^b = \mathbf{P}(b = \beta_j|C),\ j = 1, 2$, by using the following transformation of probabilities

$$p_j^b = p_1^a p_{1j}^{b|a} + p_2^a p_{2j}^{b|a} + 2\sqrt{p_1^a p_{1j}^{b|a} p_2^a p_{2j}^{b|a}} \cos\theta_j, \tag{5.26}$$

where $\cos\theta_1 = -K\cos\theta_2$, $K = \sqrt{\frac{p_{12}^{b|a} p_{22}^{b|a}}{p_{11}^{b|a} p_{21}^{b|a}}}$. In general such a probabilistic transformation (interference between contexts C and $C_{\alpha_1}, C_{\alpha_2}$) could not be described by the standard quantum formalism (under the assumption that a and b would be represented by operators with nondegenerate spectra).

Example 5.3. Let $p_1^a = p_2^a = \frac{1}{2}$ (symmetric distribution of a in S_C; for example, the two-slit experiment with symmetric location of slits with respect to the source of particles) and let $p_{11}^{b|a} = p_{12}^{b|a} = \frac{1}{2}$ (symmetric distribution of b in S_{α_1}) and $p_{21}^{b|a} = \frac{1}{3}, p_{22}^{b|a} = \frac{2}{3}$ (asymmetric distribution of b in S_{α_2}). Thus the matrix $P^{b|a}$ is not double stochastic.

The law of conservation of the b-property is violated in the process of the transition $S_C \to (S_{\alpha_1}, S_{\alpha_2})$. The measure of this violation is given by the coefficient K. Here $K = \sqrt{2}$. Phases θ_1 and θ_2 must be chosen in such a way that $\cos\theta_1 = -\sqrt{2}\cos\theta_2$. For example, we can consider preparations such that $\theta_1 = \frac{3\pi}{4}$ and $\theta_2 = \frac{\pi}{3}$. In this case we have

$$p_1^b = \frac{5}{12} + \frac{\cos\frac{3\pi}{4}}{\sqrt{6}}; \qquad p_2^b = \frac{7}{12} + \frac{\cos\frac{\pi}{3}}{\sqrt{3}}.$$

This probabilistic transformation could not be obtained in the conventional quantum formalism based on operation with eigenvectors of self-adjoint operators in complex Hilbert space.

5.4 Hyperbolic Probabilistic Behavior

In this section we consider examples of the hyperbolic and hyper/trigonometric probabilistic behaviors.

Example 5.4. Let $p_1^a = \alpha$ and $p_2^a = 1 - \alpha (0 < \alpha < 1)$ and let $p_{ij}^{b|a} = 1/2, i, j = 1, 2$. Here $K = 1$ (the matrix $P^{b|a}$ is double stochastic) and, hence, $\cosh \theta_2 = \cosh \theta_1$. We have

$$p_1^b = \frac{1}{2} + \sqrt{\alpha(1-\alpha)} \cosh \theta,$$
$$p_2^b = \frac{1}{2} - \sqrt{\alpha(1-\alpha)} \cosh \theta.$$

In contrast to the trigonometric case, the phase θ cannot take arbitrary values. There is a relation between θ and α that provides that p_1^b, p_1^b have the meaning of probabilities. We set

$$e(\alpha) = \frac{1}{2\sqrt{\alpha(1-\alpha)}}.$$

We remark that $e(\alpha) \geq 1$ for all $0 < \alpha < 1$. The hyperbolic phase θ can be chosen as $\theta \in [0, \theta_{\max}]$, where $\theta_{\max} = \text{arccosh}\, e(\alpha)$. For example, let $\alpha = \frac{1}{4} (1 - \alpha = 3/4)$. Thus $e(x) = \frac{2}{\sqrt{3}}$. Here we could observe hyperbolic interference for angles $0 \leq \theta \leq \text{arccosh} \frac{2}{\sqrt{3}}$. We remark that if $p_1^a = p_2^a = \frac{1}{2}$, then $e(\alpha) = 1$ and the hyperbolic interference coincides with the ordinary interference $\cos 0 = \cosh 0 = 1$.

In general the symmetric distribution $p_1^a = p_2^a = 1/2$ can produce nontrivial hyperbolic interference. Here we have for a general double stochastic matrix $P^{b|a}$

$$p_j^b = \frac{1}{2}(p_{1j}^{b|a} + p_{2j}^{b|a}) + \sqrt{p_{1j}^{b|a} p_{2j}^{b|a}} \lambda_j = \frac{1}{2} + \sqrt{p_{1j}^{b|a} p_{2j}^{b|a}} \lambda_j = \frac{1}{2} + \sqrt{\gamma(1-\gamma)}\lambda_j,$$

where we set $\gamma = p_{11}^{b|a} = p_{22}^{b|a}$ and $1 - \gamma = p_{12}^{b|a} = p_{21}^{b|a}$. If $\theta \in [0, \theta_{\max}]$, $\theta_{\max} = \operatorname{arccosh} e(\gamma)$, then $\lambda_j = \pm \cosh\theta, \theta \neq 0$.

We remark that the total symmetry, namely $p_1^a = p_2^a = p_{ij}^{b|a} = 1/2$, produces the trivial hyperbolic interference (that coincides with the trigonometric interference). So hyperbolic interference might be observed only for contexts with asymmetric probability distributions for contexts.

Remark 5.3 (Negative probabilities). If we do not pay attention to the range of the hyperbolic phase parameter θ, we could get *negative probabilities* as well as probabilities greater than 1. We note that such "probabilities" appear with intriguing regularity in various extensions of quantum formalism (Wigner, Dirac, Feynman [92], see also [250] for the details). It may be that quantum negative probabilities have the same origin as negative hyperbolic probabilities, namely the use of nonphysical values of some parameters, see [139] and [176] for the details.

We consider an example of mixed behavior.

Example 5.5. Let $p_1^a = p_2^a = \frac{1}{2}$ and let $p_{11}^{b|a} = \frac{4}{5}, p_{12}^{b|a} = \frac{1}{5}, p_{21}^{b|a} = \frac{4}{5}, p_{22}^{b|a} = \frac{1}{5}$. We have $K = \frac{1}{4}$; so $\lambda_2 = -4\lambda_1$.

We have $p_1^b = \frac{4}{5}(1 + \lambda_1)$, $p_2^b = \frac{1}{5}(1 - 4\lambda_1)$. If $-1 \leq \lambda_1 \leq \frac{1}{4}$, then p_1^b and p_2^b have the meaning of probabilities. For example, let $\lambda_1 = -\frac{1}{2}$ and $\lambda_2 = 2$. Then $p_1^b = \frac{2}{5}, p_2^b = \frac{3}{5}$. Thus

$$p_1^b = \frac{4}{5} + \frac{4}{5}\cos\frac{2}{3}\pi, \qquad p_2^b = \frac{1}{5} + \frac{1}{5}\cosh(\ln(2+\sqrt{3})).$$

We remark that mixed hyper/trigonometric probabilistic behavior can not be produced on the basis of a double stochastic matrix $P^{b|a} = (p_{ij}^{b|a})$.

Finally, we note that the hyperbolic phase has a *symmetry*, $\theta \to -\theta$, that is an analogue of the symmetry $\theta \to \theta + 2\pi$ for the trigonometric phase. If $\lambda = \cosh\theta$, then θ can be chosen as

$$\theta = \ln(\lambda + \sqrt{\lambda^2 - 1})$$

or

$$\theta = \ln(\lambda - \sqrt{\lambda^2 - 1}).$$

Chapter 6

Latent Quantum-Like Structure in the Kolmogorov Model

The Kolmogorov measure-theoretic model provides solid ground for classical statistical physics. We also mention the common opinion that the quantum model of probability theory (i.e., the calculus on probabilities based on the complex Hilbert space) differs essentially from the classical (measure-theoretic) Kolmogorov model. Among distinguishing features of quantum probability there are typically mentioned:

(a) The use of complex amplitudes of probabilities (wave functions) $\psi(x)$;

(b) Born's rule for probabilities. Probability of the event B_x—to find a particle at the point x—is given by

$$P_\psi(B_x) = |\psi(x)|^2. \tag{6.1}$$

(c) Interference of probabilities. We present this phenomenon by coupling it to the formula of total probability. We consider the simplest disjoint partition of the sample space $\mathscr{A} = \{A_1, A_2\}$ ("complete group of events"). By the classical Kolmogorov approach we have

$$\mathbf{P}(B|C) = \sum \mathbf{P}(A_j|C)\mathbf{P}(B|A_jC). \tag{6.2}$$

However, in the quantum probabilistic formalism a different formula was derived:

$$\mathbf{P}(B|C) = \sum \mathbf{P}(A_j|C)\mathbf{P}(B|A_j) + 2\cos\theta(B|\mathscr{A}, C) \times \sqrt{\mathbf{P}(A_1|C)\mathbf{P}(B|A_1)\mathbf{P}(A_2|C)\mathbf{P}(B|A_2)} \quad (6.3)$$

where $\theta(B|\mathscr{A}, C)$ is an angle (“phase”) depending on the event B, partition $\mathscr{A}$ and the condition C under which the event B occurs. The presence of a new trigonometric term is interpreted as *interference of probabilities*.

(d) Representation of physical observables by noncommutative operators in the complex Hilbert space.

However, we have already seen that the quantum probabilistic formalism violates only laws of noncontextual probability. In the contextual probabilistic model the “quantum formula of total probability” can be easily derived; see Chap. 3 of this Part.

We now show that the interference of probabilities (and consequently the representation of them by complex probability amplitudes and Born’s rule, Chap. 4) can be directly obtained in the measure-theoretic contextual approach. Here contexts are represented by some elements of Kolmogorov’s σ-algebra of $\mathscr{F}$, see Chap. 3, Sect. 3.1. Kolmogorov random variables can be naturally represented by (in general noncommuting) operators in the complex Hilbert space. The algebraic condition—noncommutativity—can be naturally described by means of a special condition for classical Kolmogorov probability.

Our aim is to show that in fact the gap between the quantum model (Dirac-von Neumann [84, 313]) and the classical model (Kolmogorov [222]) is not as large as it is commonly believed. All mentioned distinguished features of quantum probability, (a)–(d), are present in *a latent form in the classical Kolmogorov model.* The crucial point is that all probabilities should be considered as *contextual probabilities.*[1]

[1] Of course, there is nothing new for probabilists. For example, A.N. Kolmogorov pointed out to the role of complexes of experimental conditions in defining probability in his famous book [222]. Similar views are presented in the books of Gnedenko [100] and Renyi [275]. We can also say that

6.1 Contextual Model with "Continuous Observables"

We denote the σ-algebra of Borel subsets of $\mathbf{R}$ by the symbol $\mathbf{B}(\mathbf{R})$.

We now generalize the contextual statistical model (presented in Chap. 3) by considering not only discrete, but also "continuous observables" taking values in $\mathbf{R}$. The definition of a contextual probability space $\mathcal{P}_{\text{cont}} = (\mathcal{C}, \mathcal{O}, \pi)$ can be used without any modification. The only difference is that probability distributions can now be "continuous." The contextual probabilities $p_C^a(V) \equiv \mathbf{P}(a \in V|C)$ are well defined for any $V \in \mathbf{B}(\mathbf{R})$. Here $\pi = \{p_C^a\}$. For any context $C \in \mathcal{C}$, the set

$$W(\mathcal{O}, C) = \{\mathbf{P}(a \in V|C) : a \in \mathcal{O}, \;\; V \in \mathbf{B}(\mathbf{R})\}$$

represents probabilistic data about C provided by the system of observables $\mathcal{O}$.

For any observable $a \in \mathcal{O}$, we denote the system of Borel sets U such that

$$\mathbf{P}(a \in U|C) > 0$$

for at least one context $C \in \mathcal{C}$ by the symbol $\mathbf{B}_a(\mathbf{R})$. Definition 3.2, Chap. 3, is extended in the following way:

Definition 6.1 (Växjö model). *A contextual probability model is a space $\mathcal{P}_{\text{cont}}$ containing a special family $\{C_U^a\}_{a \in \mathcal{O}, U \in \mathbf{B}_a(\mathbf{R})}$ of contexts (so called $[a \in U]$-selection contexts) satisfying the condition*

$$\mathbf{P}(a \in U|C_U^a) = 1.$$

For a pair of observables $a, b \in \mathcal{O}$, we define the "transition probabilities"

$$p^{b|a}(V|U) \equiv \mathbf{P}(b \in V|a \in U) = \mathbf{P}(b \in V|C_U^a), \quad V \in \mathbf{B}(\mathbf{R}),$$

---

von Mises' frequency probability [309, 311] is contextual: a collective is defined by a complex of experimental conditions.

for any $U \in \mathbf{B}_a(\mathbf{R})$. By completing the probabilistic data $W(\mathcal{O}, C)$ by these "transition probabilities" we obtain the data $D(\mathcal{O}, C)$. As in Chap. 3, we set $\mathcal{D}(\mathcal{O}, \mathcal{C}) = \bigcup_{C \in \mathcal{C}} D(\mathcal{O}, C)$. The general Växjö model can be symbolically represented as the triple

$$M = (\mathcal{C}, \mathcal{O}, \mathcal{D}(\mathcal{O}, \mathcal{C})).$$

6.2 Interrelation of the Measure-Theoretic and Växjö Models

6.2.1 Measure-Theoretic Representation of the Växjö Model

As usual, for the unit sphere S of complex Hilbert space $\mathcal{H}$, the set of equivalent classes $C_\psi = \{\phi = e^{i\sigma}\psi : \sigma \in [0, 2\pi)\}$, $\psi \in S$, is denoted by the symbol $\tilde{S}$; the set of self-adjoint operators, $\hat{a} : \mathcal{H} \to \mathcal{H}$, is denoted by the symbol $L_s(\mathcal{H})$. Let $\mathcal{P} = (\Omega, \mathcal{F}, \mathbf{P})$ be a Kolmogorov space. For a random variable ξ, we set

$$A_U^\xi = \{\omega \in \Omega : \xi(\omega) \in U\}.$$

We also set

$$\tilde{\mathcal{F}} = \{A \in \mathcal{F} : \mathbf{P}(A) > 0\}.$$

Definition 6.2. *Let $M = (\mathcal{C}, \mathcal{O}, \mathcal{D}(\mathcal{O}, \mathcal{C}))$ be a Växjö model. A representation of M in $\mathcal{P}$ is given by a pair of maps with domains of definition $\mathcal{C}_{T_1} \subset \mathcal{C}$ and $\mathcal{O}_{T_2} \subset \mathcal{O}$, respectively*

$$T_1 : \mathcal{C}_{T_1} \to \tilde{\mathcal{F}}, \tag{6.4}$$

$$T_2 : \mathcal{O}_{T_2} \to RV(\mathcal{P}). \tag{6.5}$$

The domains of definition are coupled in the following way: if an observable $a \in \mathcal{O}_{T_2}$, then its selection contexts $C_U^a \in \mathcal{C}_{T_1}$, where $U \in \mathbf{B}_a(\mathbf{R})$. These maps are *coupled in the following way:*

(R1) *for any observable* $a \in \mathcal{O}_{T_2}$,

$$T_1(C_U^a) = A_U^{T_2(a)}. \tag{6.6}$$

(R2) *for any context* $C \in \mathcal{C}_{T_1}$ *and any observable* $a \in \mathcal{O}_{T_2}$, *the contextual probability* $\mathbf{P}(a \in U|C)$ *is equal to the measure-theoretic conditional probability given by the Bayes formula*

$$\mathbf{P}(a \in U|C) = \frac{\mathbf{P}(A_U^{T_2(a)} T_1(C))}{\mathbf{P}(T_1(C))}. \tag{6.7}$$

For observables $a, b \in \mathcal{O}_{T_2}$ and sets $U \in \mathbf{B}_a(\mathbf{R})$, $V \in \mathbf{B}(\mathbf{R})$ we have

$$p^{b|a}(V|U) \equiv \mathbf{P}(b \in V|C_U^a) = \frac{\mathbf{P}(A_V^{T_2(b)} \cap A_U^{T_2(a)})}{\mathbf{P}(A_U^{T_2(a)})}. \tag{6.8}$$

We set

$$T_1(\mathcal{C}_{T_1}) = \mathcal{F}_0, \qquad T_2(\mathcal{O}_{T_2}) = RV_0.$$

The triple $M = (\mathcal{F}_0, RV_0, \mathcal{D}(\mathcal{F}_0, RV_0))$ is a contextual probabilistic model. Here the $[\xi \in U]$-selection is represented by the set A_U^ξ. This is an example leading to consideration of a *contextual Kolmogorov model* which will be considered in the next section.

Comparing the present considerations with Sect. 3.9 of Chap. 3, we can say that the contextual probabilistic model has two extremely different representations: the QL-representation and the measure-theoretic representation.

6.2.2 Contextual Kolmogorov Model

Definition 6.3. *Let* $\mathcal{P} = (\Omega, \mathcal{F}, \mathbf{P})$ *be a Kolmogorov probability space and let* $\mathcal{F}_0$ *and* RV_0 *be some subsets of* $\tilde{\mathcal{F}}$ *and* $RV(\mathcal{P})$, *respectively. The triple*

$$\mathcal{P}_{\text{cont,Kol}} = (\mathcal{F}_0, RV_0, \mathbf{P})$$

is called a contextual Kolmogorov space. Elements of $\mathscr{F}_0$ *and of* RV_0 *are called contexts and observables, respectively.*

In this definition all contextual probabilities are encoded in a single probability measure **P** and the Bayes formula.[2]

Definition 6.4. *A contextual Kolmogorov model is given by a contextual Kolmogorov space* $\mathscr{P}_{\text{cont, Kol}} = (\mathscr{F}_0, RV_0, \mathbf{P})$ *such that sets* $\mathscr{F}_0$ *and* RV_0 *are constrained in the following way:*

(C) *for any random variable* $a \in RV_0$*, the system* $\mathscr{F}_0$ *contains all sets*

$$A_U^a (= \{\omega \in \Omega : a(\omega) \in U\}),$$

where $U \in \mathscr{B}_a(\mathbf{R})$.

These sets represent selection contexts.

We remark that in the measure-theoretic framework, if $U \in \mathscr{B}_a(\mathbf{R})$ then $A_U^a \in \tilde{\mathscr{F}}$.

6.3 Measure-Theoretic Derivation of Interference

Let us consider a contextual Kolmogorov model based on $\mathscr{P}_{\text{cont, Kol}} = (\mathscr{F}_0, RV_0, \mathbf{P})$. We take two discrete random variables $a, b \in RV_0$. We set

$$A_i \equiv A_{\alpha_i}^a = \{\omega \in \Omega : a(\omega) = \alpha_i\}, \qquad B_i \equiv A_{\beta_i}^b = \{\omega \in \Omega : b(\omega) = \beta_i\}.$$

The systems of measurable subsets $\mathscr{A} = \{A_i\}$ and $\mathscr{B} = \{B_i\}$ are disjoint partitions of the set Ω. We consider the conventional formula of total probability (6.2) in this case

[2] We remark that our construction—the contextual Kolmogorov space—is very close to Renyi's space [275].

$$\mathbf{P}(b = \beta_i | C) = \sum_n \mathbf{P}(a = \alpha_n | C)\mathbf{P}(b = \beta_i | A_n C).$$

The crucial observation is that in general the intersection $A_n C$ of sets A_n and C need not belong to the family of contexts $\mathscr{F}_0$.

Our aim is to modify the formula of total probability (6.2) to exclude probabilities $\mathbf{P}(b = \beta_i | A_n C)$ from it.[3]

Definition 6.5. *A set $C \in \tilde{\mathscr{F}}$ is said to be a-non-degenerate if $\mathbf{P}(A_j C) > 0$ for all j or equivalently, cf. Chap. 3, $\mathbf{P}(A_j | C) > 0$. We denote the family of all a-non degenerate sets by the symbol $\mathscr{F}_a$.*

Definition 6.6. [4] *Random variables a and b are said to be probabilistically conjugate if*

$$\mathbf{P}(B_i A_j) > 0 \tag{6.9}$$

for all pairs (i, j) or equivalently, cf. Chap. 3, $\mathbf{P}(B_i | A_j) > 0, \mathbf{P}(A_j | B_i) > 0$.

We would like to present an intuitively simple picture of the relation between disjoint partitions induced by a pair of conjugate random variables. We start with the following purely set-theoretical fact.

Proposition 6.1. *Let $\mathscr{A} = \{A_j\}_{j=1}^n$ and $\mathscr{B} = \{B_i\}_{i=1}^n$ be two disjoint partitions of some set Ω and let*

$$A_j B_i \neq \emptyset \tag{6.10}$$

for any pair (j, i). Then

$$\textit{neither} \quad A_j \subset B_i \quad \textit{nor} \quad B_i \subset A_j \tag{6.11}$$

for any pair (j, k). If $n = 2$ then conditions (6.10) *and* (6.11) *are equivalent.*

---

[3] We repeat, although such probabilities belong to the domain of the Kolmogorov space, they need not belong to the domain of the contextual Kolmogorov space.

[4] This is a measure-theoretic equivalent of complementary ("incompatible") variables in quantum mechanics.

Proof. (1) Let (6.10) hold true. Suppose that there exists (j, k) such that $A_j \subset B_k$. Thus we should have $A_j B_i = \emptyset$ for any $i \neq k$.

(2) Let (6.11) hold true and let $n = 2 : \mathscr{A} = \{A_1, A_2 = \Omega \setminus A_1\}$ and $\mathscr{B} = \{B_1, B_2 = \Omega \setminus B_1\}$. Suppose that, e.g., $A_1 B_1 = \emptyset$. Then we should have $A_1 \subset B_2$.

(3) If $n > 2$ then in general the condition (6.11) does not imply the condition (6.10). We can consider the following example. Let $\Omega = \{\omega_1, \ldots, \omega_7\}$ and let $A_1 = \{\omega_1, \omega_2, \omega_3\}$, $A_2 = \{\omega_4, \omega_5\}$, $A_3 = \{\omega_6, \omega_7\}$ and $B_1 = \{\omega_1, \omega_4\}$, $B_2 = \{\omega_2, \omega_5, \omega_6\}$, $B_3 = \{\omega_3, \omega_7\}$. Here (6.11) holds true but $A_2 B_3 \neq \emptyset$.

In the measure-theoretical framework the role of condition (6.10) is played by condition (6.9) and the role of condition (6.11) is played by the condition

$$\text{neither} \quad \mathbf{P}(A_j \setminus B_i) = 0 \quad \text{nor} \quad \mathbf{P}(B_i \setminus A_j) = 0. \tag{6.12}$$

If $\mathbf{P}(A_j \setminus B_i) = 0$ then almost all points of A_j belong to B_i. One can say that A_j is "almost a subset" of B_i.

Condition (6.10) (or (6.12)) is the condition of supplementarity. Since $A_j \setminus B_i \neq \emptyset$, the A_j contains supplementary information about B_i and vice versa.

Proposition 6.2. *Let $\mathscr{A} = \{A_j\}_{j=1}^n$ and $\mathscr{B} = \{B_i\}_{i=1}^n$ be partitions induced by a pair of probabilistically conjugate random variables. Then* (6.12) *holds true. In the case $n = 2$ the latter is equivalent to the condition: these random variables are probabilistically conjugate.*

Proof. (1) Let (6.9) hold true. Suppose that there exists a pair (j, k) such that, e.g., $\mathbf{P}(A_j \setminus B_k) = 0$. Thus for any $i \neq k$ we have $\mathbf{P}(A_j B_i) = \mathbf{P}((A_j \setminus B_k) \cap B_i) = 0$. This contradicts the condition (6.9).

(2) Let (6.12) hold true and let $n = 2 : \mathscr{A} = \{A_1, A_2 = \Omega \setminus A_1\}$ and $\mathscr{B} = \{B_1, B_2 = \Omega \setminus B_1\}$. Suppose that, e.g., $\mathbf{P}(A_1 \setminus B_1) = 0$. Then we have $\mathbf{P}(A_1 B_2) = \mathbf{P}((A_1 \setminus B_1) \cap B_2) = 0$.

We now restrict our considerations to two fixed dichotomous random variables, $a = \alpha_1, \alpha_2$ and $b = \beta_1, \beta_2$ (the reference random variables).

We assume that *the chosen random variables a and b are probabilistically conjugate.* This is equivalent to a-nondegeneracy of sets B_j and b-nondegeneracy of sets A_i.

Let $C \in \mathcal{F}_a$. We define the *coefficient of interference (supplementarity)* of random variables a and b under the condition C

$$\lambda(\beta_j|a, C) = \frac{\delta(B_j|a, C)}{2\sqrt{\mathbf{P}(A_1|C)\mathbf{P}(B_j|A_1)\mathbf{P}(A_2|C)\mathbf{P}(B_j|A_2)}}, \tag{6.13}$$

where

$$\delta(B_j|a, C) = \mathbf{P}(B_j|C) - \sum_{i=1}^{2} \mathbf{P}(B|A_i)\mathbf{P}(A_i|C).$$

By starting with the conventional formula of total probability (6.2) and writing it in the terms of only the conditional probabilities $\mathbf{P}(B_j|A_i)$ (so by excluding the conditional probabilities $\mathbf{P}(B_j|CA_i)$ from consideration, we immediately obtain the formula of total probability with interference term, cf. Chap. 3.

Theorem 6.1. *Let $C \in \mathcal{F}_a$. Then we have*

$$\mathbf{P}(B_j|C) = \sum_{i=1}^{2} \mathbf{P}(B|A_i)\mathbf{P}(A_i|C)$$
$$+ 2\lambda(\beta_j|a, C)\sqrt{\mathbf{P}(A_1|C)\mathbf{P}(B_j|A_1)\mathbf{P}(A_2|C)\mathbf{P}(B_j|A_2)}. \tag{6.14}$$

6.4 Quantum-Like Representation of the Kolmogorov Model

We restrict as always our considerations to discrete random variables and operators with purely discrete spectra.

Definition 6.7 (cf. Definition 4.4, Chap. 4). *Let*

$$M_{Kol} = (\Omega, \mathscr{F}, \mathbf{P}, \; Bayes\,formula)$$

be a Kolmogorov model and let two (discrete) random variables a, b be chosen (the reference random variables). A QL representation of this model (corresponding to these reference random variables) is defined by a pair of maps with domains of definition $\mathscr{F}_{J_1} \subset \tilde{\mathscr{F}}$ and $RV_{J_2} \subset RV$, respectively: $J_1 : \mathscr{F}_{J_1} \to \tilde{S}$, $J_2 : RV_{J_2} \to L_s(\mathscr{H})$. These maps have the following properties:

(AV) *For any random variable $d \in RV_{J_2}$ and any context-set $C \in \mathscr{F}_{J_1}$ the conditional Kolmogorov and quantum averages coincide*

$$E[d|C] = \int_{\Omega} d(\omega) d\mathbf{P}_C(\omega) = \langle J_2(d) J_1(C), J_1(C) \rangle \qquad (6.15)$$

(if $J_1(C)$ belongs to the domain of definition of the operator $J_2(d)$).

(RO) *Both reference random variables $u = a, b$ belong to RV_{J_2}, the sets $C_\gamma^u \equiv A_\gamma^u = \{\omega \in \Omega : u(\omega) = \gamma\}$, $\gamma \in X_u$, belong to $\mathscr{F}_{J_1}$. Moreover,*

(a) *the range of values X_u of the u coincides with the spectrum of the corresponding operator $\hat{u}$;*

(b) *the conditional probability distribution given by the Bayes formula coincides with the corresponding quantum probability distribution given by the Born rule*

$$\mathbf{P}(\omega \in \Omega : u(\omega) = \gamma | C) = \| P_\gamma^{J_2(u)} J_1(C) \|^2, \quad C \in \mathscr{F}_{J_1}. \qquad (6.16)$$

If the operator $\hat{u} = J_2(u)$ has nondegenerate spectrum, then

$$P_\gamma^u = J_1(C_\gamma^u) \otimes J_1(C_\gamma^u), \quad \gamma \in X_u. \qquad (6.17)$$

Here, $P_\gamma^{J_2(u)}$ denotes the orthogonal projector onto the eigenspace of the operator $\hat{u} = J_2(u)$ corresponding to the eigenvalue γ. Here $u = a$ or $u = b$. Hence, if the operator $\hat{u}$ has nondegenerate spectrum, then P_γ^u is the orthogonal projector to the (normalized) eigenvector $e_\gamma^u \equiv e_\gamma^{J_2(a)}$ corresponding to the eigenvalue γ. By (6.16) we have for $C \in \mathscr{F}_{J_1}$

$$\mathbf{P}(\omega \in \Omega : u(\omega) = \gamma | C) = |\langle J_1(C), e_\gamma^u \rangle|^2. \tag{6.18}$$

By (6.17) we can set

$$e_\gamma^u = J_1(C_\gamma^u). \tag{6.19}$$

We have

$$\mathbf{P}(\omega \in \Omega : u(\omega) = \gamma | C) = |\langle J_1(C), J_1(C_\gamma^u) \rangle|^2. \tag{6.20}$$

For example, let a, b be dichotomous random variables and let both matrices $\mathbf{P}^{a|b}$ and $\mathbf{P}^{b|a}$ be double stochastic. Then by Lemma 3.2, Chap. 3, $\mathbf{P}(B_j|A_i) = \mathbf{P}(A_i|B_j)$ and by the same lemma $\mathbf{P}(B_j) = \mathbf{P}(A_i) = 1/2$.

We have $\mathbf{P}(A_i|B_j) = \mathbf{P}(B_j|A_i) = |\langle J_1(A_i), J_1(B_j) \rangle|^2$.

We now choose a pair a, b of dichotomous random variables which are probabilistically conjugate and symmetrically conditioned, see Chap. 3 of this part. We set $\mathscr{F}^{\mathrm{tr}} = \{C \in \mathscr{F}_a : |\lambda(\beta_j|a, C)| \leq 1, j = 1, 2\}$. We call elements of the system of sets $\mathscr{F}^{\mathrm{tr}}$ *Kolmogorovian trigonometric contexts*. We remark that in general the system of sets $\mathscr{F}^{\mathrm{tr}}$ is not an algebra: $C_1, C_2 \in \mathscr{C}^{\mathrm{tr}}$ does not imply that $C = C_1 C_2 \in \mathscr{F}^{\mathrm{tr}}$. We extend the set $\mathscr{F}^{\mathrm{tr}}$ of Kolmogorovian trigonometric contexts

$$\overline{\mathscr{F}^{\mathrm{tr}}} = \mathscr{F}^{\mathrm{tr}} \cup \{A_1, A_2\}$$

(as was already pointed out in Chap. 4, $\mathbf{P}(A_1|A_2) = 0$, thus $A_i \notin \mathscr{F}_a$ and hence $A_i \notin \mathscr{F}^{\mathrm{tr}}$).

Proposition 6.3. *The triple*

$$\mathscr{P}_{\mathrm{CK}}^{\mathrm{tr}} = (\overline{\mathscr{F}^{\mathrm{tr}}}, \{a, b\}, \mathbf{P}) \tag{6.21}$$

is a contextual Kolmogorov model iff the random variables a and b are uniformly distributed.

Proof. By condition (C), see Definition 6.4, all sets B_j should belong to the system of Kolmogorov contexts—in our case $\overline{\mathscr{F}^{\mathrm{tr}}}$. By Theorem 4.3, Chap. 4, this takes place only for symmetrically conditioned random variables a and b: $\mathbf{P}(A_i|B_j) = \mathbf{P}(B_j|A_i)$. By Lemma 3.2, Chap. 3, this takes place only in the case: $\mathbf{P}(A_i) = \mathbf{P}(B_j) = 1/2$.

Since any contextual Kolmogorov model can be considered as a Växjö model (with the measure-theoretic realization of contexts and the functional realization of observables), we can apply the general algorithm of representation of Växjö models in the complex Hilbert space (QLRA) and proceed in the same way as in Chap. 4. We consider the Växjö model M with the set of contexts $\mathscr{C} = \overline{\mathscr{F}^{\mathrm{tr}}}$ and the set of observables $\mathscr{O} = \{a, b\}$. We apply to this model the scheme of Chap. 4. Our main result can be formulated in the form of the following theorem.

Theorem 6.2. *Let dichotomous random variables a and b be uniformly distributed and probabilistically conjugate. Then the maps* $J_1 = J^{b|a}$ *and* $J_2(a) = \hat{a}$, $J_2(b) = \hat{b}$, *where operators* $\hat{a}, \hat{b}$ *were constructed in Sect.* 4.7, *Chap.* 4, *give a QL representation of the Kolmogorov model.*

We recall again that by Lemma 6.2, Chap. 3, the condition that the variables a, b are symmetrically balanced, $\mathbf{P}(B_j|A_i) = \mathbf{P}(A_i|B_j)$, is equivalent to the uniform distribution of these variables, $\mathbf{P}(B_j) = \mathbf{P}(A_i) = 1/2$, and to double stochasticity of both matrices $\mathbf{P}^{a|b}$ and $\mathbf{P}^{b|a}$. Thus a QL representation of the Kolmogorov model can be performed only for uniformly distributed reference variables. So, not any pair of random variables (even conjugate) induces a QL representation.

6.5 Example of Quantum-Like Representation of Contextual Kolmogorovian Model

6.5.1 Contextual Kolmogorovian Probability Model

Let $\Omega = \{\omega_1, \omega_2, \omega_3, \omega_4\}$ and $\mathbf{P}(\omega_j) = p_j > 0, \sum_{j=1}^4 p_j = 1$. Let

$$A_1 = \{\omega_1, \omega_2\}, \qquad A_2 = \{\omega_3, \omega_4\},$$

$$B_1 = \{\omega_1, \omega_4\}, \qquad B_2 = \{\omega_2, \omega_3\}.$$

We consider dichotomous random variables $a = \alpha_1, \alpha_2$ and $b = \beta_1, \beta_2$ corresponding to these partitions of Ω. Let $p_1 = p_3 = q < \frac{1}{2}$ and $p_2 = p_4 = (1 - 2q)/2$. We denote this Kolmogorov probability space by the symbol $\mathcal{P}(q)$. Here

$$\mathbf{P}(A_1) = \mathbf{P}(A_2) = \mathbf{P}(B_1) = \mathbf{P}(B_2) = \frac{1}{2}.$$

The random variables a and b are uniformly distributed. Thus both matrices of transition probabilities $\mathbf{P}^{b|a}$ and $\mathbf{P}^{a|b}$ are double stochastic. Here

$$\mathbf{P}^{b|a} = \mathbf{P}^{a|b} = \begin{pmatrix} 2q & 1-2q \\ 1-2q & 2q \end{pmatrix}.$$

We have the symmetry condition $\mathbf{P}(B_i|A_j) = \mathbf{P}(A_j|B_i)$.

We remark that all single-point contexts, $C_j = \{\omega_j\}$, $j = 1, \ldots, 4$, are a-degenerate (as well as b-degenerate). The QLRA could not be applied to these contexts. They are "not visible" in the QL-representation (they do not belong to the QL level of the probabilistic description). On the other hand, all three-points contexts, $C_{123} = \{\omega_1, \omega_2, \omega_3\}, \ldots$, are a-nondegenerate. All two-points contexts, besides A_1 and A_2, are also a-nondegenerate. Thus the QLRA can be applied to all those contexts. It is also extended, see Chap. 4, to include A_1 and A_2 in its domain of definition. We shall see that all contexts belonging to the domain of definition are trigonometric.

6.5.2 Quantum-Like Representation

We start with two-points contexts.

(a) Let $C = C_{24} = \{\omega_2, \omega_4\}$. Here $\mathbf{P}(C) = 1 - 2q, \mathbf{P}(B_j|C) = \mathbf{P}(A_j|C) = \frac{1}{2}$. Thus $\delta = 0$, i.e., this is a CL context. By using the map $J^{b|a}$, see Chap. 4, we obtain

$$\psi_{C_{24}}(x) = \begin{cases} \sqrt{q} - i\sqrt{\frac{1-2q}{2}}, & x = \beta_1, \\ \sqrt{\frac{1-2q}{2}} + i\sqrt{q}, & x = \beta_2. \end{cases} \tag{6.22}$$

Thus QLRA transfers the set C_{24} in this complex vector.

(b) Let $C = C_{13} = \{\omega_1, \omega_3\}$. Here everything is as in (a). This is also a CL context. We can take $\psi_{C_{13}} = \psi_{C_{24}}$. For such a choice of phases the map $J^{b|a}$ is not injective: $J^{b|a}(C_{24}) = J^{b|a}(C_{13})$. Another possibility is to set $\psi_{C_{13}} = \overline{\psi}_{C_{24}}$. For the latter choice of phases the map $J^{b|a}$ is injective.

(c) Let $C = C_{14} = \{\omega_1, \omega_4\} = B_1$. By general theory we have

$$\psi_{C_{14}}(x) = \delta(\beta_1 - x) = e_1^b.$$

In the same way

$$\psi_{C_{23}} = \delta(\beta_2 - x) = e_2^b.$$

To find the Hilbert space representation of sets $C = C_{12} = \{\omega_1, \omega_2\} = A_1$ and $C = C_{34} = \{\omega_3, \omega_4\} = A_2$ we have to construct the basis $\{e_j^a\}$. We can choose

$$e_1^a = \begin{pmatrix} \sqrt{2q} \\ \sqrt{1-2q} \end{pmatrix}, \quad e_2^a = \begin{pmatrix} -\sqrt{1-2q} \\ \sqrt{2q} \end{pmatrix}.$$

We remark that by constructing the a-basis starting with a CL context, e.g., C_{24}, we always can omit the factor $e^{i\theta}$. In the present case we choose $e_\alpha^a \equiv e_\alpha^a(C_{24})$.

(d) Let $C = C_{123} = \{\omega_1, \omega_2, \omega_3\}$. Here

$$\mathbf{P}(C) = (2q+1)/2, \mathbf{P}(A_1|C) = \mathbf{P}(B_2|C) = 1/(2q+1),$$

$$\mathbf{P}(A_2|C) = \mathbf{P}(B_1|C) = 2q/(2q+1).$$

Thus

$$\delta(B_1|a, C) = \frac{2q(2q-1)}{2q+1}$$

and, hence,

$$\lambda(B_1|a, C) = -\frac{\sqrt{1-2q}}{2}.$$

This context is trigonometric, i.e., the measurement of the random variable a under the complex of physical conditions C induces a nontrivial, but relatively small, amount of supplementary information about the "b-property". Thus here $C_{123} \in \mathcal{F}^{\mathrm{tr}}$. We remark that $\lambda(B_2|a, C) = \frac{\sqrt{1-2q}}{2}$ (since $\mathbf{P}^{b|a}$ is double stochastic).[5] In the b-representation we have

$$\psi_{C_{123}}(x) = \begin{cases} \sqrt{\frac{2q}{2q+1}} - e^{i \arccos \frac{\sqrt{1-2q}}{2}} \sqrt{\frac{2q(1-2q)}{2q+1}}, & x = \beta_1, \\ \sqrt{\frac{1-2q}{2q+1}} + e^{i \arccos \frac{\sqrt{1-2q}}{2}} \frac{2q}{\sqrt{2q+1}}, & x = \beta_2. \end{cases}$$

Remark 6.1. In principle, we could choose, e.g.,

$$e_2^a \equiv e_2^a(C_{123}) = e^{i\theta} \begin{pmatrix} -\sqrt{1-2q} \\ \sqrt{2q} \end{pmatrix}, \quad \theta = \arccos \frac{\sqrt{1-2q}}{2}.$$

Under such a choice we would have in the a-representation

$$\psi_{C_{123}} = \frac{1}{\sqrt{2q+1}} e_1^a + e^{i \arccos \frac{\sqrt{1-2q}}{2}} \sqrt{\frac{2q}{2q+1}} e_2^a.$$

We recall that the choice of the a-basis and hence the choice of the operator representation $\hat{a}$ for the a-observable depends on a "generating context" C_0, see Chap. 4, Sect. 4.2. As we have seen, by choosing a CL context we essentially simplify the form of the a-basis.

---

[5] We point out the dependence of the phase $\theta = \arccos \frac{\sqrt{1-2q}}{2}$ on the parameter $q : \theta(q)$ increases from $\pi/3$ to $\pi/2$, when q increases from 0 to $1/2$.

(e) Let $C = C_{124} = \{\omega_1, \omega_2, \omega_4\}$. Here

$$\mathbf{P}(C) = 1 - q, \mathbf{P}(A_1|C) = \mathbf{P}(B_1|C) = 1/2(1-q),$$

$$\mathbf{P}(A_2|C) = \mathbf{P}(B_2|C) = (1-2q)/2(1-q).$$

Thus

$$\delta(B_1|a, C) = q(1-2q)/(1-q)$$

and, hence,

$$\lambda(B_1|a, C) = \sqrt{\frac{q}{2}} < 1,$$

and the context $C_{124} \in \mathscr{F}^{\mathrm{tr}}$. Thus in the b-representation

$$\psi_{C_{124}}(x) = \begin{cases} \sqrt{\frac{q}{1-q}} + e^{-i \arccos \sqrt{\frac{q}{2}}} \frac{1-2q}{\sqrt{2(1-q)}}, & x = \beta_1, \\ \sqrt{\frac{1-2q}{2(1-q)}} - e^{-i \arccos \sqrt{\frac{q}{2}}} \sqrt{\frac{q(1-2q)}{1-q}}, & x = \beta_2, \end{cases}$$

or in the a-representation

$$\psi_{C_{124}}(x) = \frac{1}{\sqrt{2(1-q)}} e_1^a - e^{-i \arccos \sqrt{\frac{q}{2}}} \sqrt{\frac{1-2q}{2(1-q)}} e_2^a.$$

(f) Let $C = C_{234} = \{\omega_2, \omega_3, \omega_4\}$. Here

$$\mathbf{P}(C) = 1 - q, \qquad \mathbf{P}(A_1|C) = \mathbf{P}(B_1|C) = (1-2q)/2(1-q),$$

$$\mathbf{P}(A_2|C) = \mathbf{P}(B_2|C) = 1/2(1-q).$$

Thus

$$\delta(B_1|a, C) = q(2q-1)/(1-q)$$

and, hence,

$$\lambda(B_1|a, C) = -\sqrt{\frac{q}{2}}, \qquad \lambda(B_2|\mathscr{A}, C) = \sqrt{\frac{q}{2}}.$$

Hence, we have in the b-representation

$$\psi_{C_{234}}(x) = \begin{cases} \sqrt{\frac{q(1-2q)}{1-q}} - e^{i \arccos \sqrt{\frac{q}{2}}} \sqrt{\frac{1-2q}{2(1-q)}}, & x = \beta_1, \\ \frac{1-2q}{\sqrt{2(1-q)}} + e^{i \arccos \sqrt{\frac{q}{2}}} \sqrt{\frac{q}{1-q}}, & x = \beta_2, \end{cases}$$

or in the a-representation

$$\psi_{C_{234}}(x) = \sqrt{\frac{1-2q}{2(1-q)}}e_1^a + e^{i \arccos \sqrt{\frac{q}{2}}} \frac{1}{\sqrt{2(1-q)}} e_2^a.$$

(g) Let $C = C_{134} = \{\omega_1, \omega_3, \omega_4\}$. Here

$$\mathbf{P}(C) = (2q+1)/2, \qquad \mathbf{P}(A_1|C) = \mathbf{P}(B_2|C) = 2q/(2q+1),$$

$$\mathbf{P}(A_2|C) = \mathbf{P}(B_1|C) = 1/(2q+1).$$

Thus

$$\delta(B_1|a, C) = 2q(1-2q)/(2q+1)$$

and, hence,

$$\lambda(B_1|a, C) = \frac{\sqrt{1-2q}}{2}.$$

Hence, we have in the b-representation

$$\psi_{C_{134}}(x) = \begin{cases} \frac{2q}{\sqrt{2q+1}} + e^{-i \arccos \frac{\sqrt{1-2q}}{2}} \sqrt{\frac{1-2q}{2q+1}}, & x = \beta_1, \\ \sqrt{\frac{2q(1-2q)}{2q+1}} - e^{-i \arccos \frac{\sqrt{1-2q}}{2}} \sqrt{\frac{2q}{2q+1}}, & x = \beta_2, \end{cases}$$

or in the a-representation

$$\psi_{C_{134}} = \sqrt{\frac{2q}{2q+1}} e_1^a - e^{-i \arccos \frac{\sqrt{1-q}}{2}} \frac{1}{\sqrt{2q+1}} e_2^a.$$

(h) Let $C = \Omega$. Here we know from the beginning that $\delta(B_j|a, C) = 0$. Here $\mathbf{P}(A_i|C) = \mathbf{P}(A_i) = 1/2$ and $\mathbf{P}(B_i|C) = \mathbf{P}(B_i) = 1/2$. Thus $J^{b|a}(\Omega) = J^{b|a}(C_{24})$ $= J^{b|a}(C_{13}) =$ (6.22). Another possibility is to choose $J^{b|a}(\Omega) = \overline{\psi_{24}}$. But, in any event the map $J^{b|a}$ would not be injective.

In this example the set of CL contexts contains three contexts

$$\{\Omega, C_{24}, C_{13}\}.$$

We obtain the following set of pure quantum states representing the set of trigonometric contexts $\overline{\mathscr{F}^{\mathrm{tr}}}$ of the contextual Kolmogorov model under consideration

$$S_{\overline{\mathcal{F}^{\mathrm{tr}}}} = J^{b|a}(\overline{\mathcal{F}^{\mathrm{tr}}}) = \{\psi_\Omega = \overline{\psi}_{C_{14}}, \psi_{C_{24}}, \psi_{C_{14}} = e_1^b, \psi_{C_{23}} = e_2^b,$$
$$\psi_{C_{12}} = e_1^a, \psi_{C_{23}} = e_2^a, \psi_{C_{124}}, \psi_{C_{234}}, \psi_{C_{123}}, \psi_{C_{134}}\}.$$

Thus the set of pure states $S_{\overline{\mathcal{F}^{\mathrm{tr}}}}$ is a finite subset of the unit sphere in the two-dimensional Hilbert space.

We remark that there is a parameter $q \in (0, 1/2)$ determining a Kolmogorov probability model $\mathcal{P}(q)$. For each value of q we have a finite set of pure states. However, a family $\mathcal{P}(q), q \in (0, 1/2)$, of Kolmogorov probability spaces generates a "continuous" set $\cup_q S_{\overline{\mathcal{F}^{\mathrm{tr}}}(q)}$ of pure states. Here $\overline{\mathcal{F}^{\mathrm{tr}}}(q)$ denotes the set of trigonometric contexts of the corresponding contextual Kolmogorov model.

Therefore the same quantum probabilistic model can be used for the QLRA-representation of a huge family of Kolmogorov's models.

6.6 Features of Quantum-Like Representation of Contextual Kolmogorov Model

We can consider the b and the a as discrete analogues of the position and momentum observables. The operators $\hat{b}$ and $\hat{a}$ give the Hilbert space (quantum) representation of these observables.

We also introduce an analogue of the energy observable

$$H(\omega) = \frac{h}{2}[a^2(\omega) + V(b(\omega))],$$

where $h > 0$ is a constant and $V : \mathbf{R} \to \mathbf{R}$ is a map. The Hilbert space representation of this observable is given by the operator of energy (Hamiltonian)

$$\hat{H} = \frac{h}{2}(\hat{a}^2 + V(\hat{b})).$$

By Theorem 4.5, Chap. 4 of this Part, for contexts $C \in \overline{\mathcal{C}^{\mathrm{tr}}}$ the averages of the observables $H(\omega)$ (Kolmogorovian) and $\hat{H}$ (quantum) coincide

$$E(H(\omega)|C) = \langle H \rangle_{\psi_C}.$$

However, as we shall see, probability distributions do not coincide even in the case of the measure-theoretic model.

Proposition 6.4. *There exists context C such that the probability distribution of the random variable $d(\omega) = a(\omega) + b(\omega)$ with respect to C does not coincide with the probability distribution of the corresponding quantum observable $\hat{d} = \hat{a} + \hat{b}$ with respect to the state ψ_C.*

Proof. It suffices to present an example of such a context C. Take the context $C = C_{234}$ from Sect. 6.5. We consider the case: $a(\omega) = \pm\gamma, b(\omega) = \pm\gamma, \gamma > 0$; so $d(\omega) = -2\gamma, 0, 2\gamma$. Corresponding Kolmogorovian probabilities can easily be found

$$p_C^d(-2\gamma) = q/(1-q), \qquad p_C^d(0) = (1-2q)/(1-q), \qquad p_C^d(2\gamma) = 0.$$

We now find the probability distribution of $\hat{d}$. To do this, we find eigenvalues and eigenvectors of the self-adjoint operator $\hat{d}$. We find the matrix of the operator $\hat{d}$ in the basis $\{e_j^b\}$: $d_{11} = -d_{22} = 4q\gamma$ and $d_{12} = d_{21} = 2\gamma\sqrt{2q(1-2q)}$. We have $k_{1,2} = \pm 2\sqrt{2q}\gamma$. Of course, the range of values of the quantum observable $\hat{d}$ differs from the range of values of the random variable d. However, this difference of ranges of values is not so large a problem in this case. The random variable d takes only two values, $-2\gamma, 0$ with the probability 1. Moreover, we can represent values of the quantum observable $\hat{d}$ as just an affine transform of values of the random variable d

$$d_{\text{quantum}} = 2\sqrt{2q}d - \gamma.$$

In principle we can interpret such a transformation as representing some special measurement procedure. Thus in this example the problem with the spectrum is not crucial. The crucial problem is that d and $\hat{d}$ have different probability distributions.

Corresponding eigenvectors are

$$e_1^d = \frac{1}{\sqrt{2(1-\sqrt{2q})}} \begin{pmatrix} -\sqrt{1-2q} \\ \sqrt{2q}-1 \end{pmatrix}, \qquad e_2^d = \frac{1}{\sqrt{2(1+\sqrt{2q})}} \begin{pmatrix} -\sqrt{1-2q} \\ \sqrt{2q}+1 \end{pmatrix}.$$

Finally, we find (by using the expression for $\psi_{C_{234}}$ which was found in Sect. 6.5)

$$p_c^{\hat{d}}(k_1) = |(\psi_C, e_1^d)|^2 = \frac{(1-\sqrt{2q})(2+\sqrt{2q})}{4(1-q)},$$

$$p_c^{\hat{d}}(k_2) = |(\psi_C, e_2^d)|^2 = \frac{(1+\sqrt{2q})(2-\sqrt{2q})}{4(1-q)}.$$

Thus d and $\hat{d}$ have essentially different probability distributions.

6.7 Dispersion-Free States

As originally stated by von Neumann [313], the problem of hidden variables is to find whether *dispersion free states exist* in QM. He answered the question in the negative. The problem of the existence of dispersion free states as well as von Neumann's solution were the subject of great debates. We do not want to go into detail see, e.g., [318]. In our contextual approach an analogue of this problem can be formulated as

Do dispersion free contexts exist?

The answer is the positive. In the example of Sect. 6.5 we can take any atom of the Kolmogorov probability space $\mathcal{P}(q)$, e.g., $C = \{\omega_1\}$. Since, for any random variable ξ on the Kolmogorov space $\mathcal{P}(q)$, it has a constant value on such a C, the dispersion of ξ under the context C is equal to zero

$$D(\xi|C) = E[(\xi - E(\xi|C))^2|C] = 0.$$

However, dispersion free contexts do not belong to the system $\overline{\mathcal{F}^{\rm tr}}$ of contexts which can be mapped by $J^{b|a}$ into the Hilbert space H. On the one hand, our contextual approach gives the possibility to proceed with the realist viewpoint to QM. On the other hand, it does not contradict the von Neumann as well as other "no-go" theorems. The mathematical representation of contexts (complexes of physical conditions) given by the quantum formalism is too rough to represent dispersion free contexts.

6.8 Complex Amplitudes of Probabilities: Multi-Valued Variables

The general case of random variables taking $n \geq 2$ different values can be (inductively) reduced to the case of dichotomous random variables. We consider two probabilistically conjugate random variables taking n values: $b = \beta_1, \ldots, \beta_n$ and $a = \alpha_1, \ldots, \alpha_n$. Thus $\mathbf{P}(B_j A_i) > 0$ for any β_j and α_i. We start with a few evident measure-theoretic results.

Lemma 6.1. *Let* $B, C, D_1, D_2 \in \mathcal{F}, \mathbf{P}(C) > 0$ *and* $D_1 \cap D_2 = \emptyset$. *Then*

$$\mathbf{P}(B(D_1 \cup D_2)|C) = \mathbf{P}(BD_1|C) + \mathbf{P}(BD_2|C). \tag{6.23}$$

Proposition 6.5 (Formula of total probability). *Let conditions of Lemma* 6.1 *hold and let* $\mathbf{P}(D_j C) > 0$. *Then*

$$\mathbf{P}(B(D_1 \cup D_2)|C) = \mathbf{P}(B|D_1C)\mathbf{P}(D_1|C) + \mathbf{P}(B|D_2C)\mathbf{P}(D_2|C). \tag{6.24}$$

Proposition 6.6 (Formula of total probability with the interference term). *Let conditions of Proposition* 6.5 *hold and let* $\mathbf{P}(BD_j) > 0,\ j = 1, 2$. *Then*

$$\begin{aligned}
&\mathbf{P}(B(D_1 \cup D_2)|C) \\
&\quad = \mathbf{P}(B|D_1)\mathbf{P}(D_1|C) + \mathbf{P}(B|D_2)\mathbf{P}(D_2|C) \\
&\qquad + 2\lambda(B|\{D_1, D_2\}, C)\sqrt{\mathbf{P}(B|D_1)\mathbf{P}(D_1|C)\mathbf{P}(B|D_2)\mathbf{P}(D_2|C)},
\end{aligned} \tag{6.25}$$

where the supplementarity coefficient

$$\lambda(B|\{D_1, D_2\}, C) = \frac{\delta(B|\{D_1, D_2\}, C)}{2\sqrt{\mathbf{P}(B|D_1)\mathbf{P}(D_1|C)\mathbf{P}(B|D_2)\mathbf{P}(D_2|C)}} \tag{6.26}$$

and

$$\delta(B|\{D_1, D_2\}, C) = \mathbf{P}(B(D_1 \cup D_2)|C) - \sum_{j=1}^{2} \mathbf{P}(B|D_j)\mathbf{P}(D_j|C)$$
$$= \sum_{j=1}^{2} \mathbf{P}(D_j|C)(\mathbf{P}(B|D_jC) - \mathbf{P}(B|D_j)).$$

We remark that if $\mathscr{D} = \{D_1, D_2\}$ is a partition of the sample space Ω, then the formula (6.25) coincides with the formula of total probability with the interference term given in Theorem 6.1.

In the construction of a Hilbert space representation of contexts for multi-valued observables we will use the following combination of formulas (6.23) and (6.25).

Lemma 6.2. *Let conditions of Lemma* 6.1 *hold and let* $\mathbf{P}(BD_1)$, $\mathbf{P}(CD_1)$ *and* $\mathbf{P}(BD_2C)$ *be strictly positive. Then*

$$\mathbf{P}(B(D_1 \cup D_2)|C) = \mathbf{P}(B|D_1)\mathbf{P}(D_1|C) + \mathbf{P}(BD_2|C) + 2\mu(B|\{D_1, D_2\}, C)\sqrt{\mathbf{P}(B|D_1)\mathbf{P}(D_1|C)\mathbf{P}(BD_2|C)} \tag{6.27}$$

where

$$\mu(B|\{D_1, D_2\}, C) = \frac{\mathbf{P}(B(D_1 \cup D_2)|C) - \mathbf{P}(B|D_1)\mathbf{P}(D_1|C) - \mathbf{P}(BD_2|C)}{2\sqrt{\mathbf{P}(B|D_1)\mathbf{P}(D_1|C)\mathbf{P}(BD_2|C)}}.$$

Suppose that both coefficients of supplementarity μ and λ are bounded by 1. Then we can represent them in the trigonometric form

$$\lambda(B|\{D_1, D_2\}, C) = \cos\theta(B|\{D_1, D_2\}, C),$$

$$\mu(B|\{D_1, D_2\}, C) = \cos\gamma(B|\{D_1, D_2\}, C).$$

By inserting these cos-expressions in (6.25) and (6.27) we obtain trigonometric transformations of probabilities. We have (by Lemma 6.2)

$$\begin{aligned}\mathbf{P}(B_j|C) &= \mathbf{P}(B_j(A_1 \cup \cdots \cup A_n)|C)\\ &= \mathbf{P}(B_j|A_1)\mathbf{P}(A_1|C) + \mathbf{P}(B_j(A_2 \cup \cdots \cup A_n)|C)\\ &\quad + 2\mu(B_j|\{A_1, A_2 \cup \cdots \cup A_n\}, C)\\ &\quad \times \sqrt{\mathbf{P}(B_j|A_1)\mathbf{P}(A_1|C)\mathbf{P}(B_j(A_2 \cup \cdots \cup A_n)|C)},\end{aligned}$$

where

$$\begin{aligned}&\mu(B_j|\{A_1, A_2 \cup \cdots \cup A_n\}, C)\\ &= \frac{\mathbf{P}(B_j(A_1 \cup \cdots \cup A_n)|C) - \mathbf{P}(B_j|A_1)\mathbf{P}(A_1|C) - \mathbf{P}(B_j(A_2 \cup \cdots \cup A_n)|C)}{2\sqrt{\mathbf{P}(B_j|A_1)\mathbf{P}(A_1|C)\mathbf{P}(B_j(A_2 \cup \cdots \cup A_n)|C))}}.\end{aligned}$$

Suppose that the coefficients are relatively small for all $\beta \in X$

$$|\mu(B_j|\{A_1, A_2 \cup \cdots \cup A_n\}, C)| \leq 1.$$

Then we can represent these coefficients as

$$\mu(B_j|\{A_1, A_2 \cup \cdots \cup A_n\}, C) = \cos \gamma(B_j|\{A_1, A_2 \cup \cdots \cup A_n\}, C).$$

Thus the probability $\mathbf{P}(B_j|C) = \mathbf{P}(B_j(A_1 \cup \cdots \cup A_n)|C)$ can be represented as the square of the absolute value of the complex amplitude

$$\psi_C(\beta) \equiv \psi_C^{(1)}(\beta) = \sqrt{\mathbf{P}(B_j|A_1)\mathbf{P}(A_1|C)} + e^{i\gamma_C^{(1)}(\beta)}\sqrt{\mathbf{P}(B_j(A_2 \cup \cdots \cup A_n)|C)},$$

where the phase $\gamma_C^{(1)}(\beta) \equiv \gamma(B_j|\{A_1, A_2 \cup \cdots \cup A_n\}, C)$. In the same way the probability in the second summand can be represented as

$$\begin{aligned}&\mathbf{P}(B_j(A_2 \cup \cdots \cup A_n)|C)\\ &\quad = \mathbf{P}(B_j|A_2)\mathbf{P}(A_2|C) + \mathbf{P}(B_j(A_3 \cup \cdots \cup A_n)|C)\end{aligned}$$

$$+ 2\mu(B_j|\{A_2, A_3 \cup \cdots \cup A_n\}, C)$$
$$\times \sqrt{\mathbf{P}(B_j|A_2)\mathbf{P}(A_2|C)\mathbf{P}(B_j(A_3 \cup \cdots \cup A_n)|C)},$$

where

$$\mu(B_j|\{A_2, A_3 \cup \cdots \cup A_n\}, C) = \frac{\mathbf{P}(B_j(A_2 \cup \cdots \cup A_n)|C) - \mathbf{P}(B_j|A_2)\mathbf{P}(A_2|C) - \mathbf{P}(B_j(A_3 \cup \cdots \cup A_n)|C)}{2\sqrt{\mathbf{P}(B_j|A_2)\mathbf{P}(A_2|C)\mathbf{P}(B_j(A_3 \cup \cdots \cup A_n)|C)}}.$$

By supposing that these coefficients of supplementarity are also bounded by 1 we represent the probability as the square of the absolute value of the complex amplitude

$$\psi_C^{(2)}(\beta) = \sqrt{\mathbf{P}(B_j|A_2)\mathbf{P}(A_2|C)} + e^{i\gamma_C^{(2)}(\beta)}\sqrt{\mathbf{P}(B_j(A_3 \cup \cdots \cup A_n)|C)},$$

where $\gamma_C^{(2)}(\beta) = \pm \arccos \mu(B_j|\{A_2, A_3, \cup \cdots \cup A_n\}, C)$

$$\mathbf{P}(B_j(A_2 \cup \cdots \cup A_n)|C) = |\psi_C^{(2)}(\beta)|^2.$$

On the jth step we represent $\mathbf{P}(B_j(A_j \cup \cdots \cup A_n)|C)$ as the square of the absolute value of the complex amplitude

$$\psi_C^{(j)}(\beta) = \sqrt{\mathbf{P}(B_j|A_j)\mathbf{P}(A_j|C)} + e^{i\gamma_C^{(j)}(\beta)}\sqrt{\mathbf{P}(B_j(A_{j+1} \cup \cdots \cup A_n)|C)},$$

where $\gamma_C^{(j)}(\beta)$ is the phase of the coefficient

$$\mu(B_j|\{A_j, A_{j+1} \cup \cdots \cup A_n\}, C) = \frac{\mathbf{P}(B_j(A_j \cup \cdots \cup A_n)|C) - \mathbf{P}(B_j|A_j)\mathbf{P}(A_j|C) - \mathbf{P}(B_j(A_{j+1} \cup \cdots \cup A_n)|C)}{2\sqrt{\mathbf{P}(B_j|A_j)\mathbf{P}(A_j|C)\mathbf{P}(B_j(A_{j+1} \cup \cdots \cup A_n)|C)}}.$$

It is supposed that at each step we obtain coefficients $|\mu|$ bounded by 1. At the step $j = n - 1$ we should represent the probability $\mathbf{P}(B_j(A_{n-1} \cup A_n)|C)$. At the last step we apply the formula (6.25) from Proposition 6.6.[6] We obtain

[6] We emphasize that the multi-dimensional algorithm of representation of contexts by complex probability amplitudes is based on two formulas: inductional steps, besides the final one, are based

$$\begin{aligned}
&\mathbf{P}(B_j(A_{n-1} \cup A_n)|C) \\
&\quad = \mathbf{P}(B_j|A_{n-1})\mathbf{P}(A_{n-1}|C) + \mathbf{P}(B_j|A_n)\mathbf{P}(A_n|C) \\
&\qquad + 2\lambda(B_j|\{A_{n-1}, A_n\})\sqrt{\mathbf{P}(B_j|A_{n-1})\mathbf{P}(A_{n-1}|C)\mathbf{P}(B_j|A_n)\mathbf{P}(A_n|C)},
\end{aligned}$$

where the coefficient of statistical disturbance λ was defined by (6.26). And if $|\lambda|$ is bounded by 1, then we can represent the probability as the square of the absolute value of the complex amplitude

$$\psi_C^{(n-1)}(\beta) = \sqrt{\mathbf{P}(B_j|A_{n-1})\mathbf{P}(A_{n-1}|C)} + e^{i\theta_C(\beta)}\sqrt{\mathbf{P}(B_j|A_n)\mathbf{P}(A_n|C)},$$

where $\theta_C(\beta) = \pm \arccos \lambda(\beta|\{A_{n-1}, A_n\}, C)$.

We have

$$\psi_C^{(j)}(\beta) = \sqrt{\mathbf{P}(B_j(A_j \cup \cdots \cup A_n)|C)}e^{i\alpha_C^{(j)}(\beta)},$$

where

$$\alpha_C^{(j)}(\beta) = \arg \psi_C^{(j)}(\beta) = \arccos \frac{M_j}{N_j},$$

and

$$\begin{aligned}
M_j &= \sqrt{\mathbf{P}(B_j|A_j)\mathbf{P}(A_j|C)} \\
&\quad + \mu(B_j|\{A_j, A_{j+1} \cup \cdots \cup A_n\}, C)\sqrt{\mathbf{P}(B_j(A_{j+1} \cup \cdots \cup A_n)|C)},
\end{aligned}$$

$$N_j = \sqrt{\mathbf{P}(B_j(A_j \cup \cdots \cup A_n)|C)}.$$

Finally, we have

$$\begin{aligned}
&\alpha_C^{(n-1)}(\beta) \\
&\quad = \arg \psi_C^{(n-1)}(\beta) \\
&\quad = \arccos \frac{\sqrt{\mathbf{P}(B_j|A_{n-1})\mathbf{P}(A_{n-1}|C)} + \lambda(B_j|\{A_{n-1}, A_n\}, C)\sqrt{\mathbf{P}(B_j|A_n)\mathbf{P}(A_n|C)}}{\sqrt{\mathbf{P}(B_j(A_{n-1} \cup A_n)|C)}}.
\end{aligned}$$

---

on the formula (6.27), Lemma 6.2, but the final step—on the formula (6.25). Only through this application of (6.25) do we obtain matching with the algorithm for the dichotomous case, see Chap. 4 and Sect. 6.3 of the present chapter.

Thus we have

$$\psi_C(\beta) = \sqrt{\mathbf{P}(B_j|A_1)\mathbf{P}(A_1|C)} + e^{i[\gamma_C^{(1)}(\beta)-\alpha_C^{(2)}(\beta)]}\psi_C^{(2)}(\beta)$$
$$= \sqrt{\mathbf{P}(B_j|A_1)\mathbf{P}(A_1|C)} + e^{i\beta_C^{(2)}(\beta)}\sqrt{\mathbf{P}(B_j|A_2)\mathbf{P}(A_2|C)}$$
$$+ e^{i\beta_C^{(3)}(\beta)}\psi_C^{(3)}(\beta),$$

where

$$\beta_C^{(2)}(\beta) = \gamma_C^{(1)}(\beta) - \alpha_C^{(2)}(\beta), \qquad \beta_C^{(3)}(\beta) = \beta_C^2(\beta) + \gamma_C^{(2)}(\beta) - \alpha_C^{(3)}(\beta).$$

Finally, we obtain

$$\psi_C(\beta) = \sum_{j=1}^{n} e^{i\beta_C^{(j)}(\beta)}\sqrt{\mathbf{P}(B_j|A_j)\mathbf{P}(A_j|C)}$$

with $\beta_C^{(1)}(\beta) = 0$ (this is just due to our special choice of a representation) and $\beta_C^{(n)}(\beta) = \beta_C^{(n-1)}(\beta) + \theta_C(\beta)$.

Thus by inductive splitting of multi-valued variables into dichotomous variables we represented contextual probabilities by complex amplitudes $\psi_C(\beta)$. Here Born rule holds true.

By using the notation

$$p^{b|a}(\beta_j|\alpha_i) = \mathbf{P}(B_j|A_i), \qquad p_C^b(\beta_j) = \mathbf{P}(B_j|C), \qquad p_C^a(\alpha) = \mathbf{P}(A_i|C)$$

we write

$$\psi_C(\beta) = \sum_{\alpha} e^{i\beta_C^{(\alpha)}(\beta)}\sqrt{p_C^a(\alpha)p^{b|a}(\beta|\alpha)}.$$

In particular, for $n = 3$ we have

$$\psi_C(\beta) = \sqrt{p_C^a(\alpha_1)p^{b|a}(\beta|\alpha_1)} + e^{i\beta_C^{(2)}(\beta)}\sqrt{p_C^a(\alpha_2)p^{b|a}(\beta|\alpha_2)}$$
$$+ e^{i\beta_C^{(3)}(\beta)}\sqrt{p_C^a(\alpha_3)p^{b|a}(\beta|\alpha_3)},$$

where

$$\beta_C^{(2)}(\beta) = \gamma_C^{(1)}(\beta) - \alpha_C^{(2)}(\beta), \qquad \beta_C^{(3)}(\beta) = \beta_C^{(2)}(\beta) + \theta_C(\beta).$$

We remark that each phase $\beta_C^{(j)}(\beta)$ depends on all three a-contexts, A_1, A_2, A_3. So we cannot use the symbol $\beta_C(\beta|\alpha)$. In $\beta_C^{(\alpha)}(\beta)$ the α is just the summation index; in fact, $\beta_C^{(\alpha)}(\beta) \equiv \beta_C^{(\alpha)}(\beta|A_1, A_2, A_3)$. We remark that the probability $p_C^b(\beta)$ can be represented as

$$\begin{aligned} p_C^b(\beta) &= |\psi_C(\beta)|^2 \\ &= \sum_\alpha p_C^a(\alpha) p^{b|a}(\beta|\alpha) \\ &\quad + 2 \sum_{\alpha_1 < \alpha_2} \cos[\beta_C^{(\alpha_2)}(\beta) - \beta_C^{(\alpha_1)}(\beta)] \\ &\quad \times \sqrt{p_C^a(\alpha_1) p_C^a(\alpha_2) p^{b|a}(\beta|\alpha_1) p^{b|a}(\beta|\alpha_2)}. \end{aligned}$$

6.9 Växjö Models with Multi-Kolmogorovian Structure

Let $(\Omega, \mathscr{F})$ be a measurable space (i.e. Ω is a set and $\mathscr{F}$ is a σ-algebra of its subsets). Consider the space of all measurable functions $\xi : \Omega \to \mathbf{R}$, random variables. We denote this space by the symbol RV.

Definition 6.8. *Let $\mathscr{F}_0$ and RV_0 be subsets of $\mathscr{F}$ and RV, respectively, and let $\pi = \{\mathbf{P}_C\}_{C \in \mathscr{F}_0}$ be a family of probability measures on $\mathscr{F}$. The triple*

$$\mathscr{P}_{\text{cont,multi-Kol}} = (\mathscr{F}_0, RV_0, \pi)$$

is called a contextual multi-Kolmogorov space. Elements of $\mathscr{F}_0$ and RV are called contexts and observables, respectively.

Thus for any context $C \in \mathscr{F}_0$ and any observable $a \in RV_0$ there are well-defined contextual probabilities $\mathbf{P}(a \in V|C) = \mathbf{P}_C(\omega \in \Omega : a(\omega) \in V)$, $V \in \mathscr{B}(\mathbf{R})$. The main difference from the definition of the contextual Kolmogorov space, see Definition 6.3 of this chapter, is that in general probabilities $\{\mathbf{P}_C\}$ need not be based on a single probability $\mathbf{P}$ (via the Bayes formula: $\mathbf{P}_C(A) = \mathbf{P}(A|C) = \mathbf{P}(AC)/\mathbf{P}(C)$).

Definition 6.9. *A contextual multi-Kolmogorov model is given by a contextual multi-Kolmogorov space* $\mathcal{P}_{\text{cont,multi-Kol}} = (\mathcal{F}_0, RV_0, \pi)$ *such that sets* $\mathcal{F}_0$ *and* RV_0 *are constrained in the following way:*

(C) *for any* $a \in RV_0$, *the system* $\mathcal{F}_0$ *contains all sets* $A_U^a (= \{\omega \in \Omega : a(\omega) \in U\})$, *where* $U \in \mathcal{B}_a(\mathbf{R})$.

These sets represent selection contexts.

We remark that in the multi-Kolmogorov framework a set $U \in \mathcal{B}_a(\mathbf{R})$ iff there exists $C \in \mathcal{F}_0$ such that $\mathbf{P}_C(A_U^a) > 0$.

Chapter 7

Interference of Probabilities from Law of Large Numbers

In previous chapters we derived the interference of probabilities in frequency (von Mises) and measure-theoretic (Kolmogorov) frameworks. In this chapter we would like to combine previous considerations and derive interference by using the *law of large numbers* in the Kolmogorov model [43]. Thus we would like to present the frequency derivation of the interference of probabilities in the measure-theoretic framework.

Let us remark once more that besides the probability space $(\Omega, \mathcal{F}, \mathbf{P})$ the essential role in construction of probabilistic models is played by families of random variables used for description of studied phenomena. Of course, having specified the model, one can consider the new (*canonical*) probability space with coordinate random variables. This is the cornerstone of considerations in this chapter: context transformation can be represented by transformation of random variables, cf. Chap. 4.

The main theorem of this chapter was obtained in collaboration with Alexander Bulinskii (the chair of probability, department of Mechanics and Mathematics, Moscow State University).

7.1 Kolmogorovian Description of Quantum Measurements

In the measure-theoretic framework one operates with events and events-conditioning. The formula of total probability with interference term should be interpreted in the following way.

Set $B = A_1 \cup A_2$ where $A_1 \cap A_2 = \emptyset$ and rewrite the formula of total probability with interference term as

$$\mathbf{P}(B) = \mathbf{P}(B|A_1)\mathbf{P}(A_1) + \mathbf{P}(B|A_2)\mathbf{P}(A_2) + 2\sqrt{\mathbf{P}(B|A_1)\mathbf{P}(A_1)\mathbf{P}(B|A_2)\mathbf{P}(A_2)} \cos\theta, \tag{7.1}$$

where, as usual,

$$\mathbf{P}(B|A_i) = \mathbf{P}(BA_i)/\mathbf{P}(A_i)$$

and $\mathbf{P}(A_i) > 0$, $i = 1, 2$. The trivial case $\mathbf{P}(A_i) = 0$ for some i is henceforth excluded.

For the sake of simplicity consider some physical system s possessing only properties a and b described by a pair of two-valued random variables (denoted by the same symbols a and b). Suppose that there is a procedure permitting one to reproduce the copies $s_1, s_2, \ldots$ of s to get a *statistical ensemble* of s in the following sense. Let $(a^{(1)}, b^{(1)}), (a^{(2)}, b^{(2)}), \ldots$ be a sequence of independent equally distributed random vectors defined on some probability space $(\Omega, \mathcal{F}, \mathbf{P})$. Assume that for some $\alpha_i, \beta_i \in \mathbf{R}$, $i = 1, 2$,

$$\mathbf{P}(a^{(1)} = \alpha_1) = p_1^a, \qquad \mathbf{P}(a^{(1)} = \alpha_2) = p_2^a, \tag{7.2}$$

$$\mathbf{P}(b^{(1)} = \beta_1) = p_1^b, \qquad \mathbf{P}(b^{(1)} = \beta_2) = p_2^b, \tag{7.3}$$

where $p_1^a + p_2^a = 1$ and $p_1^b + p_2^b = 1$.

For $j \in \mathbf{N}$ and elementary event $\omega \in \Omega$ the vector $(a^{(j)}(\omega), b^{(j)}(\omega))$ characterizes the properties a and b of the system s_j. (Thus these two properties are simultaneously defined. We can call them objective properties.) The independence assumption for the sequence $\{(a^{(j)}, b^{(j)})\}_{j\geq 1}$ reflects the usual non-interaction hypothesis for elements of a stochastic ensemble. Further we can use more general assumptions as well. For each $j \in \mathbf{N}$ no conditions are imposed on the joint distribution of random variables $a^{(j)}$ and $b^{(j)}$.[1] Moreover, note that for any $\omega \in \Omega$ the sequence $\{(a^{(j)}(\omega), b^{(j)}(\omega)\}_{j\geq 1}$ is not, in general, available. In fact, to get the data certain measurement procedures should be used.

Suppose that it is possible to apply to systems s_j a measurement procedure $\mathscr{M}_a$ permitting us for each $\omega \in \Omega$ to fix the value $a^{(j)}(\omega)$ of the characteristic describing property a. In other words, for every $\omega \in \Omega$ we have integers $1 \leq k_1(\omega) < k_2(\omega) < \ldots$ such that

$$a^{(k_1(\omega))}(\omega) = \alpha_1, \qquad a^{(k_2(\omega))}(\omega) = \alpha_1, \qquad \ldots$$

and, respectively,

$$a^{(m_j(\omega))}(\omega) = \alpha_2 \quad \text{for } m_1(\omega) < m_2(\omega) < \cdots$$

where $m_j(\omega) \in \mathbf{N} \setminus \bigcup_i \{k_i(\omega)\}$.

It is worth noting that, after the measurement procedure $\mathscr{M}_a$, the property b of initial systems s_j in general can be changed. Thus, new systems s'_j $(j \in \mathbf{N})$ arise. To describe the property b of these systems, it is natural to use random sequences $\{\bar{b}^{(j)}\}_{j\geq 1}$ and $\{\breve{b}^{(j)}\}_{j\geq 1}$ such that

$$\mathbf{P}(\bar{b}^{(j)} = \beta_r) = p_r^{\bar{b}}, \qquad \mathbf{P}(\breve{b}^{(j)} = \beta_r) = p_r^{\breve{b}}, \quad j, r = 1, 2, \tag{7.4}$$

where $p_1^{\bar{b}} + p_2^{\bar{b}} = 1$, $p_1^{\breve{b}} + p_2^{\breve{b}} = 1$.

Assume that for each $\omega \in \Omega$ and $i = i(\omega) \in \mathbf{N}$ such that $a^{(i)}(\omega) = \alpha_1$ the value $\bar{b}^{(i)}(\omega)$ appears instead of $b^{(i)}(\omega)$. Other values $b^{(i)}(\omega)$ (for $i = i(\omega) \in \mathbf{N} \setminus \bigcup_i \{k_i(\omega)\}$) are replaced by $\breve{b}^{(i)}(\omega)$.

[1] This is very natural from the quantum viewpoint.

So, to describe the properties a and b for the ensemble of systems s'_j ($j \in \mathbf{N}$) we obtain two-dimensional sequences

$$\{(\alpha_1, \bar{b}^{(k_j(\omega))}(\omega))\}_{j\geq 1} \quad \text{and} \quad \{(\alpha_2, \breve{b}^{(m_j(\omega))}(\omega))\}_{j\geq 1}.$$

Note that since $k_j = k_j(\omega)$ and $m_j = m_j(\omega)$ are random variables defined on probability space $(\Omega, \mathscr{F}, \mathbf{P})$, the same is true of $\bar{b}^{(k_j)} = \bar{b}^{(k_j(\omega))}(\omega)$ and $\breve{b}^{(m_j)} = \breve{b}^{(m_j(\omega))}(\omega)$.

It is always possible to construct a probability space $(\Omega, \mathscr{F}, \mathbf{P})$ where stochastic sequences $\{(a^{(j)}, b^{(j)}\}_{j\geq 1}$, $\{\bar{b}^{(j)}\}_{j\geq 1}$, $\{\breve{b}^{(j)}\}_{j\geq 1}$, with the above-mentioned properties, are defined. Moreover, using the standard product of probability spaces it is easy to guarantee the independence of terms of these sequences and independence of the collection of three sequences as well. However, we do not use the last opportunity.

Suppose also that a measurement procedure $\mathscr{M}_b$ can be applied to systems s'_j ($j \in \mathbf{N}$) to fix the values of the characteristics of property b. Thus we get systems s''_i ($i \in \mathbf{N}$) having $\bar{b}^{(k_j(\omega))}(\omega)$ or $\breve{b}^{(m_j(\omega))}(\omega)$ as b-characteristics for $\omega \in \Omega$ and $i = k_j(\omega)$ or $i = m_j(\omega)$. Of course, the a-characteristics of systems s'_j ($j \in \mathbf{N}$) in general can not be conserved.

7.2 Limit Theorems and Formula of Total Probability with Interference Term

Now we consider from the view-point of limit theorems for random variables defined on $(\Omega, \mathscr{F}, \mathbf{P})$ the approach used in previous chapters. Namely, we discuss statistical properties of b-characteristics for ensembles of systems $\{s_j\}_{j\geq 1}$ and $\{s''_j\}_{j\geq 1}$.

Let $|\cdot|$ be a number of elements of a finite set. For $\omega \in \Omega$, $N \in \mathbf{N}$ and $j = 1, 2$ let

$$q_{jN}(\omega) = |\{i \in \{1, \ldots, N\} : b^{(i)}(\omega) = \beta_j\}|/N. \tag{7.5}$$

Set $N_1(\omega) = \max\{i : k_i(\omega) \leq N\}$, $N_2(\omega) = \max\{i : m_i(\omega) \leq N\}$. For $\omega \in \Omega$ and $j = 1, 2$ put

$$n_{1j}(\omega) = |\{i \in \{1, \ldots, N_1(\omega)\} : b^{(k_i(\omega))}(\omega) = \beta_j\}|,$$

$$n_{2j}(\omega) = |\{i \in \{1, \ldots, N_2(\omega)\} : b^{(m_i(\omega))}(\omega) = \beta_j\}|,$$

$$m_{1j}(\omega) = |\{i \in \{1, \ldots, N_1(\omega)\} : \bar{b}^{(k_i(\omega))}(\omega) = \beta_j\}|,$$

$$m_{2j}(\omega) = |\{i \in \{1, \ldots, N_2(\omega)\} : \breve{b}^{(m_i(\omega))}(\omega) = \beta_j\}|.$$

To simplify the notation (for $j, r = 1, 2$ and $\omega \in \Omega$) we omit the dependence of $n_{rj}(\omega)$ and $m_{rj}(\omega)$ on N.

Evidently, for all $\omega \in \Omega$, $j = 1, 2$ and $N \in \mathbf{N}$ one has

$$q_{jN}(\omega) = \frac{m_{1j}(\omega)}{N} + \frac{m_{2j}(\omega)}{N} + \delta_j^N(\omega) \tag{7.6}$$

where

$$\delta_j^N(\omega) = \frac{n_{1j}(\omega) - m_{1j}(\omega)}{N} + \frac{n_{2j}(\omega) - m_{2j}(\omega)}{N}.$$

Theorem 7.1. *Let* $\{(a^{(j)}, b^{(j)})\}_{j\geq 1}$, $\{(a^{(j)}, \bar{b}^{(j)})\}_{j\geq 1}$ *and* $\{(a^{(j)}, \breve{b}^{(j)})\}_{j\geq 1}$ *be sequences of pairwise independent random vectors identically distributed within each sequence, i.e., for all* $j \in \mathbf{N}$,

$$(a^{(j)}, b^{(j)}) \stackrel{\mathcal{D}}{=} (a^{(1)}, b^{(1)}), \qquad (a^{(j)}, \bar{b}^{(j)}) \stackrel{\mathcal{D}}{=} (a^{(1)}, \bar{b}^{(1)}),$$

$$(a^{(j)}, \breve{b}^{(j)}) \stackrel{\mathcal{D}}{=} (a^{(1)}, \breve{b}^{(1)});$$

here $\mathcal{D}$ *means the equality of distributions. Assume that* (7.3)–(7.4) *hold true and* $a^{(1)}$ *is not degenerate. Then for all* $j, r = 1, 2$ *and* $\mathbf{P}$*-a.e.* $\omega \in \Omega$ *there exist nonrandom limits*

$$p_{rj}^{b|a} := \lim_{N\to\infty} \frac{m_{rj}(\omega)}{N_r(\omega)}, \qquad \delta_j := \lim_{N\to\infty} \delta_j^N(\omega) \tag{7.7}$$

and besides

$$p_j^b = p_1^a p_{1j}^{b|a} + p_2^a p_{2j}^{b|a} + \delta_j. \tag{7.8}$$

Proof. Borelean functions of pairwise independent random vectors are pairwise independent. Therefore, by the Etemadi SLLN (Strong Law of Large Numbers) one has for **P**-a.e. $\omega \in \Omega$ and $j = 1, 2$,

$$q_{jN}(\omega) = \frac{1}{N} \sum_{i=1}^{n} I_{\{b^{(i)}(\omega)=\beta_j\}}(\omega) \to E\, I_{\{b^{(1)}=\beta_j\}} = p_j^b \quad \text{as } N \to \infty, \tag{7.9}$$

where the indicator function of an event D

$$I_D(\omega) = \begin{cases} 1, & \omega \in D, \\ 0, & \omega \notin D. \end{cases}$$

Analogously with probability 1,

$$\frac{N_1(\omega)}{N} \to p_1^a \quad \text{and} \quad \frac{N_2(\omega)}{N} \to p_2^a \quad \text{as } N \to \infty. \tag{7.10}$$

It is clear that

$$\frac{m_{1j}(\omega)}{N} = \frac{1}{N} \sum_{i=1}^{N} I_{\{\bar{b}^{(i)}(\omega)=\beta_j\}}(\omega) I_{\{a^{(i)}(\omega)=\alpha_1\}}(\omega),$$
$$\frac{m_{2j}(\omega)}{N} = \frac{1}{N} \sum_{i=1}^{N} I_{\{\breve{b}^{(i)}(\omega)=\beta_j\}}(\omega) I_{\{a^{(i)}(\omega)=\alpha_2\}}(\omega).$$

The same reasoning shows that for **P**-a.e. $\omega \in \Omega$ there exist

$$\lim_{N\to\infty} \frac{m_{1j}(\omega)}{N} = \mathbf{P}(\bar{b}^{(1)} = \beta_j, a^{(1)} = \alpha_1), \tag{7.11}$$
$$\lim_{N\to\infty} \frac{m_{2j}(\omega)}{N} = \mathbf{P}(\breve{b}^{(1)} = \beta_j, a^{(1)} = \alpha_2). \tag{7.12}$$

The event consisting of those ω rendering $a^{(i)}(\omega) = \alpha_1$ for all $i \in \mathbf{N}$ or $a^{(i)}(\omega) = \alpha_2$ for all $i \in \mathbf{N}$ has **P**-measure zero. So, there exists an event U with $\mathbf{P}(U) = 1$ such that both $N_1(\omega) > 0$ and $N_2(\omega) > 0$ for all $\omega \in \Omega$ and $N > M(\omega)$. Hence for $\omega \in U$ and $N > M(\omega)$ we have

$$\frac{m_{rj}(\omega)}{N} = \frac{m_{rj}(\omega)}{N_r(\omega)} \frac{N_r(\omega)}{N}, \quad j, r = 1, 2.$$

Taking into account (7.10)–(7.12) we establish the existence of limits $p_{rj}^{b|a}$ in (7.7) and due to (7.9) the validity of the second relation in (7.7) follows. Now (7.8) is obvious and the proof is complete.

Remark 7.1. Successive employment of procedures $\mathscr{M}_a$ and $\mathscr{M}_b$ for systems s_j and s'_j $(j \in \mathbf{N})$ permits us to determine $N_1(\omega)$, $N_2(\omega)$ and $m_{jr}(\omega)$ for $j, r = 1, 2$. Thus, according to (7.7) and (7.10) there are strongly consistent estimates for $p_{rj}^{b|a}$ and p_r^a $(j, r = 1, 2)$. To determine $q_{jN}(\omega)$ one should apply the procedure $\mathscr{M}_b$ to systems $s_1, \ldots, s_N$. However, it would change the (distribution of) a-characteristics of these systems. Therefore, to obtain the strongly consistent estimates for parameters p_j^b $(j = 1, 2)$ we can proceed in the following manner. Invoking the principle of reproducibility we can construct besides the copies s_j $(j \in \mathbf{N})$ of a system s the copies $\tilde{s}_j$ $(j \in \mathbf{N})$ described by a sequence of random vectors $\{(\tilde{a}^{(j)}, \tilde{b}^{(j)})\}_{j\geq 1}$. The standard extension of a probability space $(\Omega, \mathscr{F}, \mathbf{P})$ can be used to define this auxiliary random sequence. For an extension we keep the same notation $(\Omega, \mathscr{F}, \mathbf{P})$. Applying the procedure $\mathscr{M}_b$ to systems $\tilde{s}_j$ $(j \in \mathbf{N})$ and defining $\tilde{q}_{jN}(\omega)$ analogously to (7.5) we see that for $\mathbf{P}$-a.e. $\omega \in \Omega$ and $j = 1, 2$,

$$\tilde{q}_{jN}(\omega)/N \to p_j^b \quad \text{as } N \to \infty.$$

Thus under the conditions of Theorem 7.1 one can provide the strongly consistent estimates for the values δ_j $(j = 1, 2)$ appearing in formula (7.8).

Remark 7.2. The notation $p_{rj}^{b|a}$ for limits in (7.7) seems natural as $m_{rj}(\omega)/N_r(\omega)$ is the relative frequency of appearing value β_j for the characteristics of property b in a sequence of those systems which have α_r value for the characteristics of property a. However, the sense of formula (7.8) is demonstrated much better by relations (7.11) and (7.12) showing that

$$p_{1j}^{b|a} = \mathbf{P}(\bar{b}^{(1)} = \beta_j | a^{(1)} = \alpha_1), \qquad p_{2j}^{b|a} = \mathbf{P}(\breve{b}^{(1)} = \beta_j | a^{(1)} = \alpha_2). \tag{7.13}$$

Thus the additional (with respect to the classical total probability formula) term δ_j appears in (7.8) because the following equalities need not hold true

$$\mathbf{P}(\bar{b}^{(1)} = \beta_j | a^{(1)} = \alpha_1) = \mathbf{P}(b^{(1)} = \beta_j | a^{(1)} = \alpha_1),$$
$$\mathbf{P}(\breve{b}^{(1)} = \beta_j | a^{(1)} = \alpha_2) = \mathbf{P}(b^{(1)} = \beta_j | a^{(1)} = \alpha_2).$$

In such a way it is more informative to rewrite (7.8) in the form

$$p_j^b = p_1^a p_{1j}^{\bar{b}|a} + p_2^a p_{2j}^{\breve{b}|a} + \delta_j, \tag{7.14}$$

setting now

$$\begin{aligned} p_{1j}^{\bar{b}|a} &= \mathbf{P}(\bar{b}^{(1)} = \beta_j | a^{(1)} = \alpha_1), \\ p_{2j}^{\breve{b}|a} &= \mathbf{P}(\breve{b}^{(1)} = \beta_j | a^{(1)} = \alpha_2), \quad j = 1, 2. \end{aligned} \tag{7.15}$$

Formula (7.14) clarifies as well that simultaneous measurement of characteristics a and b in general is not supposed.

Remark 7.3. Due to (7.14) and (7.15) one has

$$|\delta_j| \leq 1, \quad j = 1, 2.$$

In the particular case when for each $j \in \mathbf{N}$, random variables $\bar{b}^{(j)}$ and $a^{(j)}$ are independent and the same is true for $\breve{b}^{(j)}$ and $a^{(j)}$, Theorem 7.1 implies that

$$p_j^b = p_1^a p_j^{\bar{b}} + p_2^a p_j^{\breve{b}} + \delta_j. \tag{7.16}$$

So, the equality $\delta_j = 0$ (if $\delta_1 = 0$ then $\delta_2 = 0$ and vice versa) is satisfied if and only if the point p_j^b is appropriately located on the segment with end points $p_j^{\bar{b}}$ and $p_j^{\breve{b}}$. Evidently, for any given p_j^a, p_j^b it is always possible to indicate $p_j^{\bar{b}}$ and $p_j^{\breve{b}}$ $(j = 1, 2)$ to guarantee the relation $\delta_j = 0$. Namely, one takes $p_j^{\bar{b}}$ in such a way that

$$\max\{0, (p_j^b - p_2^a)/p_1^a\} \leq p_j^{\bar{b}} \leq \min\{1, p_j^b/p_1^a\}$$

and after that chooses $p_j^{\breve{b}} = (p_j^b - p_1^a p_j^{\bar{b}})/p_2^a$, $j = 1, 2$.

Remark 7.4. It is possible to write formula (7.14) analogously to (7.1) using the auxiliary parameters λ_j determined by relation

$$\delta_j = 2\sqrt{p_{1j}^{\bar{b}|a} p_1^a p_{2j}^{\breve{b}|a} p_2^a}\lambda_j, \quad j = 1, 2.$$

Note that in general λ_j can take any real values. In the particular case described by formula (7.16) we establish that

$$\lambda_j \le 1 \Longleftrightarrow t_{1j} + t_{2j} \ge \sqrt{p_j^b},$$
$$\lambda_j \ge -1 \Longleftrightarrow |t_{1j} - t_{2j}| \le \sqrt{p_j^b}$$

where $t_{1j} = \sqrt{p_j^{\bar{b}} p_1^a}$, $t_{2j} = \sqrt{p_j^{\breve{b}} p_2^a}$ $(j = 1, 2)$. If, moreover, $p_j^{\bar{b}} = p_j^{\breve{b}}$ then

$$|\lambda_j| \le 1 \Longleftrightarrow \frac{p_j^b}{1 + 2\sqrt{p_1^a p_2^a}} \le p_j^{\bar{b}} \le \frac{p_j^b}{1 - 2\sqrt{p_1^a p_2^a}}. \tag{7.17}$$

The last relation shows that the possibility to represent λ_j as the cosine of some angle in this special case means that the distribution of $\bar{b}^{(1)}$ is obtained from that of $b^{(1)}$ by a "bounded perturbation."

Remark 7.5. It is not difficult to complicate the scheme of measuring characteristics of properties a and b. For instance one can assume that procedures $\mathcal{M}_a$ and $\mathcal{M}_b$ do not permit us to fix exactly in all experiments (i.e. for all systems) the values of characteristics for properties a and b. Then for validity of Theorem 7.1 it is sufficient to suppose that the number of faults (or "nonsuccessful") measurements is **P**-a.e. $o(N)$ (as $N \to \infty$) among the systems with labels $1, \dots, N$.

Part III
Bell's Inequality

Bell's inequality really beats all records on publications, citations, discussions and controversies. As an organizer of a few international conferences on foundations of quantum theory, I was really disappointed by the stormy debates on Bell's inequality. I could not find any problem in quantum foundations which can be compared with Bell's inequality by intensiveness of discussions and strength of reactions to opponent's views, I asked myself many times:

"What is so special in Bell's inequality?"

We shall try to find a proper answer to this question in this part of the book.

Chapter 8
Probabilistic Analysis of Bell's Argument

It seems that the main source of the debate mentioned above has been Bell's mathematical imprecision (and not at all physical importance of the problem). J. Bell formulated precisely only his aim: to prove nonlocality of quantum mechanics.[1] However, he did not precisely determine mathematical probabilistic rules for formalization of the problem, [34]. This absence of rigor in mathematical formulation gives a great chance for numerous speculations. Each year people present their own (very different) views on the probabilistic structure of Bell's arguments and, consequently, they come to conclusions which are totally different from Bell's conclusions.

I have had numerous conversations with outstanding physicists. A rather common opinion is that it is totally meaningless to pay attention to foundations of probability theory. Typically such a viewpoint is motivated by considering probability as a "*physically well-defined quantity*." Therefore one need not seek a mathematically rigorous probabilistic formalization.

I totally disagree with such a viewpoint. In the same way one might say that *physicists need not take note of mathematical models of space and geometries*. For-

[1] J. Bell was a "nonlocal realist." The main aim of his investigations was to find arguments supporting the Bohmian model of quantum mechanics. Thus he wanted to save realism even at the cost of locality [34].

tunately, nobody would speak such nonsense nowadays. Suppose for the moment that we try to work in physics without determining precisely mathematical models of space. It is clear that such an activity would induce perpetual debates and a variety of paradoxes. I think that a similar thing happened with Bell's inequality (with probability, instead of space). Thus the intensity of debates on Bell's inequality is not completely determined by physical importance of the problem. The fact that the problem was not formulated in a rigorous mathematical framework could not be neglected.

Nevertheless, during the last 40 years physicists have been trying to proceed with Bell's inequalities without describing precisely the probabilistic rules that they used. Moreover, the common opinion is that one could work with only frequencies. Hence the rigorous mathematical description of a corresponding probability model is unimportant.

In this chapter we shall present a mathematical formalization of Bell's [34] (as well as von Neumann's [313]) no-go arguments. If one proceeds in such a rigorous mathematical framework, then the common conclusion about disagreement of quantum formalism with classical statistical description would not be totally justified.

Since this chapter includes analysis of standard proofs of the Bell-type inequalities [34, 59], we will reproduce these well-known proofs.

8.1 Measure-Theoretic Derivations of Bell-Type Inequalities

8.1.1 Bell's Inequality

Let $\mathscr{P} = (\Omega, \mathscr{F}, \mathbf{P})$ be a Kolmogorov probability space. For any pair of random variables $u(\omega)$, $v(\omega)$, their covariation is defined by

$$\langle u, v\rangle = \mathrm{cov}(u, v) = \int_\Omega u(\omega)v(\omega)d\mathbf{P}(\omega).$$

We reproduce the proof of Bell's inequality in the measure-theoretic framework.

Theorem 8.1 (Bell's inequality for covariations). *Let $\xi_a, \xi_b, \xi_c = \pm 1$ be random variables on $\mathcal{P}$. Then Bell's inequality*

$$|\langle \xi_a, \xi_b\rangle - \langle \xi_c, \xi_b\rangle| \leq 1 - \langle \xi_a, \xi_c\rangle \tag{8.1}$$

holds.

Proof. Set $\Delta = \langle \xi_a, \xi_b\rangle - \langle \xi_c, \xi_b\rangle$. By linearity of Lebesgue integral we obtain

$$\begin{aligned}\Delta &= \int_\Omega \xi_a(\omega)\xi_b(\omega)d\mathbf{P}(\omega) - \int_\Omega \xi_c(\omega)\xi_b(\omega)d\mathbf{P}(\omega)\\ &= \int_\Omega [\xi_a(\omega) - \xi_c(\omega)]b(\omega)d\mathbf{P}(\omega).\end{aligned} \tag{8.2}$$

As

$$\xi_a(\omega)^2 = 1, \tag{8.3}$$

we have

$$\begin{aligned}|\Delta| &= \left|\int_\Omega [1 - \xi_a(\omega)\xi_c(\omega)]\xi_a(\omega)\xi_b(\omega)d\mathbf{P}(\omega)\right|\\ &\leq \int_\Omega [1 - \xi_a(\omega)\xi_c(\omega)]d\mathbf{P}(\omega).\end{aligned} \tag{8.4}$$

8.1.2 Wigner's Inequality

We recall the following simple mathematical result, see Wigner [319].

Theorem 8.2 (Wigner inequality). *Let $\xi_a, \xi_b, \xi_c = \pm 1$ be arbitrary random variables on a Kolmogorov space $\mathcal{P}$. Then the following inequality holds*

$$\mathbf{P}(\xi_a = +1, \xi_b = +1) + \mathbf{P}(\xi_b = -1, \xi_c = +1) \geq \mathbf{P}(\xi_a = +1, \xi_c = +1). \tag{8.5}$$

Proof. We have

$$\mathbf{P}(\omega \in \Omega : \xi_a(\omega) = +1, \xi_b(\omega) = +1)$$
$$= \mathbf{P}(\omega \in \Omega : \xi_a(\omega) = +1, \xi_b(\omega) = +1, c(\omega) = +1)$$
$$+ \mathbf{P}(\omega \in \Omega : \xi_a(\omega) = +1, \xi_b(\omega) = +1, \xi_c(\omega) = -1), \tag{8.6}$$

$$\mathbf{P}(\omega \in \Omega : \xi_b(\omega) = -1, \xi_c(\omega) = +1)$$
$$= \mathbf{P}(\omega \in \Omega : \xi_a(\omega) = +1, \xi_b(\omega) = -1, c(\omega) = +1)$$
$$+ \mathbf{P}(\omega \in \Omega : \xi_a(\omega) = -1, \xi_b(\omega) = -1, \xi_c(\omega) = +1), \tag{8.7}$$

and

$$\mathbf{P}(\omega \in \Omega : \xi_a(\omega) = +1, \xi_c(\omega) = +1)$$
$$= \mathbf{P}(\omega \in \Omega : \xi_a(\omega) = +1, \xi_b(\omega) = +1, \xi_c(\omega) = +1)$$
$$+ \mathbf{P}(\omega \in \Omega : \xi_a(\omega) = +1, \xi_b(\omega) = -1, \xi_c(\omega) = +1). \tag{8.8}$$

If we add together equations (8.6) and (8.7) we obtain

$$\mathbf{P}(\omega \in \Omega : \xi_a(\omega) = +1, \xi_b(\omega) = +1) + \mathbf{P}(\omega \in \Omega : \xi_b(\omega) = -1, \xi_c(\omega) = +1)$$
$$= \mathbf{P}(\omega \in \Omega : \xi_a(\omega) = +1, \xi_b(\omega) = +1, \xi_c(\omega) = +1)$$
$$+ \mathbf{P}(\omega \in \Omega : \xi_a(\omega) = +1, \xi_b(\omega) = +1, \xi_c(\omega) = -1)$$
$$+ \mathbf{P}(\omega \in \Omega : \xi_a(\omega) = +1, \xi_b(\omega) = -1, \xi_c(\omega) = +1)$$
$$+ \mathbf{P}(\omega \in \Omega : \xi_a(\omega) = -1, \xi_b(\omega) = -1, \xi_c(\omega) = +1). \tag{8.9}$$

But the first and the third terms on the right-hand side of this equation are just those which when added together make up the term $\mathbf{P}(\omega \in \Omega : \xi_a(\omega) = +1, \xi_c(\omega) = +1)$ (Kolmogorov probability is additive). It therefore follows that

$$\mathbf{P}(\omega \in \Omega : \xi_a(\omega) = +1, \xi_b(\omega) = +1) + \mathbf{P}(\omega \in \Omega : \xi_b(\omega) = -1, \xi_c(\omega) = +1)$$
$$= \mathbf{P}(\omega \in \Omega : \xi_a(\omega) = +1, \xi_c(\omega) = +1)$$

$$+ \mathbf{P}(\omega \in \Omega : \xi_a(\omega) = +1, \xi_b(\omega) = +1, \xi_c(\omega) = -1)$$

$$+ \mathbf{P}(\omega \in \Omega : \xi_a(\omega) = -1, \xi_b(\omega) = -1, \xi_c(\omega) = +1). \tag{8.10}$$

By using non-negativity of probability we obtain the inequality

$$\mathbf{P}(\omega \in \Omega : \xi_a(\omega) = +1, \xi_b(\omega) = +1) + \mathbf{P}(\omega \in \Omega : \xi_b(\omega) = -1, \xi_c(\omega) = +1)$$

$$\geq \mathbf{P}(\omega \in \Omega : \xi_a(\omega) = +1, \xi_c(\omega) = +1). \tag{8.11}$$

8.1.3 Clauser-Horne-Shimony-Holt's Inequality

Finally, we derive the Clauser-Horne-Shimony-Holt (CHSH) inequality.

Theorem 8.3 (CHSH-inequality). *Let $\mathscr{P} = (\Omega, \mathscr{F}, \mathbf{P})$ be a Kolmogorov probability space and let $\xi_j(\omega)$ and $\xi'_j(\omega)$, $j = 1, 2$, be random variables such that*

$$|\xi_j(\omega)| \leq 1, \qquad |\xi'_j(\omega)| \leq 1, \quad \text{for almost all } \omega \in \Omega. \tag{8.12}$$

Then the following inequality for correlations holds

$$\langle \xi_1, \xi'_1 \rangle + \langle \xi_1, \xi'_2 \rangle + \langle \xi_2, \xi'_1 \rangle - \langle \xi_2, \xi'_2 \rangle \leq 2. \tag{8.13}$$

Proof. It is easy to prove the following elementary algebraic inequality for numbers bounded by 1

$$u_1 v_1 + u_1 v_2 + u_2 v_1 - u_2 v_2 \leq 2.$$

Thus

$$\xi_1(\omega)\xi'_1(\omega) + \xi_1(\omega)\xi'_2(\omega) + \xi_2(\omega)\xi'_1(\omega) - \xi_2(\omega)\xi'_2(\omega) \leq 2. \tag{8.14}$$

Finally, we integrate this inequality with respect to the measure $\mathbf{P}$.

8.2 Correspondence between Classical and Quantum Statistical Models

Typically one assumes existence of a *space of hidden variables* Ω representing states of individual physical systems. To proceed in the measure-theoretic framework, one also should consider a σ-algebra $\mathscr{F}$ of subsets of Ω (physicists never do this). On this space there are defined *classical quantities.* These are measurable functions $\xi : \Omega \to \mathbf{R}$—random variables on the measurable space $(\Omega, \mathscr{F})$.

In this framework "classical" is equivalent to existence of a *functional representation.* Denote the space of classical quantities by the symbol $V(\Omega)$. This is some space of real-valued (measurable) functions on Ω. The choice of this functional space depends on a model under consideration. For a system whose state is given by the hidden variable ω, the value $\xi(\omega)$ of a classical quantity ξ gives the objective property ξ of this system. We shall not distinguish a classical (physical) quantity and its representation by random variable.

One also considers a space of physical observables O. In the quantum model they are represented by self-adjoint operators, e.g., we can represent O by $\mathscr{L}_s(\mathscr{H})$—the space of bounded self-adjoint operators[2] in Hilbert space $\mathscr{H}$. We shall distinguish a physical observable and its operator-representative by using symbols: a and $\hat{a}$. The term "quantum observable" will be reserved for $\hat{a}$.

The main question is about existence of a correspondence between the space of random variables $V(\Omega)$ and the space of quantum observables $\mathscr{L}_s(\mathscr{H})$—about the possibility to construct a map

$$j : V(\Omega) \to \mathscr{L}_s(\mathscr{H}) \tag{8.15}$$

[2] Formally, by considering only bounded operators we restrict the formalism of quantum mechanics. However, as was emphasized by von Neumann [313], only bounded observables can be really measured.

or a map

$$i : \mathscr{L}_s(\mathscr{H}) \to V(\Omega) \tag{8.16}$$

which have "natural probabilistic properties" (j may be not a one-to-one map; its existence does not imply existence of i and vice versa). The main problem of modern theoretical quantum physics is that experimental quantum physics does not tell us much about features which such maps should have. There is a huge place for fantasies in mathematical physics. Some of them were presented in the form of various no-go theorems. We now recall the history of this problem.

8.3 Von Neumann Postulates on Classical → Quantum Correspondence

J. von Neumann was the first to try to present a list of possible features of the classical → quantum map j, see [313]. Unfortunately, von Neumann was extremely imprecise in his considerations. Nevertheless, we can try to formalize them.

(VN1) $j : V(\Omega) \to \mathscr{L}_s(\mathscr{H})$ is a one-to-one map.[3]

(VN2) For any Borel function $f : \mathbf{R} \to \mathbf{R}$ and random variable $\xi \in V(\Omega)$, the $f(\xi) \in V(\Omega)$ and we have $j(f(\xi)) = f(j(\xi)), \xi \in V(\Omega)$.

(VN3) $j(\xi_1 + \xi_2 + \cdots) = j(\xi_1) + j(\xi_2) + \cdots$ for any sequence $\xi_k \in V(\Omega)$.[4]

---

[3] Different random variables from the space $V(\Omega)$ are mapped into different quantum observables (injectivity) and any quantum observable corresponds to some random variable belonging to $V(\Omega)$ (surjectivity). By the rules of this "no-go" game, the image-space $\mathscr{L}_s(\mathscr{H})$ is given by the formalism of quantum mechanics. However, the preimage-space $V(\Omega)$ is unknown. Thus von Neumann just considered a possibility that such a classical functional space might exist.

[4] As J. von Neumann remarked: "the simultaneous measurability of $j(\xi_1), j(\xi_2), \ldots$ is not assumed", see [313], p. 314.

Any statistical model contains a space of *statistical states.* In a prequantum statistical model (which we are looking for) statistical states are represented by probability measures on the space of hidden variables Ω. Denote such a space of probabilities by $S(\Omega)$. This space is chosen depending on a classical statistical model under consideration.[5] In the quantum model statistical states are represented by von Neumann *density operators.* This space is denoted by $\mathscr{D}(\mathscr{H})$.

Roughly speaking, J. von Neumann proved that under conditions VN1–VN3 every operation of statistical averaging on $V(\Omega)$ can be represented as the quantum trace-average on $\mathscr{L}_s(\mathscr{H})$ corresponding to a quantum state $\rho \in \mathscr{D}(\mathscr{H})$. By using the language of probability measures we can say that every probability measure $\mathbf{P}$ on Ω can be represented by a quantum state ρ and vice versa. Thus we have the following "theorem" (von Neumann did not proceed in the rigorous mathematical framework).

Theorem 8.4 (von Neumann). *Under conditions VN1–VN3 (and some additional technical conditions) there are well-defined maps* $j : V(\Omega) \to \mathscr{L}_s(\mathscr{H})$ *and* $j : S(\Omega) \to \mathscr{D}(\mathscr{H})$ *such that*

$$\int_\Omega \xi(\omega) d\mathbf{P}(\omega) = \operatorname{Tr} \rho \hat{a}, \quad \textit{where } \rho = j(\mathbf{P}),\ \hat{a} = j(\xi). \tag{8.17}$$

As was pointed out, von Neumann did not specify the space $V(\Omega)$. The crucial point is that its image coincides with the space of quantum observables $\mathscr{L}_s(\mathscr{H})$. Moreover, $V(\Omega)$ is not "larger than" $\mathscr{L}_s(\mathscr{H})$: correspondence should be one-to-one. Finally, point wise addition of functions in $V(\Omega)$ should be transformed in addition of operators in $\mathscr{L}(\mathscr{H})$.

[5] For example, in classical statistical mechanics $S(\Omega)$ is the space of *all probability measures* on phase space $\Omega = \mathbf{R}^{2n}$. In a prequantum classical statistical field theory which was developed in a series of works [200, 201, 203–206, 208, 210], the space of hidden variables Ω is infinite-dimensional phase space, space of classical fields, and the space of statistical states $S(\Omega)$ consists of Gaussian measures having very small dispersion.

We say that a probability measure $\mathbf{P}$ is dispersion free (with respect to the space of random variables $V(\Omega)$) if for any $\xi \in V(\Omega) : D_{\mathbf{P}}\xi = E_{\mathbf{P}}(\xi - E_{\mathbf{P}}\xi)^2 = 0$, where $E_{\mathbf{P}}\xi = \int_\Omega \xi(\omega) d\mathbf{P}(\omega)$ is the mathematical expectation with respect to the probability measure $\mathbf{P}$. For example, let $\mathbf{P} = \delta_{\omega_0}$ be the Dirac δ-measure which is concentrated in the point $\omega_0 \in \Omega$. Then it is dispersion free with respect to the maximal space $V(\Omega)$—consisting of all random variables.

By using Theorem 8.4, J. von Neumann "proved" (he did not formulate a theorem, but just "ansatz") [313].

Theorem 8.5 (Von Neumann). *Let the space of statistical states $S(\Omega)$ contain a probability measure having zero dispersion (with respect to $V(\Omega)$). A correspondence map j between a classical statistical model*

$$M_{\mathrm{cl}} = (S(\Omega), V(\Omega))$$

and the quantum statistical model

$$N_{\mathrm{quant}} = (\mathcal{D}(\mathcal{H}), \mathcal{L}_s(\mathcal{H}))$$

satisfying the postulates VN1–VN3 (and some additional technical conditions) does not exist.

8.4 Bell-Type No-Go Theorems

As was pointed out by many outstanding physicists (e.g., by J. Bell [34] and L. Ballentine [29]), some of von Neumann's postulates [313] of classical $\rightarrow$ quantum correspondence are nonphysical. Opposite to von Neumann, in Bell-type no-go theorems different classical quantities can correspond to the same quantum observable. It is not assumed that every self-adjoint operator corresponds to some classical quantity. It might be that some self-adjoint operators have no classical counterpart (or even physical meaning). The postulate VN1 was deleted from the list for classical $\rightarrow$ quantum correspondence. The most doubtful postulate VN3 was also excluded from considerations. It was not assumed that the VN2 holds.

To proceed, we consider a family of spin operators

$$\hat{\sigma}(\theta) = \cos\theta\hat{\sigma}_z + \sin\theta\hat{\sigma}_x,$$

where $\hat{\sigma}_x, \hat{\sigma}_z$ are Pauli matrices, $\theta \in [0, 2\pi)$. These operators act in the two-dimensional state space $\mathscr{H} = \mathbf{C}^2$. We also consider spin operators for pairs of 1/2-spin particles: $\hat{\sigma}(\theta) \otimes I$ and $I \otimes \hat{\sigma}(\theta)$. They act in the four-dimensional state space $\mathscr{H} = \mathbf{C}^2 \otimes \mathbf{C}^2$. Bell's list of postulates on classical $\rightarrow$ quantum correspondence can be described as

(B1) The image $j(V(\Omega))$ contains spectral projectors for operators $\hat{\sigma}(\theta) \otimes I$ and $I \otimes \hat{\sigma}(\theta)$.

(B2) For any random variable $\xi \in V(\Omega)$, its range of values $\xi(\Omega)$ coincides with the spectrum of the operator $\hat{a} = j(\xi)$.[6]

(B3) The image $j(S(\Omega))$ contains the singlet spin state[7]

$$\psi = \frac{1}{\sqrt{2}}(|+\rangle|-\rangle - |-\rangle|+\rangle).$$

Starting with any classical $\rightarrow$ quantum mapping j we can construct a map i from quantum observables to classical random variables by setting for $\hat{a} \in j(V(\Omega))$,

$$i(\hat{a}) = \xi_a,$$

[6] Since both classical and quantum models are probabilistic, this postulate should be understood in probabilistic terms. Let $\mathbf{P} \in S(\Omega)$, $\xi \in V(\Omega)$ and let $\rho = j(\mathbf{P})$, $\hat{a} = j(\xi)$. Then, for a Borel set A, $\mathbf{P}(\omega \in \Omega : \xi(\omega) \in A) > 0$ iff $\mathbf{P}_\rho(a \in A) \equiv \mathrm{Tr}\,\rho P_A^a > 0$, where P_A^a is the spectral projector of $\hat{a}$ corresponding to the set A.

[7] This state belongs to the four-dimensional state space $\mathscr{H} = \mathbf{C}^2 \otimes \mathbf{C}^2$. By using the tensor notations we can write $\psi = \frac{1}{\sqrt{2}}[|+\rangle \otimes |-\rangle - |-\rangle \otimes |+\rangle]$.

where ξ_a is chosen from the set of random variables $j^{-1}(\hat{a})$. We also construct a map (denoted by the same symbol i) from the space of von Neumann density operators into the space of classical probability measures by choosing a probability measure $\mathbf{P}_\rho$ belonging the set $j^{-1}(\rho)$. We emphasize that such maps

$$i : \mathscr{L}_s(\mathscr{H}) \to V(\Omega)$$

and

$$i : \mathscr{D}(\mathscr{H}) \to S(\Omega)$$

are not uniquely defined! We set

$$\xi_\theta = i(\hat{\sigma}(\theta) \otimes I), \qquad \xi'_\theta = i(I \otimes \hat{\sigma}(\theta)).$$

These are classical pre-images of the spin operators for pairs of 1/2-spin particles. J. Bell also used the following postulates:

(B4) For any quantum state ρ and commuting operators $\hat{a}, \hat{b}$, the classical and quantum correlations coincide

$$\langle \xi_a, \xi_b \rangle_{\mathbf{P}_\rho} \equiv \int_\Omega \xi_a(\omega)\xi_b(\omega) d\mathbf{P}_\rho(\omega) = \langle \hat{a}\hat{b} \rangle_\rho \equiv \mathrm{Tr}\, \rho \hat{a}\hat{b}.$$

(B5) For the singlet state ψ and any θ, random variables ξ_θ and ξ'_θ are anti-correlated

$$\xi_\theta(\omega) = -\xi'_\theta(\omega) \tag{8.18}$$

almost everywhere with respect to the probability $\mathbf{P}_\psi$.

Theorem 8.6 (Bell). *Let* $\dim \mathscr{H} = 4$. *Correspondence maps*

$$j : V(\Omega) \to \mathscr{L}_s(\mathscr{H}) \quad \textit{and} \quad j : S(\Omega) \to \mathscr{D}(\mathscr{H}) \tag{8.19}$$

satisfying the postulates B1–B5 do not exist.

Proof. We apply Bell's inequality, Theorem 8.1, to random variables $\xi_\theta = i(\hat{\sigma}(\theta) \otimes I), \xi'_\theta = i(I \otimes \hat{\sigma}(\theta))$ and to a probability measure $\mathbf{P}_\psi$ corresponding to the singlet state ψ

$$|\langle \xi_{\theta_1}, \xi_{\theta_2} \rangle_{\mathbf{P}_\psi} - \langle \xi_{\theta_3}, \xi_{\theta_2} \rangle_{\mathbf{P}_\psi}| \leq 1 - \langle \xi_{\theta_1}, \xi_{\theta_3} \rangle_{\mathbf{P}_\psi}.$$

We remark that the postulate B2 was used here. To prove Bell's inequality, we used that random variables $\xi_\theta(\omega) = \pm 1$. We now apply the anti-correlation postulate B5 and rewrite the Bell's inequality

$$|\langle \xi_{\theta_1}, \xi'_{\theta_2} \rangle_{\mathbf{P}_\psi} - \langle \xi_{\theta_3}, \xi'_{\theta_2} \rangle_{\mathbf{P}_\psi}| \leq 1 + \langle \xi_{\theta_1}, \xi'_{\theta_3} \rangle_{\mathbf{P}_\psi}.$$

Finally, we apply the postulate B4 and write quantum covariations, instead of classical

$$| \operatorname{Tr}(\psi \otimes \psi)(\hat{\sigma}(\theta_1) \otimes I)(I \otimes \hat{\sigma}(\theta_2)) - \operatorname{Tr}(\psi \otimes \psi)(\hat{\sigma}(\theta_3) \otimes I)(I \otimes \hat{\sigma}(\theta_2))|$$
$$\leq 1 + \operatorname{Tr}(\psi \otimes \psi)(\hat{\sigma}(\theta_1) \otimes I)(I \otimes \hat{\sigma}(\theta_3)).$$

But this inequality is violated for a special choice of angles $\theta_1, \theta_2, \theta_3$, see the proof of Theorem 8.7 for details.

The postulate B4 on the correspondence between classical and quantum correlations can be changed to a postulate about the correspondence between classical and quantum probabilities.

(W) For any quantum state ρ and commuting operators $\hat{a}, \hat{b}$, the quantum and classical joint probability distributions coincide.

At first sight this postulate is essentially stronger than B4. However, W does not imply any coupling between the algebraic structures on the classical functional space and the quantum operator space. In contrast, by B4 the point-wise multiplication of functions should be transformed in composition of operators (in the commutative case).

Theorem 8.7 (Wigner). *Let* $\dim \mathscr{H} = 4$. *Correspondence maps*

$$j : V(\Omega) \to \mathscr{L}_s(\mathscr{H}) \quad and \quad j : S(\Omega) \to \mathscr{D}(\mathscr{H}) \tag{8.20}$$

satisfying the postulates B1–B3, W, B5 do not exist.

Proof. We shall apply Wigner's inequality for probabilities, Theorem 8.2

$$\mathbf{P}_\psi(\xi_{\theta_1} = +1, \xi_{\theta_2} = +1) + \mathbf{P}_\psi(\xi_{\theta_2} = -1, \xi_{\theta_3} = +1) \geq \mathbf{P}_\psi(\xi_{\theta_1} = +1, \xi_{\theta_3} = +1). \tag{8.21}$$

We remark that the postulate B2 was used here. We now apply the anti-correlation postulate B5 and rewrite the Wigner inequality

$$\mathbf{P}_\psi(\xi_{\theta_1} = +1, \xi'_{\theta_2} = -1) + \mathbf{P}_\psi(\xi_{\theta_2} = -1, \xi'_{\theta_3} = -1) \geq \mathbf{P}_\psi(\xi_{\theta_1} = +1, \xi'_{\theta_3} = -1). \tag{8.22}$$

We apply the postulate W and write quantum probabilities, instead of classical

$$\begin{aligned}&\mathrm{Tr}(\psi \otimes \psi) P_+(\theta_1) P'_-(\theta_2) + \mathrm{Tr}(\psi \otimes \psi) P_-(\theta_2) P'_-(\theta_3)\\ &\geq \mathrm{Tr}(\psi \otimes \psi) P_+(\theta_1) P'_-(\theta_3),\end{aligned}$$

where $P_\pm(\theta)$ and $P'_\pm(\theta)$ are spectral projectors for operators $\hat{\sigma}(\theta) \otimes I$ and $I \otimes \hat{\sigma}(\theta)$. For spin observables and the singlet state ψ we have

$$\begin{aligned}\mathbf{P}_\psi(\sigma(\theta_1) = +1, \sigma(\theta_2) = +1) &= \cos^2 \frac{\theta_1 - \theta_2}{2},\\ \mathbf{P}_\psi(\sigma(\theta_3) = +1 \sigma(\theta_2) = -1) &= \sin^2 \frac{\theta_3 - \theta_2}{2},\\ \mathbf{P}_\psi(\sigma(\theta_1) = +1 \sigma(\theta_3) = +1) &= \cos^2 \frac{\theta_1 - \theta_3}{2}.\end{aligned}$$

By (8.22) we have

$$\cos^2 \frac{\theta_1 - \theta_2}{2} + \sin^2 \frac{\theta_3 - \theta_2}{2} \geq \cos^2 \frac{\theta_1 - \theta_3}{2}.$$

We take $\theta_1 = 0, \theta_2 = 6\theta, \theta_3 = 2\theta$ and we get the following trigonometric inequality

$$\cos^2 3\theta + \sin^2 2\theta \geq \cos^2 \theta.$$

It is well known [319] that this trigonometric inequality is violated for sufficiently large θ.

We can also consider a weaker form of the postulate B2.

(CHSH) For any $\xi \in V(\Omega)$,

$$\sup\{|x| : x \in \xi(\Omega)\} = \sup\{|x| : x \in \text{Spectrum}(j(\xi)\}.$$

Theorem 8.8 (Clauser-Horne-Shimony-Holt). *Let* $\dim \mathscr{H} = 4$. *Correspondence maps*

$$j : V(\Omega) \to \mathscr{L}_s(\mathscr{H}) \quad and \quad j : S(\Omega) \to \mathscr{D}(\mathscr{H}) \tag{8.23}$$

satisfying the postulates B1, CHSH, B3, B4 do not exist.

The proof is based on the CHSH-inequality. In this no-go theorem we need not use the precise anti-correlations nor the precise coincidence of ranges of values for classical variables and quantum observables.

Our attitude with respect to Bell-type no-go theorems is similar to Bell's attitude with respect to others no-go theorems—von Neumann, Jauch-Piron and Gleasons theorems [34], pp. 4–9. As well as J. Bell did, we could speculate that some postulates about the correspondence between classical and quantum models (which were used in Bell-type no-go theorems) were nonphysical. There are many things which can be questioned.

8.5 Range of Values ("Spectral") Postulate

The proofs of Bell and Wigner no-go theorems were based on the postulate B2 on the *coincidence of ranges of values* for classical random variables and quantum observables. Moreover, one can easily construct examples of classical random variables reproducing the EPR-Bohm correlations in the case of violation of B2, see [154, 155, 188] and Chap. 11.

Is the postulate B2 really implied by the physical analysis of the situation? It seems that not at all! Henry Stapp pointed out [291]:

"The problem, basically, is that to apply quantum theory, one must divide the fundamentally undefined physical world into two idealized parts, the observed and observing system, but *the theory gives no adequate description of connection between these two parts.* The probability function is a function of degrees of freedom of the microscopic observed system, whereas the probabilities it defines are probabilities of responses of macroscopic measuring devices, and these responses are described in terms of quite different degrees of freedom."

In such a situation, rejection of the range of values condition is quite natural, since, as was pointed out by Stapp, a classical random variable ξ and its quantum counterpart $\hat{a} = j(\xi)$ depend on completely different degrees of freedom. Finally, we remark that a classical model reproducing quantum probabilistic description, but violating B2, was recently developed, see [7].

The derivation of the CHSH no-go theorem was based on the postulate CHSH—the weaker form of the range of values coincidence with postulate B2. The CHSH is also a postulate about the correspondence for ranges of values for classical random variables and quantum observables. And the above arguments against B2 can be applied against CHSH.

If the range of values postulates (in the forms VN2, B2, or CHSH) are rejected, then the classical probabilistic description does not contradict quantum mechanics.

8.6 Contextuality

In this section we shall present a very general viewpoint on the role of contextuality in Bell-type no-go theorems. Bell's original viewpoint [34] on contextuality will be presented in Sect. 8.7. The latter contextuality is *"simultaneous measurement contextuality"*—Bell's contextuality. We reserve the term *contextuality* for our general contextuality—dependence on the whole complex of physical conditions for preparation and measurement. We are aware that commonly in literature Bell-contextuality is called simply contextuality. However, using such a terminology is rather misleading, because dependence on the measurements of other compatible observables is just a very special case of dependence on the general physical context.

As was rightly pointed out by J. Bell, the only reasonable explanation of his contextuality is *action at a distance.* Another possibility is often called *"death of realism"* [82, 83]—denying the possibility to assign to quantum systems objective properties (such as the electron spin or the photon polarization)—does not sound natural. The observation of precise (anti-)correlations for the singlet state evidently contradicts the latter explanation.

In contrast to Bell-contextuality, in general contextuality implies neither action at distance nor "death of realism." Moreover, if one presents Bell's arguments in the general contextual approach, then in the classical (but contextual) probabilistic framework the probability of obtaining statistical data which would violate Bell's inequality equals zero, see Theorem 8.9.

8.6.1 Non-Injectivity of Classical → Quantum Correspondence

We first concentrate our considerations on the *classical variables → quantum observables* correspondence. As we remember, J. Bell (as well as L. Ballentine) criticized strongly the von Neumann postulate VN1. Both Bell and Ballentine (as well as many others) emphasized that there were no physical reasons to suppose (as von Neumann did) that for a quantum observable $\hat{a}$ its classical pre-image

$$j^{-1}(\hat{a}) = \{\xi \in V(\Omega) : j(\xi) = \hat{a}\}$$

should contain just one random variable. If one considers the quantum mechanical description as an *approximative description,* cf. [208], then it would be quite reasonable to assume that quantum mechanics cannot distinguish sharply prequantum physical variables. A few different classical random variables $\xi, \eta, \ldots$ can be identified in the quantum model with the same operator $\hat{a} = j(\xi) = j(\eta) = \cdots$. Moreover, there are no reasons to hope that degeneration of the map $j : V(\Omega) \to \mathcal{L}_s(\mathcal{H})$ should be small. The cardinality of the set $j^{-1}(\hat{a})$ might be huge (at least for some operators).

We now consider the *classical probabilities → quantum states* correspondence. In the same way as for variables and observables there are no physical reasons to assume injectivity of the map $j : S(\Omega) \to \mathcal{D}(\mathcal{H})$. By saying that we prepared an ensemble of systems with the fixed quantum state ρ, we could not guarantee that we really prepared the fixed classical probability distribution. The set

$$j^{-1}(\rho) = \{\mathbf{P} \in S(\Omega) : j(\mathbf{P}) = \rho\}$$

might have huge cardinality, cf. [193, 200, 201].

Our previous considerations would not induce any kind of protest from experts in quantum foundations (including J. Bell). However, our following conclusions might not be appreciated so much. We remark that the derivations of all Bell-type no-go theorems were based on the possibility of selecting for any quantum state (at least for the singlet state) one fixed classical probability measure $\mathbf{P}_\rho \in j^{-1}(\rho)$ and for any quantum observable (at least for spin observables) the fixed random variable $\xi \in j^{-1}(\hat{a})$.

The crucial counterargument is that, at the experimental level in all Bell-type inequalities, one should use data which is obtained in *a few different runs of measurements* (at least three, but in the real experimental framework four), see [139]. In the light of the above discussion on the non-injectivity of classical → quantum correspondence there are no physical reasons to assume that we would be able to obtain the same classical probability distribution and the same classical random variables (for example, corresponding to spin observables).[8] We are not able to guarantee that all runs of measurements are performed under the same physical conditions.

Let us consider a new random variable C describing a *complex of physical conditions* (context) during a run of measurements. And let us try to proceed as J. Bell and his followers did by proving inequalities for correlations and probabilities. Now classical probability measures corresponding to a quantum state ρ (in particular, to the singlet state) depend on runs $C : \mathbf{P}_\rho \equiv \mathbf{P}_{\rho,C}$ as well as random variables: $\xi_{a,C}(\omega), \xi_{b,C}(\omega), \xi_{c,C}(\omega)$. We start with the correlation inequality. There are three different complexes of physical conditions C_1, C_2, C_3 inducing correlations which were considered in Theorem 8.1. Here

$$\langle \xi_a, \xi_b \rangle (C_1) = \int_\Omega \xi_{a,C_1}(\omega)\xi_{b,C_1}(\omega)\mathbf{P}_{\rho,C_1}(\omega),$$

$$\langle \xi_c, \xi_b \rangle (C_2) = \int_\Omega \xi_{c,C_2}(\omega)\xi_{b,C_2}(\omega)\mathbf{P}_{\rho,C_2}(\omega).$$

If $C_1 \neq C_2$ we are not able to perform operations with integrals as we did in Theorem 8.1. We can not obtain the Bell's inequality involving the third correlation

$$\langle \xi_a, \xi_c \rangle (C_3) = \int_\Omega \xi_{a,C_3}(\omega)\xi_{c,C_3}(\omega)\mathbf{P}_{\rho,C_3}(\omega)$$

for a context C_3. To derive Bell's inequality, we should assume that

$$C_1 = C_2 = C_3. \tag{8.24}$$

The same counterarguments can be used against the derivations of Wigner's inequality and CHSH's inequality. Let us now formalize the procedure of correspondence between classical and quantum models in the case of context-dependence. Denote by $\mathscr{C}$ the set of contexts under consideration. We suppose that on $\mathscr{C}$ there is defined

[8] Even if we use the same macroscopic preparation and measurement devices, fluctuations of micro-parameters can induce different physical conditions.

a probability measure **Q**. Instead of degenerate maps $j : V(\Omega) \to \mathcal{L}_s(\mathcal{H})$ and $j : S(\Omega) \to \mathcal{D}(\mathcal{H})$, we consider random maps

$$i : \mathcal{C} \times \mathcal{L}_s(\mathcal{H}) \to V(\Omega),$$

$$i : \mathcal{C} \times \mathcal{D}(\mathcal{H}) \to S(\Omega).$$

For any context C (considered now as a random parameter) and any quantum observable $\hat{a}$ there is uniquely defined a random variable $\xi(\omega) = i(C, \hat{a})(\omega)$ and for any quantum state ρ there is uniquely defined a probability measure $\mathbf{P} = i(C, \rho)$. In this framework we can formulate an interesting problem[9]

What is the probability of obtaining statistical data which would satisfy Bell-type inequalities?

It is natural to assume that the probability that precisely the same complex of physical conditions would be realized in a few experiments is equal to zero.[10] Thus the probability **Q** is *"continuous"*

$$\mathbf{Q}(C) = 0 \tag{8.25}$$

for any single point $C \in \mathcal{C}$. Such a condition assumes that quantum mechanics provides only a rough description of the real physical situation.

Theorem 8.9. *Under the assumption* (8.25) *the probability of obtaining statistical data which would satisfy Bell-type inequalities is zero.*

Proof. Since $\mathbf{Q}(C) = 0$, the probability of obtaining in three different experiments totally identical complexes of physical conditions is zero.

[9] Thus there are two random parameters: C describing a complex of physical conditions and ω describing the hidden state of a system.

[10] Both source and measurement apparatus are macroscopic systems which states depend on a huge number of parameters.

8.6.2 Bell's Inequality and Experiment

The standard conclusion from Bell's considerations is that the experimental violation of Bell-type inequalities confirmed disagreement between classical probabilistic description and experiment. Our probabilistic analysis demonstrated that such a conclusion was not totally justified. In general, existence of a prequantum classical probabilistic model would only imply that Bell-type inequalities could hold with zero probability.

8.7 Bell-Contextuality and Action at a Distance

In Bell's approach contexts are completely determined by compatible observables. Thus all micro-conditions of preparation and measurement are ignored. The simultaneous measurement of a compatible observable is considered as the only source of contextuality. Let $\hat{a} \in \mathcal{L}_s(\mathcal{H})$. Contexts for the a-measurement are given by all observables which are compatible with $a : C = C_b$, where $\hat{b} \in \mathcal{L}_s(\mathcal{H})$ and $[\hat{a}, \hat{b}] = 0$. In contrast to our previous considerations, C_b does not depend on a quantum state ρ or a run of preparation or measurement. Nevertheless, Bell-contextuality (as any contextuality) also blocks the derivations of all Bell-type theorems.

As Bell pointed out, if measurements of a and b are performed in separated regions of space-time, then his contextuality can be interpreted as action at a distance

$$i(C_b, \hat{a})(\omega) = \xi_a(C_b, \omega).$$

Thus a classical random variable $\xi_a(C_b, \omega)$ is not determined uniquely by a and it depends on the measurement of b. The latter acts instantaneously at a distance.

In principle, *Bell's argument in favor of action at a distance might stimulate investigations on finding direct evidence of such an action.* Unfortunately, it did not happen. Bell-type no-go theorems were considered as the final proofs of "quantum nonlocality." However, as we have seen, Bell-contextuality and consequent nonlocality is only one of many possibilities to block the derivations of no-go theorems. One could not have any definite conclusion from Bell-type theorems, besides the evident remark: Since we do not know a prequantum classical statistical model and the rules of classical → quantum correspondence, we can play a lot with such hypothetical rules. The known no-go theorems are the results of such games with correspondence rules.

Personally I think that violation of the spectral postulate or (and) general preparation and measurement contextuality provides essentially more natural possibilities for blocking Bell-type theorems than nonlocality.

Conclusion. The only value of Bell's arguments was the great stimulation of experimental technologies for entangled photons.[11]

[11] Such a viewpoint on the role of Bell's arguments might be disappointing for physicists. However, my analysis demonstrated: the violation of Bell's inequality has such a huge diversity of interpretations that no definite conclusion could be made. Of course, one might (following Bell) speculate about incompatibility of the local realist description and quantum formalism. But such speculations do not have any reasonable ground. Bell's arguments, which definitely played a stimulating role for quantum foundations in 1960–70th became one of the main barriers for development of quantum physics. These arguments should be rejected as soon as possible as totally meaningless. It would open the door for new studies and new experiments. If one wants, e.g., to confirm Bohr's original position—the completeness of QM and impossibility to go beyond it, then new theoretical and experimental arguments must be presented. Bohr's original views in the form of the principle of complementarity are not so attractive in the light of modern developments of quantum technologies. If one wants to confirm that QM is really based on nonlocal reality, then again new theoretical and especially experimental arguments must be presented. The most important consequence of my analysis is that there is no risk of encountering the dreadful nonlocalty (as was claimed by Bell) on the way beyond QM.

Chapter 9

Bell's Inequality for Conditional Probabilities

In Chap. 8 we have already discussed probabilistic assumptions which were explicitly and implicitly used by Bell in his considerations. We remark that in Chap. 8 we did not discuss even once *conditional probability*. On the other hand, conditional probability plays a fundamental role both in classical and quantum probabilistic models. The classical probability model is not only the Kolmogorov probability space, but also the Bayes formula for conditional probabilities. The quantum probabilistic model is not only Born's postulate, but also the definition of quantum conditional probability based on the von Neumann projection postulate, see Part I: Chap. 1.

Contrary to rather common opinion, Bell did not compare the classical probabilistic model, Kolmogorov's model (which is the basis of classical statistical mechanics) and the quantum probabilistic model. "Bell's classical probabilistic model," see Sect. 9.1, differs crucially from Kolmogorov's model of probability. He created his own "classical probability model." The latter was compared with quantum mechanics. However, classical statistical mechanics is described by Kolmogorov's model, and not by Bell's model. Already at the classical level J. Bell lost direct connection with classical statistical mechanics. He used a rather unusual "classical probability model."

In Bell's "classical probability model" there is no definition of conditional probability. The only thing which is evident from Bell's works is that conditional probability in his model cannot be defined by Bayes' formula. We shall show this. To prove this, we use the approach based on Bell-type inequalities in the conventional probabilistic approach, Kolmogorov's model. We prove an analog of Wigner's inequality, but for conditional probabilities (under the assumption that they are defined by Bayes' formula as it should be in Kolmogorov's model). By using this inequality we show that predictions of the conventional (Kolmogorov) and quantum probability models disagree *already in the case of non-composite systems.*

Thus even the *probability of polarization of a single photon or spin of a single electron* could not be described by conventional probability theory (which is the basis of classical statistical mechanics). Hence, quantum mechanics and classical statistical mechanics differ even for noncomposite systems.

Can one prove nonlocality by showing that these theories disagree also for composite systems?

9.1 Measure-Theoretic Probability Models

9.1.1 Conventional Probability Model and Classical Statistical Mechanics

Definition 9.1. *Let* $\mathcal{O} = \{a, b, \dots c\}$ *be a system of physical observables. This system permits a classical probabilistic description if there exists a Kolmogorov probability space* $\mathcal{P} = (\Omega, \mathcal{F}, \mathbf{P})$ *such that all observables belonging to* $\mathcal{O}$ *can be represented by random variables on* $\mathcal{P}$ *and conditional probabilities are defined by the Bayes formula:* $\mathbf{P}(a = \alpha | b = \beta) = \frac{\mathbf{P}(a=\alpha, b=\beta)}{\mathbf{P}(b=\beta)}$.

Thus the possibility of a classical probabilistic description of some system of physical observables is equivalent to the possibility of using the Kolmogorov probability model. We remark that classical statistical mechanics is based on a classical probabilistic description.[1] We would like to analyze the possibility of applying a classical probabilistic description to quantum mechanics.

9.1.2 Bell's Probability Model and Classical Statistical Mechanics

Let us consider a probabilistic model which is based on the Kolmogorov probability space (as the original Kolmogorov probability model, see Part I: Chap. 2), but there is no such notion as conditional probability. In particular, there is no Bayes' formula. Such a probability model we call *Bell's model*, since J. Bell was the first who started to work in such a framework: the measure-theoretic approach without the Bayes formula (and, in fact, without any definition of conditional probability).

One should sharply distinguish the notions of the Kolmogorov probability space and the Kolmogorov probability model. We remark that both Bell's and Kolmogorov's probability models are based on a measure-theoretical approach, the Kolmogorov probability space, but they differ by their viewpoints on conditional probabilities. Bell's model is essentially wider than Kolmogorov's. Bell did not assume (as Kolmogorov did) a constraint in the form of Bayes' formula for conditional probabilities. In general Bell's classical model may contradict the classical statistical mechanical description.

9.1.3 Confronting Bell's Classical Probabilistic Model and Quantum Mechanics

One may confront even Bell's classical probabilistic model with the probabilistic structure of quantum mechanics. It is well known that J. Bell showed that his classical model contradicts quantum mechanics in the case of a composite system. How-

[1] In general a Växjö model does not permit a classical probabilistic description.

ever, as was already mentioned, such an approach has nothing to do with comparison of classical and quantum mechanics. The crucial question is *"Why do classical and quantum mechanics have different probabilistic structures even in the case of a single photon?"*—and not at all: *"Why can such a difference also be demonstrated for entangled photons?"*

9.2 Wigner-Type Inequality for Conditional Probabilities

Let $\mathscr{P} = (\Omega, \mathscr{F}, \mathbf{P})$ be a Kolmogorov probability space. As a simple consequence of Theorem 8.2, Chap. 8, we obtain

Theorem 9.1 (Wigner inequality for conditional probabilities). *Let $\xi_a, \xi_b, \xi_c = \pm 1$ be symmetrically distributed random variables on $\mathscr{P}$*

$$\mathbf{P}(\omega \in \Omega : \xi_a = +1) = \mathbf{P}(\omega \in \Omega : \xi_a = -1) = \cdots$$
$$= \mathbf{P}(\omega \in \Omega : \xi_c = -1) = 1/2.$$

Then the following inequality holds true

$$\mathbf{P}(\xi_a = +1|\xi_b = +1) + \mathbf{P}(\xi_c = +1|\xi_b = -1) \geq \mathbf{P}(\xi_a = +1|\xi_c = +1). \quad (9.1)$$

Proof. We have $\mathbf{P}(\xi_b = +1) = \mathbf{P}(\xi_b = -1) = \mathbf{P}(\xi_a = +1) = \mathbf{P}(\xi_a = -1) = \mathbf{P}(\xi_c = +1) = \mathbf{P}(\xi_c = -1) = 1/2$. Thus

$$\mathbf{P}(\xi_a = +1|\xi_b = +1) + \mathbf{P}(\xi_c = +1|\xi_b = -1)$$
$$= 2\mathbf{P}(\xi_a = +1, \xi_b = +1) + 2\mathbf{P}(\xi_c = +1, \xi_b = -1)$$

and $\mathbf{P}(\xi_a = +1|\xi_c = +1) = 2\mathbf{P}(\xi_a = +1, \xi_c = +1)$. Hence by (8.5) we get (9.1).

We emphasize again that the main distinguishing feature of (9.1) is the presence of *only conditional probabilities.* Conditional probabilities can always be calculated by using quantum formalism—even for incompatible observables! In fact, we need not consider pairs of particles, since conditional probabilities are well defined even for noncomposite quantum systems. By Theorem 9.1 we can confront the conventional classical (Kolmogorov) model of probability and the quantum model in the case of a single quantum system.

9.3 Impossibility of Classical Description of Spin of Single Electron

We consider now the following postulates P and CP on classical $\to$ quantum correspondence j (see Chap. 8)

(P) For any random variable $\eta \in V(\Omega)$ and any probability $\mathbf{P} \in S(\Omega)$, the probability distribution of η coincides with the probability distribution of $\hat{a} = j(\eta)$

$$\mathbf{P}(\omega \in \Omega : \eta(\omega) \in A) = \mathrm{Tr}\, \rho P_A^a, \quad \rho = j(\mathbf{P}), \tag{9.2}$$

for any Borel set $A \subset \mathbf{R}$.

As always, P_A^a is the spectral projector of the self-adjoint operator $\hat{a}$ corresponding to the Borel set A.

We recall that in quantum mechanics the conditional probability is defined as

$$\mathbf{P}_\rho(b \in B|a \in A) = \mathrm{Tr}\, \rho_A^a P_B^b, \quad \rho_A^a = \frac{P_A^a \rho P_A^a}{\mathrm{Tr}\, \rho P_A^a}, \tag{9.3}$$

for any pair of Borel sets A and B. It is assumed that $\mathrm{Tr}\, \rho P_A^a > 0$. Thus by the quantum formalism we have

$$\mathbf{P}_\rho(b \in B|a \in A) = \frac{\mathrm{Tr}\, P_A^a \rho P_A^a P_B^b}{\mathrm{Tr}\, \rho P_A^a}. \tag{9.4}$$

(CP) For $\xi, \eta \in V(\Omega)$ and $\mathbf{P} \in S(\Omega)$, conditional probabilities given by Bayes' formula coincide with quantum conditional probabilities for $\hat{a} = j(\eta), \hat{b} = j(\xi), \rho = j(\mathbf{P})$

$$\mathbf{P}(\xi(\omega) \in B | \eta(\omega) \in A) = \frac{\mathrm{Tr}\, P_A^a \rho P_A^a P_B^b}{\mathrm{Tr}\, \rho P_A^a} \tag{9.5}$$

for any pair of Borel sets $A (\mathbf{P}(A) > 0, \mathrm{Tr}\, \rho P_A^a > 0)$ and B.

We do not assume that the operators $\hat{a}$ and $\hat{b}$ should commute. We recall again that quantum conditional probabilities are defined even for incompatible quantum observables. We remark that if the range of values postulate B2, see Chap. 8 of this part, holds true, then for $\rho = j(\mathbf{P})$

$$\mathbf{P}(A) > 0 \quad \text{iff} \quad \mathrm{Tr}\, \rho P_A^a > 0.$$

We consider the following condition, cf. B1, on the image of the map j mapping classical variables (measurable functions on the "space of hidden parameters" Ω) in quantum observables (self-adjoint operators in complex Hilbert space):

(B1K) The image $j(V(\Omega))$ contains spectral projectors for the spin operators $\hat{\sigma}(\theta)$ for 1/2-spin particle, $\theta \in [0, 2P]$.

We obtain the following no-go theorem for a classical → quantum correspondence preserving conditional probabilities.

Theorem 9.2 (Khrennikov). *Let* $\dim \mathscr{H} = 2$ *and let* ψ *be a pure quantum state. A Kolmogorov model* $\mathscr{P}_\psi = (\Omega_\psi, \mathscr{F}_\psi, \mathbf{P}_\psi)$ *such that classical and quantum conditional probabilities coincide does not exist.*

Proof. We consider a family of spin operators

$$\hat{\sigma}(\theta) = \cos\theta \hat{\sigma}_z + \sin\theta \hat{\sigma}_x,$$

where $\hat{\sigma}_x, \hat{\sigma}_z$ are Pauli matrices, $\theta \in [0, 2P)$. We can apply the inequality (9.1) to classical random variables corresponding to any three spin operators: $\xi_{\sigma(\theta_1)}(\omega)$,

$\xi_{\sigma(\theta_2)}(\omega), \xi_{\sigma(\theta_3)}(\omega)$. We remark that injectivity of the classical $\rightarrow$ quantum correspondence map j is not supposed. In any event we can choose a random variable $\xi_{\sigma(\theta)}$ belonging to the pre-image $j^{-1}(\hat{\sigma}(\theta)) = \{\xi \in V(\Omega) : j(\xi) = \hat{\sigma}(\theta)\}$. We have

$$\mathbf{P}_\psi(\xi_{\sigma(\theta_1)} = +1|\xi_{\sigma(\theta_2)} = +1) + \mathbf{P}_\psi(\xi_{\sigma(\theta_3)} = +1|\xi_{\sigma(\theta_2)} = -1)$$
$$\geq \mathbf{P}_\psi(\xi_{\sigma(\theta_1)} = +1|\xi_{\sigma(\theta_3)} = +1). \tag{9.6}$$

We can compute conditional probabilities by using quantum formalism, see Part I: Chap. 1.[2] If we have two arbitrary dichotomous observables a and b which are represented by quantum operators $\hat{a}$ and $\hat{b}$, then

$$\mathbf{P}_\psi(a = \alpha_i|b = \beta_j) = |\langle e_i^a, e_j^b\rangle|^2,$$

where $\{e_i^a\}$ and $\{e_i^b\}$ are systems of normalized eigenvectors of operators $\hat{a}$ and $\hat{b}$, respectively: $\hat{a}e_i^a = \alpha_i e_i^a, \hat{b}e_j^b = \beta_j e_j^b$. The assumption of coincidence of classical and quantum conditional probabilities implies that

$$\mathbf{P}_\psi(\omega \in \Omega_\psi : \xi_a(\omega) = \alpha_i|\xi_b(\omega) = \beta_j) = |\langle e_i^a, e_j^b\rangle|^2,$$

where $\xi_a = i_\psi(\hat{a}), \xi_b = i_\psi(\hat{b})$. For the spin operators we have

$$\hat{\sigma}(\theta)\varphi_+(\theta) = \varphi_+(\theta), \quad \text{where } \varphi_+(\theta) = \left(\cos\frac{\theta}{2}, \sin\frac{\theta}{2}\right),$$
$$\hat{\sigma}(\theta)\varphi_-(\theta) = -\varphi_-(\theta), \quad \text{where } \varphi_-(\theta) = \left(-\sin\frac{\theta}{2}, \cos\frac{\theta}{2}\right).$$

Thus

$$\mathbf{P}_\psi(\omega \in \Omega_\psi : \xi_{\sigma(\theta_1)}(\omega) = +1|\xi_{\sigma(\theta_2)}(\omega) = +1) = \cos^2\frac{\theta_1 - \theta_2}{2},$$
$$\mathbf{P}_\psi(\omega \in \Omega_\psi : \xi_{\sigma(\theta_3)}(\omega) = +1|\xi_{\sigma(\theta_2)}(\omega) = -1) = \sin^2\frac{\theta_3 - \theta_2}{2},$$

[2] Spin-operators have nondegenerate spectra. Therefore there is no problem with the definition of conditional probabilities. This is precisely the case considered by P. Dirac and J. von Neumann.

$$\mathbf{P}_\psi(\omega \in \Omega_\psi : \xi_{\sigma(\theta_1)}(\omega) = +1 | \xi_{\sigma(\theta_3)}(\omega) = +1) = \cos^2 \frac{\theta_1 - \theta_3}{2}.$$

By (9.6) we have

$$\cos^2 \frac{\theta_1 - \theta_2}{2} + \sin^2 \frac{\theta_3 - \theta_2}{2} \geq \cos^2 \frac{\theta_1 - \theta_3}{2}.$$

We take $\theta_1 = 0, \theta_2 = 6\theta, \theta_3 = 2\theta$ and we get the following trigonometric inequality

$$\cos^2 3\theta + \sin^2 2\theta \geq \cos^2 \theta.$$

This trigonometric inequality is violated for sufficiently large θ.

Conclusion. The classical probabilistic model (Kolomogorov's model) does not exist even for quantum systems with the Hilbert space of dimension $\mathbf{d} = \mathbf{2}$.[3]

We consider the weakest possible form of Bell's postulate B3.

(B3K) The image $j(S(\Omega))$ contains at least one pure state ψ.

Theorem 9.3 (Khrennikov). *Let* $\dim \mathscr{H} = 2$. *Correspondence maps*

$$j : V(\Omega) \to \mathscr{L}_s(\mathscr{H}) \quad \text{and} \quad j : S(\Omega) \to \mathscr{D}(\mathscr{H}) \tag{9.7}$$

satisfying the postulates B1K, B2, B3K and P, CP do not exist.

We continue our no-go activity in the conditional probabilistic framework. In the classical Kolmogorov model the conditional probability distribution for $\xi|\eta$ in combination with the probability distribution for η determine the joint probability distribution

$$\begin{aligned} &\mathbf{P}(\omega \in \Omega : \xi(\omega) \in B,\ \eta(\omega) \in A) \\ &\quad = \mathbf{P}(\omega \in \Omega : \eta(\omega) \in A)\mathbf{P}(\omega \in \Omega : \xi(\omega) \in B | \eta(\omega) \in A). \end{aligned} \tag{9.8}$$

[3] We recall that J. Bell [34] considered the Hilbert space of dimension $\mathbf{d} = \mathbf{4}$ to confront the quantum and classical probabilistic models. However, Bell's classical probabilistic model was not at all the conventional Kolmogorov model, because, opposite to Kolmogorov, Bell did not use Bayes' formula to define conditional probabilities.

By postulates P and CP the following quantum representation for the joint probability distribution should take place

$$\mathbf{P}(\omega \in \Omega : \xi(\omega) \in B,\ \eta(\omega) \in A) = \mathrm{Tr}\, P_A^a \rho P_A^a P_B^b, \tag{9.9}$$

where $\rho = j(\mathbf{P}), \hat{a} = j(\eta), \hat{b} = j(\xi)$.

Lemma 9.1. *Let* $\dim \mathcal{H} = 2$. *There exist a pure state* ψ *and two spin operators such that the right-hand side of* (9.9) *does not define the joint probability distribution for any pair of Kolmogorovian random variables.*

Proof. We remark that the left-hand side of (9.9) is symmetric, but the right-hand side is not. Hence, Lemma 9.1 is a consequence of the following result:

Lemma 9.2 (Kirkpatrick, [213]). *Let* P *and* Q *be two arbitrary orthogonal projectors. If for any pure state* ψ *we have*

$$\mathrm{Tr}\, P[\psi \otimes \psi]PQ = \mathrm{Tr}\, Q[\psi \otimes \psi]QP, \tag{9.10}$$

then

$$[Q, P] = 0. \tag{9.11}$$

Proof. We have from (9.10) that $\|QP\psi\|^2 = \|PQ\psi\|^2$, or

$$\langle QP\psi, QP\psi\rangle = \langle PQP\psi, \psi\rangle = \langle PQ\psi, PQ\psi\rangle = \langle QPQ\psi, \psi\rangle.$$

If the quadratic forms of two self-adjoint operators coincide, then these operators also coincide. We remark that $(PQP)^* = P^*Q^*P^* = PQP$ as well as $(QPQ)^* = Q^*P^*Q^* = QPQ$. Thus we have $PQP = QPQ$. From this equality we obtain

$$(PQ)^2 = PQP, \qquad (QP)^2 = QPQ. \tag{9.12}$$

We set $C = [Q, P]$. We remark that if we prove that the operator $CC^* = 0$, then we get that $C = 0$. By (9.12) we have

$$CC^* = QPQ - (PQ)^2 - (QP)^2 + PQP = 0.$$

We consider the weakest possible form of the postulate B1 which is even weaker than the postulate B1K.

(B1K′) The image $j(V(\Omega))$ contains at least two noncommutative orthogonal projectors.

However, to proceed with the postulate B1K′, we should consider a stronger form of the postulate B3K.

(B3K′) The image $j(S(\Omega))$ contains all pure states.

Theorem 9.4 (Khrennikov). *Let* $\dim \mathscr{H} = 2$. *Correspondence maps*

$$j : V(\Omega) \to \mathscr{L}_s(\mathscr{H}) \quad \text{and} \quad j : S(\Omega) \to \mathscr{D}(\mathscr{H}) \tag{9.13}$$

satisfying the postulates B1K′, B2, B3K′, and P, CP do not exist.

Thus we have demonstrated that the classical probabilistic formalism (Kolmogorov's measure-theoretic model) is incompatible with the quantum probabilistic formalism (Dirac-von Neumann Hilbert space model) already in the case of noncomposite quantum systems—for systems with two quantum degrees of freedom.

By using the generalization of Wigner inequality for conditional probabilities, Theorem 9.1, we have shown that classical conditional probabilities (defined via Bayes' formula) are incompatible with quantum conditional probabilities (defined via projections of states).

The violation of Bayes' rule for quantum conditional probabilities is one of the main problems of foundations of quantum physics. We recall that the definition of quantum conditional probabilities is based (at least for observables with nondegenerate spectra) on the *von Neumann projection postulate.* Thus it seems that one of the *main mysteries of quantum theory is encoded in the von Neumann projection postulate.*

J. Bell did not pay any attention to the problem of *correspondence between classical and quantum conditional probabilities,* see L. Accardi [2, 3] for a discussion. He chose a totally different strategy by concentrating on quantum nonlocality. However, by demonstrating the violation of Bell-type inequalities for composite systems (four quantum degrees of freedom) it is impossible to clarify the violation of the conditional probabilistic version of Wigner's inequality, (9.1), which takes place already for non composite systems (two quantum degrees of freedom).

Chapter 10

Frequency Probabilistic Analysis of Bell-Type Considerations

As we have seen in two previous chapters, the derivation of Bell's inequality (and its generalizations) is based on a number of assumptions on probability distributions. However, such assumptions are always formulated in the standard measure-theoretical framework. In this chapter we perform the frequency probabilistic analysis of Bell's assumptions on probability distributions which are used to derive Bell's inequality. By operating with von Mises collectives (or in the more pragmatic approach with S-sequences) we get the possibility of finding physical and mathematical conditions which imply (or do not imply) Bell's assumptions.

The main conclusion of our frequency analysis is that

(a) either some of probability distributions do not exist at all

or

(b) so-called Bell's locality condition is merely the condition of *independence of collectives*.

We remark that our frequency analysis could not be considered as an argument against nonlocality. However, it demonstrated that Bell's conclusion (either nonlocality or death of realism[1]) cannot be considered as completely justified.

[1] Under the assumption of incompleteness of QM.

We recall, see Part I: Chap. 2, that in the von Mises framework a probability distribution is not a primary object of the mathematical description of a physical model. The probability distribution is induced by properties of a *collective* (random sequence). R. von Mises continuously emphasized: “first collective, then probability.” Therefore all properties of probability distributions are just numerical representations of properties of collectives (or more generally S-sequences). In particular, Bell’s assumptions on probability distributions can be analyzed from the viewpoint of collectives which generate these distributions. Moreover, even the assumption that hidden variables (of particles and measurement devices and their combinations) induce collectives (via the stabilization of relative frequencies) is questionable.

Our analysis demonstrates that (under the assumption that the simultaneous behavior of hidden variables and measurements produce collectives) the root of the problem is the question on the possibility of *combining* collectives corresponding to different settings of measurement devices (combining collectives corresponding to two different pairs of settings).

Here we plan to study it in more detail. The operation of combining of two different collectives (random sequences) $x = (x_1, x_2, \ldots, x_N, \ldots)$ and $y = (y_1, y_2, \ldots, y_N, \ldots)$ (to obtain a new collective $z = (z_1, z_2, \ldots, z_N, \ldots)$, where $z_j = (x_j, y_j)$) is a rather delicate operation. At the moment there are no physical reasons to suppose that EPR-like collectives are combinable. We remark that there is nothing especially “quantum” in the existence of noncombinable collectives, see [150].

Even if we suppose that collectives are combinable, then, to derive Bell’s inequality, we must also assume that these collectives are *independent*. The latter assumption is even more doubtful, see our analysis.

Finally, we discuss the problem of transmission of information with the aid of EPR pairs.

As was already remarked, our frequency analysis could not be considered as an argument against nonlocality. The quantum reality may be nonlocal. The only thing that we demonstrated is that *Bell's inequality is based on such doubtful probabilistic considerations that it could not be considered as the crucial argument against local realism.* It seems that at the present level of the development of quantum physics some other theories (for example, Bohmian mechanics) give essentially more rigorous arguments in the favor of nonlocality than Bell's inequality (cf., even with Schrödinger's original remark that the wave function of a composite system is not defined on the physical space $\mathbf{R}^3$).

10.1 Frequency Probabilistic Description of Hidden Variables

We consider the standard EPR-Bohm framework. We consider a composite system $s = (s^1, s^2)$, e.g., a pair of correlated electrons or photons. Observables (e.g., spin or polarization projections) for the subsystem s^1 are denoted by $a, b, c, \ldots$ and for the subsystem s^2 by $a', b', c', \ldots$, respectively.

HV (hidden variables) are denoted by ω; it is supposed that the set of HV is finite

$$\Omega = \{\omega_1, \omega_2, \ldots, \omega_M\}.$$

We consider a complex of physical conditions C (context). It is represented by a sequence of pairs of particles

$$\mathbf{s} = (s_j = \{s_j^1, s_j^2\},\ j = 1, 2, \ldots).$$

Let $\omega_j \in \Omega,\ j = 1, 2, \ldots$ be the value of the hidden variable for the jth pair. For settings a and b of measurement apparatuses we consider sequences of pairs

$$x_{a,\omega} = ((\alpha_1, \omega_1), \ldots, (\alpha_N, \omega_N), \ldots), \tag{10.1}$$

$$x_{b',\omega} = ((\beta_1, \omega_1), \ldots, (\beta_N, \omega_N), \ldots), \tag{10.2}$$

where α_j and β_j are the jth results for measurements of observables a and b.

This frequency approach is very general. We do not exclude the influence of the HV of measurement devices. Denote these parameters by the symbols $\omega^a, \omega^{b'}$, respectively. Then observables can be represented as functions of HV $\omega, \omega^a, \omega^{b'}$

$$a = a(\omega^a, \omega^{b'} \omega), \qquad b = b(\omega^a, \omega^{b'}, \omega). \tag{10.3}$$

We can formulate the condition of *individual locality*

$$a = a(\omega^a, \omega), \qquad b = b(\omega^{b'}, \omega). \tag{10.4}$$

So the result of a measurement of the observable a does not depend on the internal state of the measurement device for the b'-measurement and vice versa. This condition of locality is very natural from the physical viewpoint, cf. with a condition of "probabilistic locality", *Bell-Clauser-Horne factorability condition*, Sect. 10.2 of this chapter.

We emphasize that by using only the condition of individual locality (10.4) it is not possible to prove Bell's inequality. Thus the individual locality does not contradict predictions of the quantum formalism and corresponding experimental measurements.

We also remark that (10.4) is about *realistic observables*: states of a system, ω, and a corresponding measurement device, ω^a or $\omega^{b'}$, determine uniquely the result of measurement. Thus (10.4) can be considered as a definition of (individual) local realism. This is a very natural definition of local realism. And it does not contradict predictions of quantum formalism.

Thus in the local realistic framework we have

$$\alpha_j = a(\omega_j^a, \omega_j), \qquad \beta_j = b(\omega_j^{b'}, \omega_j).$$

Such a general HV-model is known as a *stochastic* HV-model. In the stochastic HV-model observables can be non-dichotomous (and even continuous) and they need not satisfy the condition of precise correlation (or anti-correlation), see [69, 283].

However, one could still prove Bell's inequality by using the condition of "probabilistic locality", Bell-Clauser-Horne factorability condition [283]. In Sect. 10.2 we shall show that this condition is totally nonphysical.

The first question about sequences (10.1), (10.2) induced by measurements of parts s^1 and s^2 of a composite system is:

Are sequences $x_{a,\omega}$ and $x_{b',\omega}$ S-sequences (collectives)?

We must recognize that the present level of the development of quantum physics is not sufficient to give a definite answer to this question. The statistical stabilization in sequences $x_{a,\omega}, x_{b',\omega}$ could not be motivated by the statistical stabilization on the macro-level, namely, for values of α, β. The average over hidden variables may compensate oscillations of relative frequencies

$$\nu_N((\alpha_l, \omega_k); x_{a,\omega}) = \frac{n_N((\alpha_l, \omega_k); x_{a,\omega})}{N},$$
$$\nu_N((\beta_l, \omega_k); x_{b,\omega}) = \frac{n_N((\beta_l, \omega_k); x_{b,\omega})}{N},$$

where $n_N((\alpha_l, \omega_k); x_{a,\omega})$ and $n_N((\beta_l, \omega_k); x_{b,\omega})$ are numbers of occurrence of the results $a = \alpha_l$ and $b' = \beta_l$, respectively, in N trials. Such a situation is quite natural (at least from the mathematical viewpoint), see Sect. 10.3. So there must be some physical reasons supporting such a stabilization at the micro-level. The authors of [59, 60, 82, 83, 283] did not provide such physical reasons. Therefore, in principle, even the possibility of operating with probabilities

$$\mathbf{P}(a = \alpha_l, \omega = \omega_k) \equiv \mathbf{P}((\alpha_l, \omega_k); x_{a,\omega}) = \lim_{N\to\infty} \nu_N((\alpha_l, \omega_k); x_{a,\omega}),$$
$$\mathbf{P}(b = \beta_l, \omega = \omega_k) \equiv \mathbf{P}((\beta_l, \omega_k); x_{b,\omega}) = \lim_{N\to\infty} \nu_N((\beta_l, \omega_k); x_{b,\omega})$$

is rather questionable.

Nevertheless, let us try to proceed as authors of [59, 60, 82, 83, 283] did and let us assume that $x_{a,\omega}$ and $x_{b',\omega}$ are S-sequences (or collectives). Thus we assumed that the frequency probabilities $\mathbf{P}(a = \alpha_l, \omega = \omega_k)$, $\mathbf{P}(b' = \beta_l, \omega = \omega_k)$ are well defined. However, the next question arises immediately:

Are S-sequences (collectives) $x_{a,\omega}$ and $x_{b',\omega}$ combinable?

There are no physical reasons to claim that they are combinable. In principle they may be non-combinable. Both measurement apparatuses are complex systems depending on a huge number of parameters, ω^a, $\omega^{b'}$. It is not clear why fluctuations of these parameters and HV ω of a microsystem must be probabilistically consistent (and produce a probability distribution).

The existence of macro-probabilities $\mathbf{P}(a = \alpha_l, b' = \beta_m)$ does not imply that the statistical stabilization on the micro-level also should take place. The macro-stabilization can be a result of the overage with respect to hidden variables, see [139].

Remark 10.1 (Time window, coincidence loophole and combining of sequences of observations). In fact, the physical assumption on the possibility of performing simultaneous measurements of the observables a and b' is encoded in the operation of combining of $x_{a,\omega}$ and $x_{b',\omega}$. In general, we should consider frequencies

$$\nu_N((a = \alpha_l, \omega = \omega_k), (b' = \beta_m, \omega = \omega_q)),$$

i.e., the systems s_1 and s_2 are assigned with the HV ω_k and ω_q, respectively. But assuming the simultaneous measurement we restrict ourselves to consideration of frequencies

$$\nu_N((a = \alpha_l, \omega = \omega_k), (b' = \beta_m, , \omega = \omega_k))$$

which we denoted by

$$\nu_N(a = \alpha_l, \omega = \omega_k, b' = \beta_m).$$

Thus by assuming that sequences $x_{a,\omega}$ and $x_{b',\omega}$ are combinable we assume that it is possible to solve all problems with the choice of time-window and coincidence loophole, see Larsson [233].

Nevertheless, to proceed we suppose that $x_{a,\omega}$ and $x_{b',\omega}$ are combinable. In such a case the frequency probabilities $\mathbf{P}(a = \alpha_l, \omega = \omega_k, b' = \beta_m)$ are well defined. However, the possibility of combining any two collectives does not imply the possibility of combining any three collectives. Of course, if one assumes (as J. Bell) the possibility of combining of any triple of collectives, e.g., $x_{a,\omega}, x_{c,\omega}, x_{b',\omega}$, then one would be able to prove Bell's inequality by using the simultaneous probability distribution[2]

$$\mathbf{P}(a = \alpha_l, c = \gamma_n, b' = \beta_m) = \sum_m \mathbf{P}(a = \alpha_l, c = \gamma_n, b' = \beta_m, \omega = \omega_k)$$

see Rastal [274], Fine [93]. However, I could not find any physical reason for existence of this distribution. It should be assigned with incompatible observables, a and c (measurements of different spin or polarization projections of the same particle)!

The common point of view presented in literature is that one can escape operating with probability distributions for triples of observables (among which at least two are incompatible) and prove Bell's inequality by operating only with probability distributions for pairs of observables, see, e.g., [283]. However, to derive Bell's inequality in such a framework, one should assume the factorization of these probabilities. We shall discuss this condition in more detail in the next section.

10.2 Bell's Locality (Bell-Clauser-Horne Factorability) Condition

Proofs [283] of Bell-type inequalities for stochastic HV-models are based on the so-called Bell's locality condition

$$\mathbf{P}(a = \alpha_l, b' = \beta_m|\omega_k) = \mathbf{P}(a = \alpha_l|\omega_k)\mathbf{P}(b' = \beta_m|\omega_k). \qquad (10.5)$$

[2] For the deterministic HV-model, all observables are dichotomous and satisfy the condition of precise correlation or anti-correlation.

This condition is also known as the Bell-Clauser-Horne factorability condition.[3] We analyse this condition in the contextual frequency probabilistic framework. We proceed under the assumption that $x_{a,\omega}$ and $x_{b',\omega}$ are combinable S-sequences (or collectives). To define the conditional probability $\mathbf{P}(a = \alpha_l, b' = \beta_m|\omega_k)$ in the von Mises framework, we consider the S-sequence (collective) $x_{a,\omega,b'}$—the result of combination of $x_{a,\omega}$ and $x_{b',\omega}$. We fix the value ω_k of HV ω and select in $x_{a,\omega,b'}$ the subsequence $x_{a,\omega_k,b'}$ consisting of triples in which $\omega = \omega_k$. Since $x_{a,\omega,b'}$ is an S-sequence, the $x_{a,\omega_k,b'}$ is also an S-sequence. Therefore the probability

$$\begin{aligned}\mathbf{P}(a = \alpha_l, b' = \beta_m|\omega_k) &\equiv \mathbf{P}_{x_{a,\omega,b'}}(a = \alpha_l, b' = \beta_m|\omega_k)\\ &= \mathbf{P}_{x_{a,\omega_k,b'}}(a = \alpha_l, b' = \beta_m).\end{aligned}$$

To define the conditional probabilities $\mathbf{P}(a = \alpha_l|\omega_k)$, $\mathbf{P}(b' = \beta_m|\omega_k)$, we consider S-sequences $x_{a,\omega}$ and $x_{b',\omega}$ and select in them the subsequences x_{a,ω_k} and x_{b',ω_k} consisting of pairs in which $\omega = \omega_k$. By definition

$$\begin{aligned}\mathbf{P}(a = \alpha_l|\omega_k) &\equiv \mathbf{P}_{x_{a,\omega}}(a = \alpha_l|\omega_k) = \mathbf{P}_{x_{a,\omega_k}}(a = \alpha_l);\\ \mathbf{P}(b' = \beta_m|\omega_k) &\equiv \mathbf{P}_{x_{b',\omega}}(b' = \beta_m|\omega_k) = \mathbf{P}_{x_{\omega_k,b'}}(b' = \beta_m).\end{aligned}$$

We remark that under the above assumptions x_{a,ω_k} and $x_{\omega_k,b'}$ are combinable and Bell's nonlocality condition (10.5) is nothing else than the *condition of independence of S-sequences (or collectives)* x_{a,ω_k} and $x_{\omega_k,b'}$

$$\mathbf{P}_{x_{a,\omega_k,b'}}(a = \alpha_l, b' = \beta_m) = \mathbf{P}_{x_{a,\omega_k}}(a = \alpha_l)\mathbf{P}_{x_{\omega_k,b'}}(b' = \beta_m). \tag{10.6}$$

Contrary to the measure-theoretic (noncontextual) framework in that the condition (10.5) can be interpreted as the condition of measurement independence, in our framework this condition, see (10.6), is not so natural from the physical viewpoint. There are no physical reasons for independence of S-sequences (collectives) x_{a,ω_k}

[3] De Muynck [79] used the term conditional statistical independence.

and $x_{\omega_k,b'}$. In general they should be dependent through the common HV parameter ω_k.

On the other hand, in the conventional approach a conditional probability is probability for events, namely $\mathbf{P}(B|A)$ has the meaning of the probability of the event B under the condition of the event A. Here $\mathbf{P}(B|A) \neq \mathbf{P}(B)$ implies that the event B depends on the event A. Specially in the EPR-Bohm framework the violation of (10.5) implies that the event $B = \{$obtain the value $b' = \beta_m$ for a particle s^2 having the state $\omega = \omega_k\}$ depends on the event $A = \{$obtain the value $a = \alpha_l$ for a particle s^1 having the state $\omega = \omega_k\}$. In principle such a dependence of events may be interpreted (as J. Bell did) as an evidence of nonlocality.

Conclusion. By operating with abstract (noncontextual) probabilities one could really interpret the factorability condition (10.5) as a locality condition. Violation of this condition could be interpreted as dependence of observations for subsystems s^1 and s^2. In the frequency contextual approach the factorability condition (10.5)–(10.6) has a totally different meaning. It has nothing to do with independence of observations for different subsystems of a composite system. This is the condition of independence of the S-sequences (or collectives) x_{a,ω_k} and $x_{\omega_k,b'}$. However, in general they should be dependent through the common parameter ω_k (the common preparation). Therefore there are no reasons to expect that the factorability condition (10.5) holds in the EPR-Bohm experiment.

10.3 Chaos of Hidden Variables and Frequencies for Macro-Observables

Let a hidden variable ω have two values, 0 and 1. Let b be a physical observable having two values, 0 and 1. Suppose that in the sequence

$$x_{b,\omega} = ((b_j, \omega_j),\ j = 1, 2, \ldots),$$

relative frequencies

$$\nu_N(b = \beta, \omega = k) \approx \frac{1}{2}\sin^2\phi_N \quad \text{for } \beta = 1,\ k = 0 \text{ and } \beta = 0,\ k = 0;$$
$$\nu_N(b = \beta, \omega = k) \approx \frac{1}{2}\cos^2\phi_N \quad \text{for } \beta = 1,\ k = 1 \text{ and } \beta = 0,\ k = 1,$$

where ϕ_N are some parameters ('phases') encoding the frequency behavior.

If 'phases' ϕ_N do not stabilize (mod 2π) when $N \to \infty$, then frequencies $\nu_N(b = \beta, \omega = k)$ fluctuate when $N \to \infty$. Hence the sequence $x_{b,\omega}$ is not a collective nor an S-sequence (the principle of the statistical stabilization is violated). However, the frequency probabilities for the physical observable b are well defined

$$\mathbf{P}(b = \beta) = \lim_{N\to\infty} \sum_{k=0,1} \nu_N(b = \beta, \omega = k) = 1/2. \tag{10.7}$$

This 'paradox' (chaotic micro-fluctuations induce statistical stabilization on the macro-level) has a simple explanation. If $x_{b,\omega}$ is a collective, then all probabilities $\mathbf{P}(b = \beta, \omega = k)$ are well defined. Hence it is possible to change the order of limit and sum

$$\lim_{N\to\infty} \sum_{k=0,1} = \sum_{k=0,1} \lim_{N\to\infty} \tag{10.8}$$

and obtain that

$$\mathbf{P}(b = \beta) = \sum_{k=0,1} \mathbf{P}(b = \beta, \omega = k). \tag{10.9}$$

If $x_{b,\omega}$ is not a collective, then, of course, the probabilities $\mathbf{P}(b = \beta, \omega = k)$ are not defined (at least some of them). Thus it is impossible to use (10.8) and (10.9). However, the nonexistence of the right-hand side of (10.8) does not imply nonexistence of the left-hand side of (10.8). The latter gives well-defined frequency probabilities for macro-variables, see [139] for details and further examples.

Let us consider the right-hand side of (10.8). Here we find the limits of relative frequencies and then we perform averaging over the set of hidden variables Ω. Of course, this is possible only if $x_{b,\omega}$ is a collective. We remark that such a procedure has *never been performed in experimental studies.* An experimenter is not able to prepare an ensemble of systems for a fixed value of the hidden variable. Therefore he does not know anything about behavior of frequencies $\nu_N(b = \beta, \omega = k)$. They really may fluctuate, for example, in quantum experiments. In the experimental framework there is performed the procedure represented by the left-hand side of (10.8). In fact, experimenters study behavior of averages of frequencies with respect to the set of hidden variables.

Remark 10.2. Of course, the sample should be representative with respect to the set of values of the hidden variable. This is an additional assumption and its validity also can be questioned. It is quite natural if the set of values of the hidden variable is finite. But if, e.g., $\Omega = \mathbf{R}^n$, where $n = 10^{100...0}$, then there are no reasons to suppose that a sample containing $N = 10^4$ particles can give us a representative sample of values of the hidden variable.

10.4 Fluctuating Distributions of Hidden Variables

In this section we consider the deterministic HV-model. Here, for each physical observable d, $d = d(\omega)$. Thus if for a system s we have $\omega(s) = \omega_0$, then $d(s) = d(\omega_0)$. HV of measurement devices, $\omega^a, \omega^{b'}$ will not be taken into account. As before, we suppose that the set of hidden variables is finite $\Omega = \{\omega_1, \ldots, \omega_M\}$.

Let us consider a family of finite ensembles $\mathbf{S} = \{S\}$ such that, for any natural number N, the family $\mathbf{S}$ contains at least one ensemble S with $|S| \geq N$ (as always $|S|$ denotes the number of elements in the ensemble S).

We introduce a probabilistic characteristic of the family $\mathbf{S}$ (the coefficient of statistical non-reproducibility) $\varepsilon_{\mathbf{S}}$. This coefficient will play an important role in further considerations. For $S \subset \mathbf{S}$, we define "ensemble probability" (ensemble proportion) by

$$\mathbf{P}_S(\omega_k) = \frac{|\{s \in S : \omega(s) = \omega_k\}|}{|S|}.$$

We set

$$W_{\mathbf{S}} = \{w = (S^N)_{N=1}^{\infty} : S^N \in \mathbf{S},\ |S^N| \geq N\}.$$

It is the set of all sequences of ensembles from the family $\mathbf{S}$ such that $|S^N| \geq N$. The main idea is to consider sequences of ensembles with an increasing number of elements in ensembles.

For two sequences of finite ensembles

$$w_1 = (S_1^N)_{N=1}^{\infty}, \qquad w_2 = (S_2^N)_{N=1}^{\infty} \in W_{\mathbf{S}},$$

we set

$$\delta(w_1, w_2) = \overline{\lim}_{N\to\infty} \sum_{k=1}^{M} |\mathbf{P}_{S_1^N}(\omega_k) - \mathbf{P}_{S_2^N}(\omega_k)|$$

and

$$\varepsilon_{\mathbf{S}} = \sup_{w_1, w_2 \in W_{\mathbf{S}}} \delta(w_1, w_2).$$

Let us consider a context C represented by a family of finite ensembles $\mathbf{S}_C = \{S\}$, where each S is an ensemble of pairs of particles.

Let $u, v' = \pm 1$ be some observables in the EPR-Bohm framework. The first one, u, corresponds to some observable for the system s^1 and the second one, v', corresponds to some observable for the system s^2 in a pair $s = (s^1, s^2)$ of correlated systems.

We start with consideration of the frequency covariation

$$\langle u, v' \rangle_{x_{uv'}}$$

with respect to a random sequence

$$x_{uv'} = (x_1, x_2, \ldots, x_N, \ldots),$$

where $x_i = (\gamma_i, \mu_i)$, which is induced by measurements of the pair of observables (u, v'). The random sequence $x_{uv'}$ is obtained by measurements for a sequence of finite ensembles

$$w_{uv'} = \{S_{uv}^N\} \in W_{S_C}$$

of physical systems (pairs of correlated quantum particles). Thus any ensemble $S_{uv'}^N$ is prepared under the complex of physical conditions C. The index uv' has no relation to the ensemble preparation. It just shows that this concrete ensemble will be used for measurements of this pair of observables.

Our aim is to represent the frequency covariation $\langle u, v' \rangle_{x_{uv'}}$ as the limit of the ensemble covariations

$$\langle u, v' \rangle_{w_{uv'}} = \lim_{N \to \infty} \langle u, v' \rangle_{w_{uv'}}^N .$$

We have

$$\langle u, v' \rangle_{x_{uv'}}^N = \sum_{k=1}^{M} \gamma_k \mu_k \mathbf{P}_{S_{uv'}^N}(\omega_k),$$

where $\gamma_k = u(\omega_k)$, $\mu_k = v'(\omega_k)$. The latter sum can be represented as

$$\frac{1}{N} \sum_{i=1}^{N} u(\omega(i)) v'(\omega(i)).$$

Thus

$$\langle u, v' \rangle_{x_{uv'}} = \lim_{N \to \infty} \langle u, v' \rangle_{x_{uv'}}^N = \lim_{N \to \infty} \sum_{k=1}^{M} \gamma_k \mu_k \mathbf{P}_{S_{uv}^N}(\omega_k) \equiv \langle u, v' \rangle_{w_{uv'}} .$$

The latter is the average with respect to the sequence of ensembles $w_{uv'} = \{S^N\}$.

We remark that in general we can not change the order of the limit and the sum; in general

$$\lim_{N \to \infty} \sum_{k=1}^{M} \neq \sum_{k=1}^{M} \lim_{N \to \infty} ,$$

because probabilities $\mathbf{p}_k^{uv'} \equiv \lim_{N \to \infty} \mathbf{P}_{S_{uv'}^N}(\omega_k)$ may be not well defined, cf. Sect. 10.6.

Now let us consider observables satisfying the condition of precise correlations (anti-correlations can be considered in the same way). Here a and a' (b and $b', \ldots$) correspond to the same setting of measurement devices

$$a(s^1) = a'(s^2), \qquad b(s^1) = b'(s^2), \qquad c(s^1) = c'(s^2), \qquad \dots,$$
$$\text{for } s = (s^1, s^2). \tag{10.10}$$

Let us now consider four observables a, c, c', b'. We would like to estimate (as J. Bell did) the following combination of covariations

$$\Delta = \langle a, b' \rangle_{x_{ab'}} - \langle c, b' \rangle_{x_{cb'}}.$$

To find each of the covariations, we should prepare its own sequence of ensembles representing the context C. For the pair of settings (or e.g. of two polarization beam splitters) a, b'

$$w_{ab'} = \{S^N_{ab'}\};$$

for the pair c, b

$$w_{cb'} = \{S^N_{cb'}\};$$

for the pair a, c'

$$w_{ac'} = \{S^N_{ac'}\}.$$

We recall that indexes $ab', \dots, ac'$ have nothing to do with the preparation of corresponding ensembles. They only denote runs of the same preparation procedure which are used for measurements of corresponding pairs of observables on two parts of a composite system, namely $(a, b'), \dots, (a, c')$. We have

$$\Delta = \langle a, b' \rangle_{w_{ab'}} - \langle c, b \rangle_{w_{cb'}} = \lim_{N\to\infty} \sum_k (\mathbf{P}_{S^N_{ab'}}(\omega_k)\alpha_k - \mathbf{P}_{S^N_{cb'}}(\omega_k)\gamma_k)\beta_k$$

and

$$\langle a, c' \rangle_{x_{ac'}} = \langle a, c' \rangle_{w_{ac'}} = \lim_{N\to\infty} \sum_k \alpha_k \gamma_k \mathbf{P}_{S^N_{ac'}}(\omega_k).$$

We now suppose that

$$\lim_{N\to\infty} \mathbf{P}_{S^N_{ab'}}(\omega_k) = \lim_{N\to\infty} \mathbf{P}_{S^N_{cb'}}(\omega_k) = \lim_{N\to\infty} \mathbf{P}_{S^N_{ac'}}(\omega_k) \equiv \mathbf{p}_k. \tag{10.11}$$

By this assumption all preparation runs produce the same probability distribution (in our case it is discrete and given by the probabilities p_k). Hence

$$\Delta = \sum_{k=1}^{M} \mathbf{p}_k(\alpha_k - \gamma_k)\beta_k$$

and

$$\langle a, c' \rangle_{x_{ac'}} = \sum_{k=1}^{M} \mathbf{p}_k \alpha_k \gamma_k.$$

We can now apply Theorem 8.1, Chap. 8, for the discrete probability distribution $\{\mathbf{p}_k\}_{k=1}^{M}$ and obtain Bell's inequality.

However, there are no physical grounds for the condition (10.11)—existence of the probability distribution of HV. The use of this assumption may induce doubts in Bell's arguments. It would be very attractive to try to proceed without this assumption. Therefore the following type of considerations is very popular, cf. [85, 87]. Suppose that in preparation runs for all three experiments (measurements of (a, b'), (c, b') and (a, c')) the experimenter was extremely lucky. He was able to prepare the sequences $w_{ab'}, w_{cb'}, w_{ac'}$ of ensembles of pairs of correlated particles having the same probability distribution of values of the hidden variable for each triple of ensembles $S_{ab'}^N, S_{cb'}^N, S_{ac'}^N$

$$\mathbf{P}_{S_{ab'}^N}(\omega_k) = \mathbf{P}_{S_{cb'}^N}(\omega_k) = \mathbf{P}_{S_{ac'}^N}(\omega_k) \equiv \mathbf{p}_k^N \tag{10.12}$$

Such a lucky experimenter would be able to prove Bell's inequality for each fixed N (by using the discrete probability distribution $\{\mathbf{p}_k^N\}_{k=1}^{M}$) and then to obtain Bell's inequality in the limit $N \to \infty$. In such an approach we need not make the assumption (10.11)—existence of the probability distribution of hidden variables. But neither is the assumption (10.12) justified. The probability of producing three such sequences of physical systems is negligibly small, see Chap. 8 of this Part. Therefore we should consider the general case when different preparation runs produce different sequences of hidden variables.

Remark 10.3 (p-adic probability models, negative probabilities and Bell's inequality). All our considerations were based on statistical stabilization with respect to the real metric. In [136, 137, 139] we considered statistical stabilization with respect to a p-adic metric. The field of p-adic numbers $\mathbf{Q}_p$, where $p > 1$ is a prime number, can be constructed (as is the field of real numbers $\mathbf{R}$) as a completion of the field of rational numbers $\mathbf{Q}$. The p-adic metric differs strongly from the real one. As for finite ensembles S, ensemble probabilities $\mathbf{P}_S(a) = \frac{n(a)}{N}$ are rational numbers, we can study their behavior not only with respect to the real metric on $\mathbf{Q}$, but also with respect to the p-adic metric. p-adic probability theory gives numerous examples of ensemble probabilities fluctuating in the real metric and stabilizing in the p-adic metric. However, the p-adic stabilization of probabilities does not imply the possibility of repeating Bell's proof for p-adic probabilities: these probabilities may be negative rational numbers, see [137, 141] (compare with Muckenheim, [250]).

10.5 Generalized Bell's Inequality

Theorem 10.1 ([139] Generalized Bell's inequality). *Under the condition of precise correlation* (10.10) *the inequality for frequency covariations*

$$|\langle a, b'\rangle - \langle c, b'\rangle| \leq (1 + 2\varepsilon_C) - \langle a, c'\rangle \tag{10.13}$$

holds true.

Proof. We have

$$\begin{aligned}
|\Delta| &= |\langle a, b'\rangle_{x_{ab'}} - \langle c, b'\rangle_{x_{cb'}}| \\
&\leq \overline{\lim}_{N\to\infty}\left|\sum_{k=1}^{M} \mathbf{P}_{S^N_{ab'}}(\omega_k)(\alpha_k - \gamma_k)\beta_k\right| \\
&\quad + \overline{\lim}_{N\to\infty}\left|\sum_{k=1}^{M} (\mathbf{P}_{S^N_{ab'}}(\omega_k) - \mathbf{P}_{S^N_{cb'}})\gamma_k\beta_k\right|
\end{aligned}$$

$$\leq \varepsilon_{S_C} + \lim_{N\to\infty} \sum_{k=1}^{M} \mathbf{P}_{S^N_{ab'}}(\omega_k)|\alpha_k\beta_k|(1-\alpha_k\gamma_k)$$

$$\leq (1+\varepsilon_{S_C}) - \langle a, c'\rangle_{w_{ac'}} + \overline{\lim}_{N\to\infty} \sum_{k=1}^{M} |\mathbf{P}_{S^N_{ac'}}(\omega_k) - \mathbf{P}_{S^N_{ab'}}(\omega_k)||\alpha_k\gamma_k|$$

$$\leq (1+2\varepsilon_{S_C}) - \langle a, c'\rangle_{w_{ac'}}.$$

In the light of this theorem, experiments to verify Bell's inequality might be interpreted as experiments to estimate a statistical measure of ensemble nonreproducibility ε_{S_C}. The violation of Bell's inequality shows that there exist contexts (represented by quantum states $\psi = \psi_C$) for that $\varepsilon_{S_C} \neq 0$. Moreover, it is possible to estimate ε_{S_C}. Such an interpretation is no less justified than a nonlocal interpretation.

10.6 Transmission of Information with the Aid of Dependent Collectives

There have been numerous discussions on the possibility of using "nonlocality condition"

$$\mathbf{P}(a=\alpha, \omega=k, b'=\beta) \neq \mathbf{P}(a=\alpha, \omega=k)\mathbf{P}(b'=\beta, \omega=k) \qquad (10.14)$$

for the transmission of information, see, for example, [283]. Typically such a transmission of information was connected with 'essentially quantum' properties (so-called entanglement). However, the standard scheme can be applied to transfer information with the aid of any two dependent combinable collectives. Let $u = (u_j)$ and $v = (v_j)$ be dependent collectives. As they are combinable, conditional probabilities

$$\mathbf{P}(v=\beta|u=\alpha) - \lim_{N\to\infty} \nu_N(v-\beta; v(\alpha)), \quad \alpha, \beta = \pm 1,$$

are well defined. Here, as usual, $v(\alpha)$ is a collective obtained from v by the choice of subsequence v_{j_k} of values of the v-observable such that $u_{j_k} = \alpha$. As collectives are dependent, we have, for example,

$$\mathbf{P}_1 = \mathbf{P}(v = 1|u = +1) \neq \mathbf{P}_2 = \mathbf{P}(v = 1|u = -1).$$

Then we can proceed as in all quantum stories, see, e.g., [283]. For example:

Bob prepares a statistical ensemble of pairs which components are described by collectives u and v respectively. He chooses sub-collective $v(+1)$ and sends it to Alice. If Alice knows the relation between probabilities, she can easily rediscover the bit of information.

Chapter 11

Original EPR-Experiment: Local Realistic Model

Einstein, Podolsky and Rosen (EPR) presented an argument to show that there are situations in which the scheme of quantum theory seems to be incomplete [89]. Though the EPR work dealt with continuous position and momentum variables, most of the further activity was concentrated almost exclusively on systems of discrete spin variables following the Bohm and Bell [34] works, see Chap. 8 of this part.

Bell's theorem, see Chap. 8 of this part, states (at least in the original Bell's interpretation) that there are quantum spin correlation functions that can not be represented as classical correlation functions of separated random variables. It has been interpreted as incompatibility of the requirement of locality with the statistical predictions of quantum mechanics. It is now widely accepted, as a result of Bell's theorem and related experiments, that local realism must be rejected and there exists the so called quantum nonlocality.

However, in Chap. 8 we performed a detailed analysis of probabilistic assumptions in theorems of J. von Neumann and J. Bell. We showed that, without a rigorous formalization of the probabilistic content of Bell's arguments, one cannot forcefully derive the fundamental dilemma that we are often being offered: that is, either nonlocality or the death of reality.

For example, if one pays attention to the problem of the *range coincidence* for prequantum classical variables and quantum observables, the derivation of Bell's inequality would be immediately blocked.

The same thing would happen if *noninjectivity of classical → quantum correspondence* would be taken into account.

Contextuality in the EPR-Bohm experiment is not reduced to contextuality of simultaneous measurement of compatible observables (Bell's contextuality). Therefore contextuality can be explained without appealing to nonlocality.

If one takes into account the *time structure* of the EPR experiment and considers the *frequency definition of probability*, then Bell's inequality could not be derived.[1]

Moreover, Igor Volovich pointed out, see in [308], that in the derivation of the conclusion about quantum nonlocality the fundamental fact that space-time exists was neglected. If we take into account the spatial dependence of the wave function, then the standard formalism of quantum mechanics might be consistent with local realism.

In this chapter we present a detailed analysis of the space-time structure of the EPR experiment and, in particular, we prove that, for the original EPR experiment for the position and momentum, the correlations have a local realistic representation, see Andrei Khrennikov and Igor Volovich (Steklov Mathematical Institute of Russian Academy of Science), [154, 155, 188].

11.1 Space and Arguments of Einstein, Podolsky, Rosen, and Bell

11.1.1 Bell's Local Realism

A mathematical formulation of Bell's local realism may be given by the following relation

$$\langle\psi|\hat{a}(\alpha)\hat{b}(\beta)|\psi\rangle = E\xi(\alpha)\eta(\beta). \tag{11.1}$$

[1] Similar conclusions were obtained by K. Hess and W. Philipp who also took into account the time structure of the EPR experiment, see [116–120].

Here $\hat{a}(\alpha)$ and $\hat{b}(\beta)$ are self-adjoint operators which commute on a natural domain and α and β are certain indices. Here E is a mathematical expectation and $\xi(\alpha)$ and $\eta(\beta)$ are two stochastic processes such that the range of $\xi(\alpha)$ is the spectrum of $\hat{a}(\alpha)$ and the range of $\eta(\beta)$ is the spectrum of $\hat{b}(\beta)$ and ψ is a normalized vector from a Hilbert space. Then we say that the triplet

$$\{\hat{a}(\alpha), \hat{b}(\beta), \psi\}$$

satisfies *Bell's local realism* (BLR) condition.

Let $\alpha = (\alpha_1, \alpha_2, \alpha_3)$ and $\beta = (\beta_1, \beta_2, \beta_3)$ be two unit vectors in three-dimensional space $\mathbf{R}^3$. We denote by (α, β) their inner product in $\mathbf{R}^3$: $(\alpha, \beta) = \sum_{i=1}^{3} \alpha_i \beta_i$.

It was proved that a two-spin quantum correlation function (which is equal to just $-(\alpha, \beta)$) can not be represented in the form (11.1)

$$D_{spin}(\alpha, \beta) = \langle \psi_{spin} | (\sigma, \alpha) \otimes (\sigma, \beta) | \psi_{spin} \rangle \neq E\xi(\alpha)\eta(\beta) \tag{11.2}$$

if one has a bound

$$|\xi(\alpha)| \leq 1, \qquad |\eta(\beta)| \leq 1. \tag{11.3}$$

Here $\sigma = (\sigma_1, \sigma_2, \sigma_3)$ are the Pauli matrices, and $(\sigma, \alpha) = \sum_{i=1}^{3} \sigma_i \alpha_i$ and

$$\psi_{spin} = \frac{1}{\sqrt{2}} \left(\begin{pmatrix} 0 \\ 1 \end{pmatrix} \otimes \begin{pmatrix} 1 \\ 0 \end{pmatrix} - \begin{pmatrix} 1 \\ 0 \end{pmatrix} \otimes \begin{pmatrix} 0 \\ 1 \end{pmatrix} \right).$$

Therefore the correlation function of two spins does not satisfy the BLR condition (11.1). In this sense sometimes one speaks about quantum non-locality.

11.1.2 Einstein's Local Realism

Note however that in the previous discussion the space-time parameters were not explicitly involved though one speaks about non-locality. Actually "local realism" in the Bell sense as it was formulated above in (11.1) is a notion which has nothing to do with the notion of locality in the ordinary three-dimensional space. Therefore we define also another notion which we will call the condition of local realism in the sense of Einstein.

In a Hilbert space H, let there be given a family of self-adjoint operators $\{\hat{a}(a, \mathcal{O})\}$ and $\{\hat{b}(b, \mathcal{O})\}$ parameterized by the regions $\mathcal{O}$ in the Minkowsky space-time. Suppose that one has a representation

$$\langle\psi|\hat{a}(\alpha, \mathcal{O}_1)\hat{b}(\beta, \mathcal{O}_2)|\psi\rangle = E\xi(\alpha, \mathcal{O}_1)\eta(\beta, \mathcal{O}_2) \tag{11.4}$$

for $\alpha, \beta, \mathcal{O}_1, \mathcal{O}_2$ for which the operators commute. Then we say that the quadruplet

$$\{\hat{a}(\alpha, \mathcal{O}_1), \hat{b}(\beta, \mathcal{O}_2), \psi\}$$

satisfies the *Einstein local realism* (ELR) condition.

11.1.3 Local Realist Representation for Quantum Spin Correlations

Quantum correlation describing the localized measurements of spins in the regions $\mathcal{O}_1$ and $\mathcal{O}_2$ includes the projection operators $P_{\mathcal{O}_1}$ and $P_{\mathcal{O}_2}$. In contrast to Bell's theorem (11.2) now there exists a local realist representation [154, 155, 188]

$$\langle\psi|(\sigma, \alpha)P_{\mathcal{O}_1} \otimes (\sigma, \beta)P_{\mathcal{O}_2}|\psi\rangle = E\xi(\mathcal{O}_1, \alpha)\eta(\mathcal{O}_2, \beta) \tag{11.5}$$

if the distance between the regions $\mathcal{O}_1$ and $\mathcal{O}_2$ is large enough. Here all classical random variables are bounded by 1.

Since detectors of particles are obviously located somewhere in space it shows that loopholes are unavoidable in experiments aimed to establish a violation of Bell's inequalities. Though there were some reports on experimental derivation of violation of Bell's inequalities, in fact such violations always were based on additional assumptions besides local realism. No genuine Bell's inequalities have been violated since always some loopholes were found in the experiments, for a review see for example [276, 277]. There were many discussions of proposals for experiments which could avoid the loopholes, however up to now a convincing proposal still has not been advanced.

One can compare the situation with attempts to measure the position and momentum of a particle in a single experiment. Also one could speak about some technical difficulties (similar to the efficiency of a detectors loophole) and hope that someone could come up with a proposal to make an experiment without loopholes. However we know from the uncertainty relation for the measurement of momentum and position that it is not possible. Similarly the formula (11.5) shows that a loophole-free experiment in which a violation of Bell's inequalities will be observed is impossible if the distance between detectors is large enough. Therefore loopholes in Bell's experiments are irreducible.

11.1.4 EPR versus Bohm and Bell

The original EPR system involving continuous variables was considered by Bell in [34]. He mentioned that if one admits "measurement" of an arbitrary "observable" at an arbitrary state, then it is easy to mimic his work on spin variables (just take a two-dimensional subspace and define an analogue of spin operators). The problem which he was discussing in [34] is a narrower problem, restricted to measurement of positions only, on two non-interacting spin-less particles in free space. Bell used the Wigner distribution approach to quantum mechanics.The original EPR state has a nonnegative Wigner distribution. Bell argues that it gives a local, classical model of hidden variables and therefore the EPR state should not violate local realism. He then considers a state with non-positive Wigner distribution and demonstrates that this state violates local realism.

Bell's proof of violation of local realism in phase space has been criticized at many occasions, because of the use of an unnormalizable Wigner distribution. Then it was demonstrated that the Wigner function of the EPR state, though positive definite, provides an evidence of the nonlocal character of this state if one measures a parity operator.

In [188] we have applied to the original EPR problem the method which was used by Bell in his well-known paper [34]. He has shown that the correlation function of two spins cannot be represented by classical correlations of separated bounded random variables. This Bell's theorem has been interpreted as incompatibility of local realism with quantum mechanics. It was shown in [188] that, in contrast to Bell's theorem for spin correlation functions, the correlation function of positions (or momenta) of two particles always admits a representation in the form of classical correlation of separated random variables.The following representation was proved:

$$\langle\psi|\hat{q}_1(\alpha_1)\hat{q}_2(\alpha_2)|\psi\rangle = E\xi_1(\alpha_1)\xi_2(\alpha_2). \tag{11.6}$$

(The notations used in this equality will be explained in the next section.)

Therefore we *obtain a local realistic (in the sense of Bell and in the sense of Einstein as well) representation for the correlation function in the original EPR model.*

This result looks rather surprising since it is the common opinion in the quantum community that the Bohm-Bell reformulation of the EPR paradox (i.e. with discrete observables—projections of spin) is equivalent to the original one (i.e. with continuous observables—position and momentum).

11.2 Bell's Theorem and Ranges of Values of Observables

The role of the range coincidence assumption in Bell's considerations was discussed in detail in Chap. 8 of this part of the book. We present here a simple local (in the sense of Bell) classical probabilistic model which reproduces the quantum mechanical correlation of two spins. Let us take as a probability space Λ just 3 points: $\Lambda = \{1, 2, 3\}$ and the expectation $Ef = \frac{1}{3}\sum_{\lambda=1}^{3} f(\lambda)$. Let the random fields be $\xi(\alpha, \lambda) = \eta(\alpha, \lambda) = \sqrt{3}\alpha_\lambda$, $\lambda = 1, 2, 3$. Then one has the relation $(\alpha, \beta) = E\xi(\alpha)\xi(\beta)$. Bell's theorem is not valid in this case because we do not have the bound (11.3). Instead we have

$$|\xi(\alpha, \lambda)| \leq \sqrt{3}.$$

This model shows that the boundedness of observables by 1 plays a crucial role in the proof of Bell's theorem. Actually to reproduce (α, β) we can use even a deterministic model: simply the first experimentalist will report on the measurement of the components of the vector $(\alpha_1, \alpha_2, \alpha_3)$ and the second on the measurement of the components of the vector $(\beta_1, \beta_2, \beta_3)$.

11.3 Correlation Functions in EPR Model

Now let us apply a similar approach to the original EPR case [89]. The Hilbert space of two one-dimensional particles is $L^2(R) \otimes L^2(R)$ and canonical coordinates and momenta q_1, q_2, p_1, p_2 are represented by the operators $\hat{q}_1, \hat{q}_2, \hat{p}_1, \hat{p}_2$ which obey thc commutation rclations

$$[\hat{q}_m, \hat{p}_n] = i\delta_{mn}, \qquad [\hat{q}_m, \hat{q}_n] = 0, \qquad [\hat{p}_m, \hat{p}_n] = 0, \quad m, n = 1, 2. \tag{11.7}$$

We introduce canonical transformations of our variables

$$\hat{q}_n(\alpha) = \hat{q}_n \cos\alpha - \hat{p}_n \sin\alpha, \qquad \hat{p}_n(\alpha) = \hat{q}_n \sin\alpha + \hat{p}_n \cos\alpha, \quad n = 1, 2. \tag{11.8}$$

Then one gets

$$[\hat{q}_m(\alpha), \hat{p}_n(\alpha)] = i\delta_{mn}, \quad n = 1, 2. \tag{11.9}$$

In particular one has $\hat{q}_n(0) = \hat{q}_n$, $\hat{q}_n(3P/2) = \hat{p}_n$, $n = 1, 2$.

Now let us consider the correlation function

$$D(\alpha, \beta) = \langle \psi | \hat{q}_1(\alpha) \otimes \hat{q}_2(\beta) | \psi \rangle. \tag{11.10}$$

The correlation function $D(\alpha, \beta)$ (11.10) is an analogue of the Bell correlation function $D_{spin}(\alpha, \beta)$. Bell in [34] suggested that we consider the correlation function of just the free evolutions of the particles at different times (see below).

We are interested in the question whether the quantum mechanical correlation function (11.10) can be represented in the form

$$\langle \psi | \hat{q}_1(\alpha) \otimes \hat{q}_2(\beta) | \psi \rangle = E\xi_1(\alpha)\xi_2(\beta). \tag{11.11}$$

Here $\xi_1(\alpha) = \xi_1(\alpha, \lambda)$, $\xi_2(\beta) = \xi_2(\beta, \lambda)$ are two real random processes, possibly unbounded. The parameters λ are interpreted as hidden variables in a realist theory.

Theorem 11.1. *For an arbitrary state $\psi \in L^2(R) \otimes L^2(R)$ on which pairwise products of operators $\hat{q}_1, \hat{q}_2, \hat{p}_1, \hat{p}_2$ are defined, there exist random processes $\xi_1(\alpha, \lambda)$ and $\xi_2(\beta, \lambda)$ such that the relation* (11.11) *is valid.*

Proof. We rewrite the correlation function $D(\alpha, \beta)$ (11.10) in the form

$$\begin{aligned} \langle \psi | \hat{q}_1(\alpha) \otimes \hat{q}_2(\beta) | \psi \rangle = {} & \langle \hat{q}_1 \hat{q}_2 \rangle \cos\alpha \cos\beta - \langle \hat{p}_1 \hat{q}_2 \rangle \sin\alpha \cos\beta \\ & - \langle \hat{q}_1 \hat{p}_2 \rangle \cos\alpha \sin\beta + \langle \hat{p}_1 \hat{p}_2 \rangle \sin\alpha \sin\beta. \end{aligned} \tag{11.12}$$

Here we use the notation

$$\langle \hat{q}_1 \hat{q}_2 \rangle = \langle \psi | \hat{q}_1 \hat{q}_2 | \psi \rangle.$$

Now let us set

$$\xi_1(\alpha, \lambda) = f_1(\lambda) \cos\alpha - g_1(\lambda) \sin\alpha,$$

$$\xi_2(\beta, \lambda) = f_2(\lambda) \cos\beta - g_2(\lambda) \sin\beta.$$

Here real functions $f_n(\lambda), g_n(\lambda),\ n = 1, 2$ are such that

$$Ef_1 f_2 = \langle \hat{q}_1 \hat{q}_2 \rangle, \qquad Eg_1 f_2 = \langle \hat{p}_1 \hat{q}_2 \rangle,$$
$$Ef_1 g_2 = \langle \hat{q}_1 \hat{p}_2 \rangle, \qquad Eg_1 g_2 = \langle \hat{p}_1 \hat{p}_2 \rangle. \tag{11.13}$$

We use for the expectation the notation $Ef_1 f_2 = \int f_1(\lambda) f_2, (\lambda) d\rho(\lambda)$. To solve the system of equations (11.13) we take

$$f_n(\lambda) = \sum_{\mu=1}^{2} F_{n\mu} \eta_\mu(\lambda), \qquad g_n(\lambda) = \sum_{\mu=1}^{2} G_{n\mu} \eta_\mu(\lambda), \tag{11.14}$$

where $F_{n\mu}, G_{n\mu}$ are constants and $E\eta_\mu \eta_\nu = \delta_{\mu\nu}$. We write

$$\langle \hat{q}_1 \hat{q}_2 \rangle = A, \qquad \langle \hat{p}_1 \hat{q}_2 \rangle = B, \qquad \langle \hat{q}_1 \hat{p}_2 \rangle = C, \qquad \langle \hat{p}_1 \hat{p}_2 \rangle = D.$$

A solution of (11.13) may be given for example by

$$f_1 = A\eta_1, \qquad f_2 = \eta_1,$$
$$g_1 = B\eta_1 + \left(D - \frac{BC}{A} \right) \eta_2, \qquad g_2 = \frac{C}{A} \eta_1 + \eta_2.$$

Hence the representation of the quantum correlation function in terms of the separated classical random processes (11.11) is proved.

Remark 11.1. We were able to solve the system of equations (11.13), because there are no bounds to the random variables f_1, f_2, g_1, g_2. In the case of the Bohm spin model one has the bound (11.3) which leads to Bell's inequality and as a result an analogue of (11.13) in the Bohm model has no solution.

Remark 11.2. The condition of reality of the functions $\xi_1(\alpha, \lambda)$ and $\xi_2(\beta, \lambda)$ is important. It means that the range of $\xi_1(\alpha, \lambda)$ and $\xi_2(\beta, \lambda)$ are the sets of eigenvalues of the operator $\hat{q}_1(\alpha)$ and $\hat{q}_2(\beta)$. If we relax this condition, then one can get a hidden variable representation just by using an expansion of unity

$$\langle \psi | \hat{q}_1(\alpha) \hat{q}_2(\beta) | \psi \rangle = \sum_\lambda \langle \psi | \hat{q}_1(\alpha) | \lambda \rangle \langle \lambda | \hat{q}_2(\beta) | \psi \rangle.$$

Similarly one can prove a representation

$$\langle\psi|\hat{q}_1(t_1)\otimes\hat{q}_2(t_2)|\psi\rangle = \int \xi_1(t_1,\lambda)\xi_2(t_2,\lambda)d\rho(\lambda), \tag{11.15}$$

where $\hat{q}_n(t) = \hat{q}_n + \hat{p}_n t$, $n = 1, 2$ is a free quantum evolution of the particles. It is enough to take

$$\xi_1(t_1,\lambda) = f_1(\lambda) + g_1(\lambda)t_1, \qquad \xi_2(t_2,\lambda) = f_2(\lambda) + g_2(\lambda)t_2.$$

Remark 11.3. In fact we can prove a more general theorem. If $f(s,t)$ is a function of two variables, then it can be represented as the expectation of two stochastic processes $f(s,t) = E\xi(s)\eta(t)$. Indeed, if $f(s,t) = \sum_n g_n(s)h_n(t)$, then we can take

$$\xi(s,\omega) = \sum_n g_n(s)x_n(\omega), \qquad \eta(t,\omega) = \sum_n h_n(s)x_n(\omega),$$

where $Ex_nx_m = \delta_{nm}$.

11.4 Space-Time Dependence of Correlation Functions and Disentanglement

11.4.1 Modified Bell's Equation

As was already mentioned in the previous sections, the space part of the wave function of particles was neglected in Bohm-Bell considerations. However, exactly the space part is relevant to the discussion of locality. The Hilbert space assigned to one particle with spin 1/2 is $\mathbf{C}^2\otimes L^2(\mathbf{R}^3)$ and the Hilbert space of two particles is $\mathbf{C}^2\otimes L^2(\mathbf{R}^3)\otimes\mathbf{C}^2\otimes L^2(\mathbf{R}^3)$. The complete wave function is $\psi = (\psi_{ij}(\mathbf{r}_1,\mathbf{r}_2,t))$ where i and j are spinor indices, t is time and $\mathbf{r}_1$ and $\mathbf{r}_2$ are vectors in three-dimensional space.

We suppose that there are two detectors (A and B) which are located in space $\mathbf{R}^3$ within the two localized regions $\mathcal{O}_1$ and $\mathcal{O}_2$ respectively, well separated from one another. If one makes a local observation in the region $\mathcal{O}_1$, then this means that one measures not only the spin observable σ_i but also some other observable which describes the localization of the particle like the energy density or the projection operator $P_{\mathcal{O}}$ to the region $\mathcal{O}$. Normally in experiments there are polarizers and detectors. We will consider here correlation functions which include the projection operators $P_{\mathcal{O}}$.

Quantum correlation describing the localized measurements of spins in the regions $\mathcal{O}_1$ and $\mathcal{O}_2$ is

$$\omega((\sigma,\alpha)P_{\mathcal{O}_1}\otimes(\sigma,\beta)P_{\mathcal{O}_2}) = \langle\psi|(\sigma,\alpha)P_{\mathcal{O}_1}\otimes(\sigma,\beta)P_{\mathcal{O}_2}|\psi\rangle. \tag{11.16}$$

Let us consider the simplest case when the wave function has the form of the product of the spin function and the spatial function $\psi = \psi_{spin}\phi(\mathbf{r}_1,\mathbf{r}_2)$. Here $\phi(\mathbf{r}_1,\mathbf{r}_2)$ is a complex-valued function. Then one has

$$\omega((\sigma,\alpha)P_{\mathcal{O}_1}\otimes(\sigma,\beta)P_{\mathcal{O}_2}) = g(\mathcal{O}_1,\mathcal{O}_2)D_{spin}(\alpha,\beta), \tag{11.17}$$

where the function

$$g(\mathcal{O}_1,\mathcal{O}_2) = \int_{\mathcal{O}_1\times\mathcal{O}_2} |\phi(\mathbf{r}_1,\mathbf{r}_2)|^2 d\mathbf{r}_1 d\mathbf{r}_2 \tag{11.18}$$

describes correlation of particles in space. It is the probability of finding one particle in the region $\mathcal{O}_1$ and another particle in the region $\mathcal{O}_2$.

One has

$$0 \le g(\mathcal{O}_1,\mathcal{O}_2) \le 1. \tag{11.19}$$

11.4.2 Disentanglement

If $\mathcal{O}_1$ is a bounded region and $\mathcal{O}_1(l)$ is a translation of $\mathcal{O}_1$ to the 3-vector l, then one can prove

$$\lim_{|l|\to\infty} g(\mathcal{O}_1(l),\mathcal{O}_2) = 0. \tag{11.20}$$

Since

$$\langle \psi_{spin} | (\sigma, \alpha) \otimes I | \psi_{spin} \rangle = 0,$$

we have

$$\omega((\sigma, \alpha) P_{\mathcal{O}_1} \otimes I) = 0.$$

Therefore we have proved the following proposition which says that the state $\psi = \psi_{spin}\phi(\mathbf{r}_1, \mathbf{r}_2)$ becomes disentangled (factorized) at large distances.

Proposition 11.1. *One has the following property of the asymptotic factorization (disentanglement) at large distances*

$$\lim_{|l| \to \infty} [\omega((\sigma, \alpha) P_{\mathcal{O}_1(l)} \otimes (\sigma, \beta) P_{\mathcal{O}_2}) - \omega((\sigma, \alpha) P_{\mathcal{O}_1(l)} \otimes I) \omega(I \otimes (\sigma, \beta) P_{\mathcal{O}_2})] = 0 \tag{11.21}$$

or

$$\lim_{|l| \to \infty} \omega((\sigma, \alpha) P_{\mathcal{O}_1(l)} \otimes (\sigma, \beta) P_{\mathcal{O}_2}) = 0.$$

Now one inquires whether one can write a representation

$$\omega((\sigma, \alpha) P_{\mathcal{O}_1} \otimes (\sigma, \beta) P_{\mathcal{O}_2}) = \int \xi_1(\alpha, \mathcal{O}_1, \lambda) \xi_2(\beta, \mathcal{O}_2, \lambda) d\rho(\lambda), \tag{11.22}$$

where $|\xi_1(\alpha, \mathcal{O}_1, \lambda)| \leq 1$, $|\xi_2(\beta, \mathcal{O}_2, \lambda)| \leq 1$.

Remark 11.4. A local modified equation reads

$$|\phi(\mathbf{r_1}, \mathbf{r_2}, t)|^2(\alpha, \beta) = E\xi(\alpha, \mathbf{r_1}, t)\eta(\beta, \mathbf{r_2}, t).$$

If we are interested in the conditional probability of finding the projection of spin along vector α for the particle 1 in the region $\mathcal{O}_1$ and the projection of spin along the vector β for the particle 2 in the region $\mathcal{O}_2$, then we have to divide both sides of (11.22) by $g(\mathcal{O}_1, \mathcal{O}_2)$.

Note that here the classical random variable $\xi_1 = \xi_1(\alpha, \mathcal{O}_1, \lambda)$ is not only separated in the sense of Bell (i.e. it depends only on α), but it is also local in three-dimensional space, since it depends only on the region $\mathcal{O}_1$. The classical random variable ξ_2 is also local in three-dimensional space, since it depends only on $\mathcal{O}_2$. Note also that since the eigenvalues of the projector $P_{\mathcal{O}}$ are 0 or 1, then one should have $|\xi_n(\alpha, \mathcal{O}_n)| \leq 1, \ n = 1, 2$.

Due to the property of the asymptotic factorization and the vanishing of the quantum correlation for large $|l|$, there exists a trivial asymptotic classical representation of the form (11.22) with $\xi = \eta = 0$.

We can do even better and find a classical representation which will be valid uniformly for large $|l|$.

Let us take now the wave function ϕ of the form $\phi = \psi_1(\mathbf{r}_1)\psi_2(\mathbf{r}_2)$ where

$$\int_{R^3} |\psi_1(\mathbf{r}_1)|^2 d\mathbf{r}_1 = 1, \qquad \int_{R^3} |\psi_2(\mathbf{r}_2)|^2 d\mathbf{r}_2 = 1.$$

In this case

$$g(\mathcal{O}_1(l), \mathcal{O}_2) = \int_{\mathcal{O}_1(l)} |\psi_1(\mathbf{r}_1)|^2 d\mathbf{r}_1 \cdot \int_{\mathcal{O}_2} |\psi_2(\mathbf{r}_2)|^2 d\mathbf{r}_2.$$

There exists such $L > 0$ that

$$\int_{B_L} |\psi_1(\mathbf{r}_1)|^2 d\mathbf{r}_1 = \varepsilon < 1/2,$$

where $B_L = \{\mathbf{r} \in R^3 : |\mathbf{r}| \geq L\}$.

We have the following

Theorem 11.2. *Under the above assumptions and for large enough $|l|$ there exists the following representation of the quantum correlation function*

$$\omega((\sigma, \alpha)P_{\mathcal{O}_1(l)} \otimes (\sigma, \beta)P_{\mathcal{O}_2}) = E\xi(\mathcal{O}_1(l), \alpha)\xi(\mathcal{O}_2, \beta),$$

where all classical random variables are bounded by 1.

11.5 Role of Space-Time in EPR Argument

Mathematical definitions of local realism in the sense of Bell and in the sense of Einstein were given in this chapter. We demonstrated that if we include into the quantum mechanical formalism the space-time structure in the standard way, then quantum mechanics might be consistent with Einstein's local realism. Thus one even should not go beyond QM, but simply apply the standard formalism of QM.

We claim, see Khrennikov and Volovich [153, 154, 188], that *loopholes are unavoidable in experiments aimed at establishing a violation of Bell's inequalities!*[2]

It is shown also that, for the original EPR correlation functions which deal with positions and momenta, one can get a local realistic representation in terms of separated random processes. The representation is obtained for any state including entangled states. Therefore the original EPR model does not lead to quantum nonlocality in the sense of Bell even for entangled states. One can get quantum nonlocality in the EPR situation only if we rather artificially restrict ourself in the measurements with a two-dimensional subspace of the infinite-dimensional Hilbert space corresponding to the position or momentum observables.[3]

An interrelation of the roles of entangled states and the bounded by 1 observables in considerations of local realism and quantum nonlocality deserves further study.

---

[2] We remark that our position with respect to loopholes in the EPR-Bohm experiment was repeated by Richard Gill who assigned it to Bell. However, Gill's publication was a result of intensive email exchange with us (directly after the first conference "Quantum Theory: Reconsideration of Foundations", [151]) in which we explicated our position. On the other hand, Bell by himself did never point to such a possibility. Hence, it was really unfair to couple the thesis that loopholes are unavoidable to Bell and not to Khrennikov and Volovich.

[3] And if one neglects all problems induced by explicit and implicit assumptions in Bell's considerations, cf. Chap. 8.

The main experimental consequence of our analysis of the space-time structure of the ERP-type experiments is that new detailed *experiments on dependence of experimental correlations on time and space should be performed as soon as possible.*

Part IV
Interrelation between Classical and Quantum Probabilities

In this part we continue our studies to minimize the gap between the classical and quantum probabilistic models. On the one hand, many distinguishing features of quantum probability are present in classical probability. Thus it is possible to do a lot in QM by using classical probability. On the other hand, classical probability has a very natural QL representation. Thus it is possible to represent purely classical (e.g., physical) models in the QL way. Such a QL representation is incomplete and can be considered as a simplification of the originally very complex classical statistical description. It is essentially easier to work in the QL representation. In this part we shall consider a few topics on quantum features of classical statistical models:

1. We shall see that diffraction and interference histograms (which are typical for QM) can be produced not only by, e.g., electrons or photons, but even by deterministic particles. The crucial point is that dynamics of particles is in *discrete time*. Thus such a distinguishing feature of QM as the interference of probabilities can be obtained in the purely classical framework under the assumption that a time quantum exists.
2. We shall show that the quantum probabilistic description can naturally appear in classical statistical mechanics for disordered systems (spin glasses).
3. We shall derive Schrödinger's equation in the contextual probabilistic framework: under some assumptions the dynamics of classical probabilities can be represented in the form of the QM-dynamics in the complex Hilbert space. Under weaker restrictions, classical probabilistic dynamics induces a nonlinear generalization of Schrödinger's equation.

Chapter 12

Discrete Time Dynamics

We recall that one of the most distinguishing features of quantum probabilistic behavior is the appearance of *interference* structures created by ensembles of elementary particles. Interference is a direct consequence of the QL formula of total probability, see Part II, Chap. 3. In this chapter we shall be interested in this formula in a special case of symmetric a-selections. We consider such a context C that probabilities $p_C(\alpha_1)$ and $p_C(\alpha_2)$ are equal: $p_C(\alpha_1) = p_C(\alpha_2) = 1/2$. (In the context of the two-slit experiment it corresponds to the situation when slits are located symmetrically with respect to the source of particles). In this case the QL-formula of total probability can be written in the form

$$P_{12} = \frac{1}{2}(P_1 + P_2) + \sqrt{P_1 P_2} \cos\theta,$$

where $P_{12} = p_C(\beta)$ and $P_j = p^{b|a}(\beta|\alpha_j)$, $j = 1, 2$; cf. the classical rule

$$P_{12} = \frac{1}{2}(P_1 + P_2).$$

In this chapter we study contextual probabilistic behavior of dynamical systems with *discrete time*. We are interested in diffraction and interference effects for deterministic dynamics. We start with the deterministic model for scattering of charged particles on the charged screen with a single slit, Sect. 12.2. The resulting diffraction picture has a nontrivial minimum-maximum distribution [152, 154, 172]. This diffraction picture has no relation to "self-diffraction" of particles, no wave-structure is involved in our considerations. The basic source of diffraction is the *discrete time scale* which is used in our mathematical model: instead of Newton's differential equations—continuous time evolution, we consider *difference equations—discrete time evolution*. The diffraction effect disappears as the time discreteness parameter τ goes to zero.

In Sects. 12.3, 12.4 we consider the deterministic model for scattering of charged particles on the charged screen with two slits. The resulting interference picture has a nontrivial minimum-maximum distribution (and the interference effect disappears as the time discreteness parameter τ goes to zero). Moreover, in the same way as in quantum mechanics (see, e.g., the book of Feynman and Hibs [91] for a detailed analysis) the classical formula (6.23) is violated (R. Feynman did not pay attention to the conditional structure of this formula and considered it as a formula of addition of probabilities and not as the formula of total probability). Instead of the classical rule (6.23), we obtain nontrivial trigonometric interference. Here we use our contextual probabilistic calculus which (opposite to the quantum Hilbert space calculus) can be applied to ensembles of deterministic particles. We extract contextual probabilities from the statistical data which is obtained from numerical simulation. Hence we find the coefficients of supplementarity λ, see Part II, Chap. 3. In our simulations these coefficients are always less than 1.[1] Therefore we can represent them in the cos-way and find the probabilistic phases θ. Finally, by using the algorithm proposed in Part II, QLRA, we reconstruct the wave function. This is the wave function of the complex of experimental physical conditions C under consideration.

[1] We do not know: "Is this the general consequence of discrete-time deterministic dynamics or just a property of those concrete dynamical systems which we studied numerically?" This is an extremely interesting problem: "Can one find discrete-time deterministic dynamics which would produce the hyperbolic interference?"

The common viewpoint might be that we study just a discrete approximation to the continuous Newton model. It is supposed that the latter model gives the right picture of "classical physical reality." Contrary to such a conventional viewpoint on the continuous Newton model, we speculate that

The continuous Newton model is just an approximation of physical reality.

The right picture is given by discrete difference equations.[2]

Our investigations can also be interpreted in the following way. Let us start with a postulate about the discreteness of time.

Postulate D. *There exists the fundamental (indivisible) quant of time* τ.

"Which features of the conventional quantum formalism could be reproduced on the basis of postulate D?"

We have demonstrated that at least diffraction and interference effects can be reproduced by Newtonian particles with discrete time dynamics.

The results of this chapter were obtained in a series of papers [152, 154, 172] of the author of this book and his graduate student Yaroslav Volovich who performed rather tricky physical and numerical modeling.

12.1 Discrete Time in Newton's Equations

Consider the well-known Newton's equation

$$\mathbf{F} = m\ddot{\mathbf{r}}. \tag{12.1}$$

Let us rewrite the second-order differential equation (12.1) as a system of first-order differential equations. We have

[2] We would like to mention an analogy between models with discrete time and p-adic models, see, e.g., [136, 137, 307].

$$\begin{aligned}\mathbf{F} &= m\dot{\mathbf{v}},\\ \mathbf{v} &= \dot{\mathbf{r}}.\end{aligned} \tag{12.2}$$

In the system (12.2) the derivatives assume the continuousness of time. Let us now introduce a discreteness parameter τ. We have

$$\begin{aligned}\mathbf{F} &= m\frac{\mathbf{v}(t+\tau)-\mathbf{v}(t)}{\tau},\\ \mathbf{v}(t+\tau) &= \frac{\mathbf{r}(t+\tau)-\mathbf{r}(t)}{\tau}.\end{aligned} \tag{12.3}$$

In the limit of $\tau \to 0$, (12.3) is equivalent to (12.1) and (12.2). Now let us rewrite the system (12.3) in a directly computable way

$$\begin{aligned}\mathbf{v}(t+\tau) &= \mathbf{v}(t) + \mathbf{F}\tau/m,\\ \mathbf{r}(t+\tau) &= \mathbf{r}(t) + \mathbf{v}(t)\tau.\end{aligned} \tag{12.4}$$

where $\mathbf{F} = \mathbf{F}(\mathbf{r}(t), \mathbf{v}(t), t)$.

Note that in our model the coordinate space is continuous.

12.2 Diffraction Pattern in a Single Slit Scattering

Consider the following experimental setup (see Fig. 12.1). A particle source e is located in front of the center of a slit in a screen S_1. Near the slit there is a nonzero force field which affects particles,

$$\mathbf{F}(x, y) \neq 0, \quad \text{for } (x, y) \in \delta$$

(x and y are coordinates along horizontal and vertical axes' respectively, the axes origin is in the center of the slit). Particles pass through the slit and concentrate on a second screen S_2. We study a particle density on the screen S_2.

Let us start from the simplest case when the force field is constant and perpendicular to the screens, $\mathbf{F}(x, y) = \mathbf{e}_x F_0$, for $x \geq 0$ and 0 otherwise. In this case

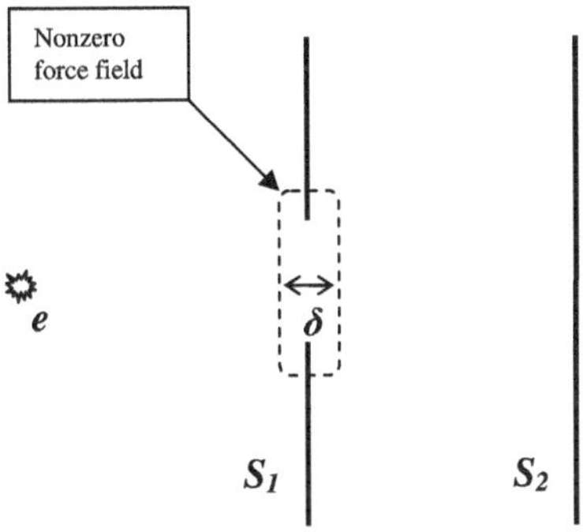

Fig. 12.1 Particle scattering on a screen with a slit. Particles are emitted from the source e, pass through a slit in screen S_1 and gather on screen S_2. In the region δ near the slit there is a nonzero force field which affects the particles

the region δ effectively covers the whole space between the screens S_1 and S_2. Although such a case is rather nonphysical, in the general case it allows us to construct a simple model of the particle dynamics. We will come below to a more physically natural model. Let the source e be point-like, emitting particles with constant velocity v_0 under random evenly distributed angles $\alpha \in [0, 2\pi)$. In this case trajectories of particles emitted by e (in discrete time dynamics) in the region $x < 0$ (i.e. before the screen S_1) form concentric circles originating from e with the radii $r_n = v_0 \tau n$, $n = 1, 2, \ldots$ (Fig. 12.2). We get the circles—which are trajectories of many particles emitted under close angles—in the region where there is no force field and particles move along straight lines, exactly as in classical dynamics. Let a_i be the points where these circles enter the region δ. The distance between the center of the slit and a_i is given by

$$a_i = \operatorname{sign}(i)\sqrt{v_0^2 \tau^2 (n_0 + |i|)^2 - d^2}, \quad i = \pm 1, \pm 2, \ldots, \tag{12.5}$$

where d is the distance between e and S_1, and n_0 is the largest integral value not greater than a fraction $d / v_0 \tau$,

$$n_0 = \left\lfloor \frac{d}{v_0 \tau} \right\rfloor. \tag{12.6}$$

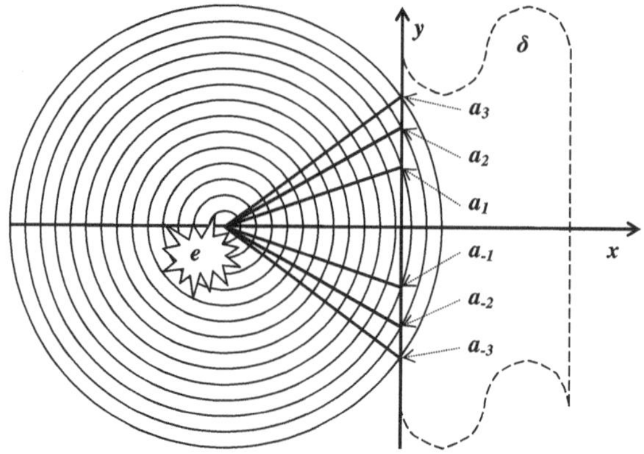

Fig. 12.2 Particles emitted from the source e enter the region δ. Points $a_{\pm 1}$, $a_{\pm 2}$, $a_{\pm 3}$, ... are origins of deviation which form a diffraction pattern

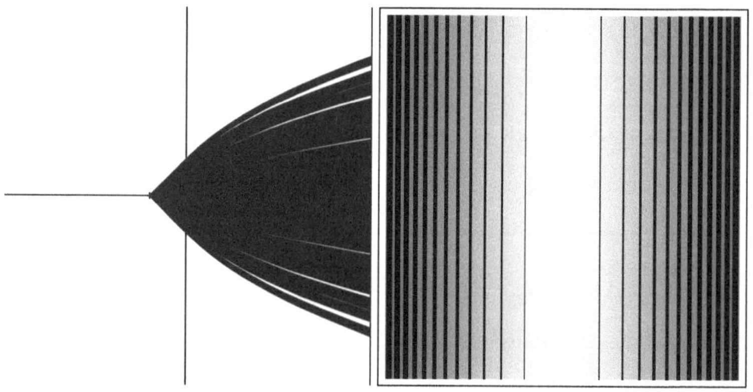

Fig. 12.3 Particle trajectories (*left*) and diffraction pattern (*right*)—*brighter area* corresponds to the higher particle density—and for a single-slit scattering computed in the discrete time formalism

This rather simple setup already produces an interesting nontrivial diffraction pattern (Fig. 12.3).

The points a_i are the origins of deviation from 'classical' (i.e. continuous time) trajectories. This deviation forms a diffraction pattern. To argue this let φ_i be an angle between the horizontal axis and a line connecting e and a_i. Particles emitted under angles less than φ_i ($i \geq 1$) become affected by the force field in the region δ one step earlier than those emitted under angles greater than φ_i. As a result there appears a 'fork'—even very close trajectories but with the angles above and below φ_i becoming separated. One could get the points of minima of the diffraction pattern by following the trajectory along the line $e \to a_i$ and the parabolic curve (movement in a constant field) in the region $x > 0$ until the trajectory hits the screen S_2.

The case described above is not completely physically reasonable, indeed it requires a force field everywhere in the half-plane $x \geq 0$, thus it could be considered only as a simplified model which still demonstrates some interesting properties of discrete time dynamics. An example of such a property is a Fresnel-like phenomenon of a black region behind the center of the slit which is discussed below.

One could see from (12.5)–(12.6) that if

$$d = v_0 \tau k + \varepsilon, \quad \text{where } k = 1, 2, \ldots$$

and $\varepsilon \to 0$, then $|a_{+1} - a_{-1}| \to 0$, This means that particle trajectories become separated, forming a region without dots within the center of the screen S_2 just behind the slit (Fig. 12.4).

Now let us consider the case when a field is localized near the slit. We performed numerical simulations taking the force field of the Gaussian form

$$\mathbf{F}(x, y) = \mathbf{e}_x \, F_0 e^{-\sigma x^2}. \tag{12.7}$$

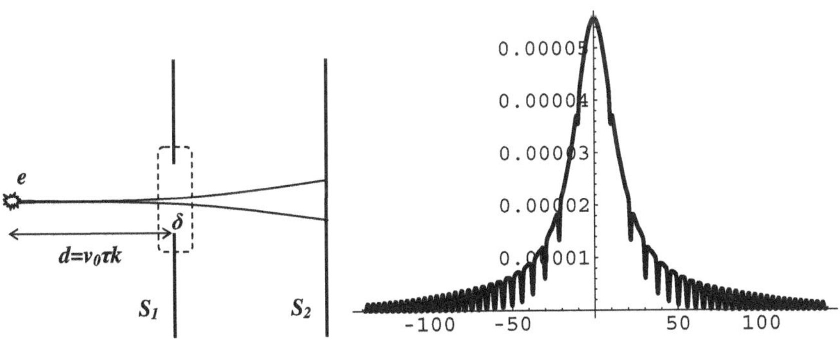

Fig. 12.4 Fresnel-like phenomenon of a black region behind the center of the slit in the discrete time formalism (*left*) and diffraction pattern for the Gaussian force (12.7)—*right*

The results of this simulation again show a diffraction-like picture. It is very stimulating that such a setup also produces nontrivial patterns, since it means that one could start thinking of the physical nature of this force field. We have also considered the case of a non-point-like Gaussian source distributed along the vertical line; results of this simulation are presented in Fig. 12.4.

Conclusion. We have shown that a QL diffraction picture could appear as a statistical effect for deterministic particles, i.e., having trajectories and obeying deterministic equations, if one introduces a discrete time. The nature of the resulting diffraction picture (particle distribution) does not follow from the geometry of a force field alone, but is strongly attached to the discreteness parameter τ.

12.3 Interference in the Two-Slit Experiment for Deterministic Particles

We now consider a classical analog of the two-slit experiment (Fig. 12.5). The uniformly charged round particles are emitted at the point e with fixed velocity with the angles evenly distributed in the range $[0, 2\pi)$. Each particle interacts with the uniformly charged flat screen S_1. The charge distributions on the particle and the screen stay unchanged even if the particle comes close to the screen. Physically this is a good approximation when the particle and the screen are both made of dielectric. There are two rectangular slits in the screen (on the Fig. 12.5 the slits are perpendicular to the plane of the picture). Particles pass through the slits in screen S_1 and gather on screen S_2.

We consider three experiments. In the first one the bottom slit is closed with the shutter, in the second the upper slit is closed, and in the third both slits are left open. The charge distribution on the shutter is the same as on the screen, i.e. in the first

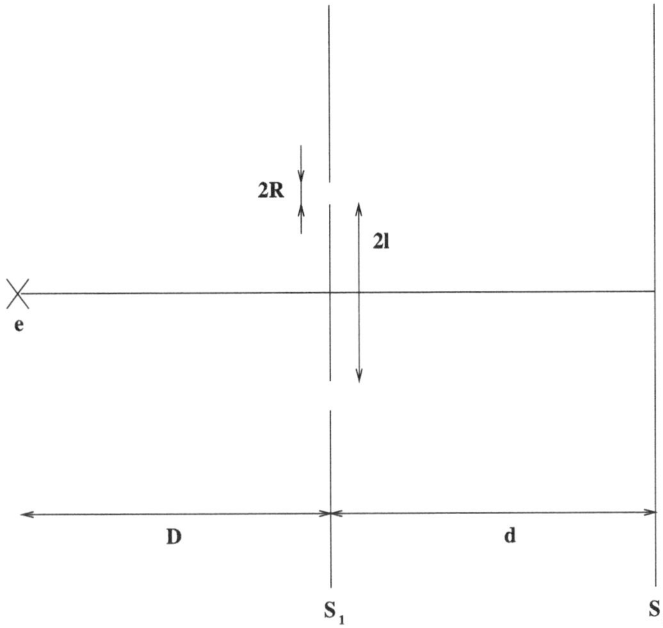

Fig. 12.5 Two-slit experiment

two experiments one can consider the uniformly charged screen to have only one slit. In this and several paragraphs below, by the screen we mean screen S_1.

Now let us write the equations of motion in each of three experiments ($i = 1, 2, 3$)

$$m\ddot{\mathbf{r}} = \mathbf{F_i} \tag{12.8}$$

where $\mathbf{r}$ determines the place of the particle. Here $\mathbf{F_i}$ is force affecting the particle in each experiment. It is given by Coulomb's law

$$\mathbf{F_i} = \int_{\mathscr{D}_i} \frac{q\sigma}{|\mathbf{r}'|^2} \cdot \frac{\mathbf{r}'}{|\mathbf{r}'|} ds \tag{12.9}$$

where $\mathbf{r}'$ is a vector from an element on the screen to the particle, q is the charge of the particle, σ is charge density on the screen, i.e. charge of a unit square. We integrate over the surface of the screen, the integration region $\mathscr{D}_i$ is the plane of the screen except the slits; as mentioned above it is different in each experiment depending on which slits are opened.

Projecting equations (12.8)–(12.9) to the xy-plane, where x and y denote horizontal and vertical coordinates of the particle respectively, we get

$$\begin{aligned} m\ddot{x} &= q\sigma \int\limits_{\Gamma_i} dy' \int\limits_{\mathbb{R}} dz' \frac{x}{(x^2 + (y - y')^2 + z'^2)^{3/2}}, \\ m\ddot{y} &= q\sigma \int\limits_{\Gamma_i} dy' \int\limits_{\mathbb{R}} dz' \frac{y - y'}{(x^2 + (y - y')^2 + z'^2)^{3/2}}, \end{aligned} \tag{12.10}$$

where Γ_i indicates the integration region for the i-th experiment. In our previous notation $\mathscr{D}_i = \Gamma_i \times \mathbb{R}$. We have

$$\begin{aligned} \Gamma_1 &= (-\infty, l) \cup (l + 2R, +\infty), \\ \Gamma_2 &= (-\infty, -l - 2R) \cup (-l, +\infty), \\ \Gamma_3 &= (-\infty, -l - 2R) \cup (-l, l) \cup (l + 2R, +\infty). \end{aligned} \tag{12.11}$$

Here $2l$ is the distance between slits and $2R$ is the height of the slit.

Integrating the rhs of (12.10) we get

$$\begin{aligned} m\ddot{x} &= q\sigma \sum_{(a,b)\subset\Gamma_i} 2\left(\arctan\frac{b - y}{x} - \arctan\frac{a - y}{x}\right), \\ m\ddot{y} &= q\sigma \sum_{(a,b)\subset\Gamma_i} \ln(x^2 + (b - y)^2) - \ln(x^2 + (a - y)^2), \end{aligned} \tag{12.12}$$

where the notation $(a, b) \subset \Gamma_i$ means that the sum extends over all subranges of Γ_i given in (12.11). For example for $i = 1$ we have two summands with $(a = -\infty, b = l)$ and $(a = l + 2R, b = +\infty)$, and (12.12) will take the form

$$\begin{aligned} m\ddot{x} &= 2q\sigma\left(\pi + \arctan\frac{l - y}{x} - \arctan\frac{l + 2R - y}{x}\right), \\ m\ddot{y} &= q\sigma \ln\frac{x^2 + (l - y)^2}{x^2 + (l + 2R - y)^2}. \end{aligned} \tag{12.13}$$

Here we took into account that $\arctan(\pm\infty) = \pm\pi/2$ and the sum of the logarithms for $a = -\infty$ and $b = +\infty$ vanishes.

We take the initial values

$$\begin{aligned} x(0) &= -D, & \dot{x}(0) &= v_0 \cos\alpha, \\ y(0) &= 0, & \dot{y}(0) &= v_0 \sin\alpha, \end{aligned} \tag{12.14}$$

where angle α is a random variable uniformly distributed in $[0, 2\pi)$. The constant parameters v_0 and D are initial velocity and distance between the emitter and the screen.

Particles are emitted at point e (see Fig. 12.5), move obeying (12.12), (12.14) passing through slit(s) in the screen S_1 and gather on the screen S_2. Having points where particles hit the screen S_2, we compute frequencies with which particles appear on screen S_2 as a function of coordinates on the screen. We interpret these frequencies as probability distributions. We are interested in computing the probability distribution over a vertical line on screen S_2 with $z = 0$. That is why we consider a motion only in the xy-plane and initial values (12.14) do not contain a z-coordinate.

We solve the equations of motion (12.12) with initial conditions (12.14) numerically. We use Runge-Kutta 4th order switching to Adams 4th order method. We used a GNU C++ (g++) compiler to realize the simulation on an Ultra-SPARC computer running Solaris. We had to explore about 10^5 trajectories and we used a 4-processor parallel computer located at Växjö University. The computation process was easy to make parallel as moving particles are not interacting, i.e. one could consider them as being emitted with long intervals. The algorithm automatically adjusted the computation precision, making shorter steps when the particle came near to the first screen or the coordinates (x and y) are changed more than minimum precision allowed. The first stage of computation was calibration when the algorithm determined the angle ranges for which the particles passed through the slits and hit the second screen. This reduced the angle range from $[0, 2\pi)$ to a set of ranges, which are different in each experiment. In fact we used symmetry of the first two emissions (when only upper

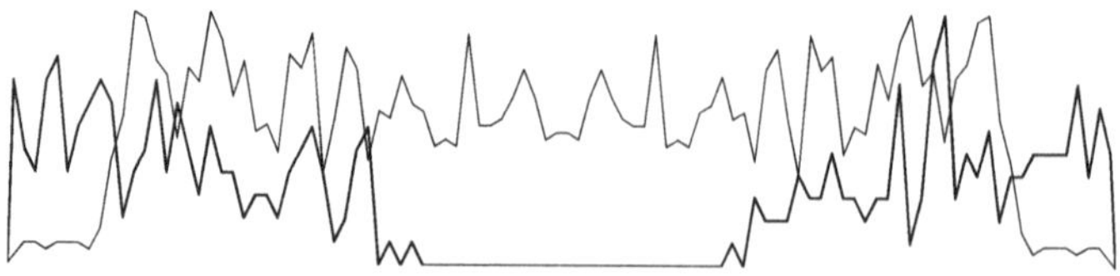

Fig. 12.6 Probability distributions. *Thin line* is $(\frac{P_1}{2} + \frac{P_2}{2})$, *thick line* is P_{12}

or lower slit is opened) making computations only for the first emission. The second screen was separated with cells of equal size, the diameter of a particle. The number of particles which hit each cell was calculated and interpreted as a probability distribution.

Remark 12.1. In fact, we present in this picture densities for numbers of particles, i.e., not normalized distributions. In particular, this implies (due to errors of statistical stabilization) asymmetry of densities. This asymmetry would disappear after normalization when the number of particles goes to infinity.

Let us denote the probability distribution in the first experiment (only the upper slit is opened) as $P_1 = P_1(y)$, in the second experiment (only the lower slit is opened) as $P_2 = P_2(y)$, and in the third experiment (both slits are opened) as $P_{12} = P_{12}(y)$. Although the force is different in each experiment, it is quite clear that (Fig. 12.6)

$$P_{12} \neq \frac{P_1}{2} + \frac{P_2}{2}. \tag{12.15}$$

To become an equality the above equation should have an extra term,

$$P_{12} = \frac{P_1}{2} + \frac{P_2}{2} + \sqrt{P_1 P_2}\cos\theta, \tag{12.16}$$

where $\sqrt{P_1 P_2}\cos\theta$ is a so-called interference term (Fig. 12.7), and $\theta = \theta(y)$ is spread along the y-axis.

The function

$$\cos\theta = \frac{2P_{12} - (P_1 + P_2)}{\sqrt{P_1 P_2}} \tag{12.17}$$

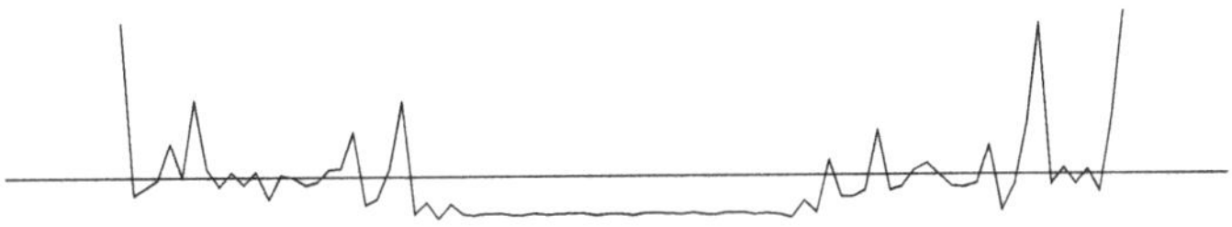

Fig. 12.7 Interference term, $\cos\theta = (2P_{12} - (P_1 + P_2))/\sqrt{P_1 P_2}$

is shown in (Fig. 12.7). Please note that as there are ranges where P_1 or P_2 are equal to zero, i.e. $P_1 P_2 = 0$, the function $\cos\theta$ is not determined and from (12.16) we see that P_{12} does not depend on it.

12.4 Physical Interpretation of Results of Computer Simulation

We provided a QL-description of the two-slit experiment with charged macroscopic bodies. In this mathematical description there are presented all features of the standard quantum mechanical description: interference, phase shift, wave function,... However, it seems that from the physical point of view our macroscopic version of the two-slit experiment crucially differs from the quantum two-slit experiment—at least if you use the conventional interpretation of quantum mechanics.

By the conventional interpretation of quantum mechanics one has the following picture (Heisenberg, Bohr, Dirac,..., Peres, Mermin, Zeilinger, Leggett,...).

"It is especially impossible in principle to predict with certainty both through which slit an elementary particle will go and where it will appear in the interference pattern."

An elementary particle does not have a trajectory... It is impossible to predict an individual event, in the sense that it is not possible, not even in principle, to arrive at an accurate and detailed prediction and description of a particular process resulting in a particular event.

In our experiment the presence of the QL-representation of statistical data does not imply the impossibility of an individual description by using Newton equations. Here it is possible to predict with certainty both through which slit a charged macro-system will go and where it will appear in the interference pattern. However, in general it is possible only theoretically. If a charged system is very sensitive to our measurement devices (e.g., it has small mass and large charge), then it would be impossible to verify experimentally through which slit the charged macro-system will go without disturbing or even destroying the interference pattern. Our experiment demonstrated that the theoretical possibility of an individual description does not imply that we can escape a purely probabilistic description. Here randomness also is a fundamental characteristic.

Conclusion. We have shown that the proposed classical model of the two-slit experiment (based on the Postulate (D)) has a context dependence and could be adequately described with contextual formalism. The QL-behavior for macro-systems is demonstrated. We simulated the QL-interference for macroscopic objects. Such a simulation essentially reduced the gap between micro- and macro-worlds.

The two-slit experiment with charged macroscopic bodies demonstrated that wavelike descriptions of experimental statistical data and experimental nonreducibility of randomness do not imply automatically the impossibility of a description of individual processes.

12.5 Discrete Time Dynamics

In classical mechanics a dynamical function $A = A(p, q)$ (here p and q are momenta and coordinates of the system) evolves according to the equation

$$D_t A = \{A, H\} \tag{12.18}$$

where $H = H(p, q)$ is a Hamiltonian of the system and in the right-hand side is a Poisson bracket, which could be presented as

$$\{A, B\} = \frac{\partial A}{\partial q}\frac{\partial B}{\partial p} - \frac{\partial A}{\partial p}\frac{\partial B}{\partial q}. \tag{12.19}$$

The left-hand side of (12.18) contains a continuous time derivative

$$D_t A = \frac{dA}{dt}.$$

As was mentioned earlier, we are interested in the study of dynamics with discrete time. This is done with the help of a *discrete derivative* which is postulated to be

$$D_t^{(\tau)} A = \frac{1}{\tau}[A(t+\tau) - A(t)],$$

where τ is the discreteness parameter. This parameter is finite and is treated in the same way as Planck constant in quantum mechanical formalism. In particular if τ is small relative to dimensions of the system, then classical approximation with a continuous derivative might work well (although this could not be the case all the time in the same sense as there are examples when quantum formalism is reasonable even for macroscopic systems, for example in superfluidity).

Summarizing, the discrete time dynamical equation is postulated to be

$$D_t^{(\tau)} A = \{A, H\}, \tag{12.20}$$

where $A(p, q)$ is a real-valued function of real-valued dynamical variables and in the right-hand side there is a classical Poisson bracket (12.19). Equation (12.20) could be solved in the sense that we can write

$$A(t+\tau) = A(t) + \tau\{A, H\} \tag{12.21}$$

thus providing the evolution of any dynamical function $A = A(p, q)$.

Note that in our model the coordinate space is continuous.

12.6 Motion in Central Potential

Here we will study the properties of motion in a central potential $U = U(r)$ in discrete time formalism. As we will see, discreteness of time enriches mechanics with some new properties which are usually thought of as having a quantum nature. In particular, as it will be shown below, in discrete time mechanics stationary orbits (i.e. finite motion) have discrete energy spectrum. We point out that the phase space is assumed here to be a continuous real manifold.

Following the general approach described in the previous section we start from the classical Hamiltonian and then write the dynamical equations. In polar coordinates (r, φ) the Hamiltonian of the system with mass m in central potential $U(r)$ is given by

$$H(r, p_r, \varphi, p_\varphi) = \frac{p_r^2}{2m} + \frac{p_\varphi^2}{2mr^2} + U(r), \tag{12.22}$$

where p_r and p_φ denote momenta corresponding to r and φ—radial and angular coordinates respectively. Using (12.21) let us write the dynamical equations. We obtain

$$r(t+\tau) = r(t) + \tau\frac{p_r}{m}, \tag{12.23}$$

$$p_r(t+\tau) = p_r(t) + \tau\left(\frac{p_\varphi^2}{mr^3} - \frac{\partial U}{\partial r}\right), \tag{12.24}$$

$$\varphi(t+\tau) = \varphi(t) + \tau\frac{p_\varphi}{mr^2}, \tag{12.25}$$

$$p_\varphi(t+\tau) = p_\varphi(t). \tag{12.26}$$

The equation (12.26) corresponds to angular momentum conservation in the central field—this is a direct analog of the angular momentum conservation law in classical (continuous time) dynamics and is a consequence of the fact that our Hamiltonian does not depend on φ, (it is a so-called *cyclic variable*).

Let us limit ourselves to circular stationary periodic orbits. In this case we should have $r(t+\tau) = r(t)$ and thus using (12.23) we see that the radial momentum should

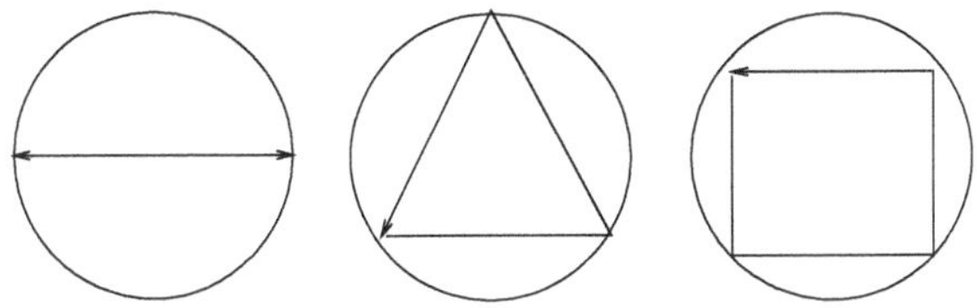

Fig. 12.8 First three trajectories

be zero,

$$p_r(t) = 0. \tag{12.27}$$

From (12.24) and (12.27) we obtain the condition

$$\frac{p_\varphi^2}{mr^3} = \frac{\partial U}{\partial r}. \tag{12.28}$$

Let us finally come to the angular coordinate φ. For stable motion the following *periodicity* condition should be satisfied (Fig. 12.8):

$$\varphi(n\tau) = \varphi(0) + 2\pi, \tag{12.29}$$

where $n = 1, 2, \ldots$ (note that discreteness of n is a consequence of discreteness of time). Using (12.25) and (12.29) we get

$$n\tau \frac{p_\varphi}{mr^2} = 2\pi,$$

or

$$p_\varphi = \frac{2\pi mr^2}{n\tau}. \tag{12.30}$$

From (12.28) and (12.30) we get the following equation for the radius of the n-th orbit

$$\frac{4\pi^2 mr}{n^2\tau^2} = \frac{\partial U}{\partial r}. \tag{12.31}$$

If potential $U(r)$ is known, then from equation (12.31) we can find $r = r_n$. If we consider physical potentials, i.e. potentials for which the force, $-U'(r)$, is smooth, negative, and is strictly monotonically decreasing in absolute value as r grows, vanishing on infinity, then a solution of (12.31) always exists and is unique. Now upon substituting r_n for the Hamiltonian (12.22) we obtain energy levels E_n.

For circular periodic orbits the original Hamiltonian (12.22) due to (12.27) and (12.28) simplifies to the form

$$H = \frac{1}{2} r \frac{\partial U}{\partial r} + U(r).$$

Thus the expression for energy E_n of the n-th stable periodic orbit in terms of its radius r_n is given by

$$E_n = \frac{1}{2} r_n \frac{\partial U}{\partial r}\bigg|_{r=r_n} + U(r_n), \quad n = 1, 2, \ldots. \tag{12.32}$$

As we see, if potential $U(r)$ allows stationary periodic motion and (12.31) has unique positive solutions r_n, then the energy spectrum is discrete. This situation is directly analogous to quantum mechanics where for finite motion we might expect discrete energy levels.

12.7 Energy Levels of the Hydrogen Atom

Our task in this section is to study whether the discrete time dynamics in a central field can lead to determination of the energy levels of the hydrogen atom. We treat energy levels as given (measured) quantities and our task is to find the corresponding potential. We restrict ourselves to the simplest case when the atom is unperturbed by external electric or magnetic fields and thus currently we do not study splitting of its energy levels (note that in order to observe "degenerate" levels in a quantum mechanical treatment one needs to somehow perturb the system in such a way that the levels split).

We start from the following energy spectrum for the hydrogen atom

$$E_n = -\frac{\gamma}{n^2}, \quad n = 1, 2, \ldots, \tag{12.33}$$

where $\gamma \approx 13.6$ eV is ionization energy of the hydrogen atom. The task is to find such $U = U(r)$ that leads to the spectrum (12.33). Using (12.31), (12.32), and (12.33) we get

$$\frac{1}{2}\xi\frac{r_n^2}{n^2} + U(r_n) = -\frac{\gamma}{n^2}, \tag{12.34}$$

where the constant ξ is given by (note that ξ depends on discreteness parameter τ)

$$\xi = \frac{4\pi^2 m}{\tau^2} \tag{12.35}$$

and $n = 1, 2, \ldots$. Let us rewrite equation (12.34) in the form

$$U(r_n) = -\frac{1}{n^2}\left(\frac{1}{2}\xi r_n^2 + \gamma\right). \tag{12.36}$$

We want to find $U(r)$. The idea is to obtain dependence of $r_n = f(n)$ on n and then, inverting it substitute $n = f^{-1}(r_n)$ in (12.36); here f^{-1} denotes the function inverse to f. As a result of this procedure we will get rid of explicit dependence of the right-hand side of (12.36) on n, it will depend only on r_n. Then we interpolate the result for any $r \geq 0$. The resulting $U(r)$ we check by substitution into (12.31)–(12.32).

Let us proceed as described above. First, we have to find dependence of r_n on n. To do this let us assume that n is continuous and take the derivative in n of both sides of (12.36). This gives

$$\left.\frac{\partial U}{\partial r}\right|_{r=r_n}\frac{dr_n}{dn} - \frac{\xi r_n^2}{n^3} - \frac{\xi r_n r_n'}{n^2} + \frac{2\gamma}{n^3} \tag{12.37}$$

Now, from (12.31) we find

$$\left.\frac{\partial U}{\partial r}\right|_{r=r_n} = \xi\frac{r_n}{n^2}. \tag{12.38}$$

Substituting (12.38) into (12.37) we obtain the following differential equation for r_n

$$2\xi n r_n r_n' - \xi r_n^2 - 2\gamma = 0. \tag{12.39}$$

Solving (12.39) and taking into consideration only positive solutions we obtain

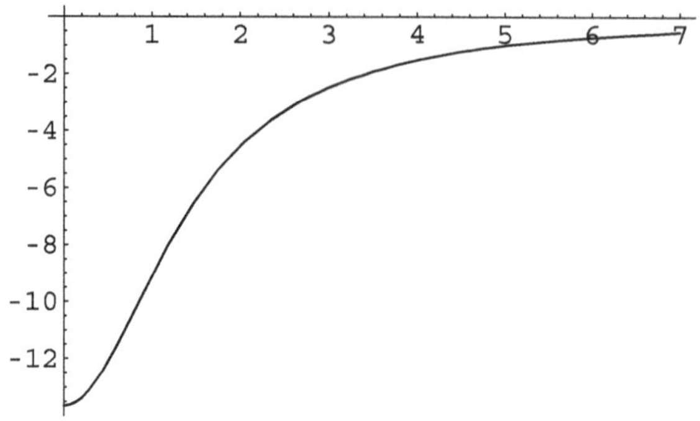

Fig. 12.9 The shape of potential (12.42) which leads to energy levels of the hydrogen atom (we put $\beta = \xi = \gamma = 1$)

$$r_n = \sqrt{\frac{-2\gamma + e^{2\beta\xi} n}{\xi}}, \tag{12.40}$$

where the constant β is due to integration. Inverting (12.40) we get

$$n = e^{-2\beta\xi}(2\gamma + r_n^2 \xi). \tag{12.41}$$

Substituting (12.41) into (12.36) we obtain

$$U(r_n) = -\frac{e^{4\beta\xi}}{4\gamma + 2r_n^2 \xi}$$

or performing interpolation, i.e. putting r in place of r_n, we finally get

$$U(r) = -\frac{e^{4\beta\xi}}{4\gamma + 2r^2 \xi}. \tag{12.42}$$

Now, substituting (12.42) into (12.31)–(12.32) we come to the expected energy levels (12.33) of the hydrogen atom.

The form of the potential (12.42) is presented on Fig. 12.9. It is interesting to note that unlike Coulomb's potential it is nonsingular at $r = 0$.

12.8 Spectrum of Harmonic Oscillator

The procedure described in the previous section could be used to obtain potentials corresponding to an arbitrary energy spectrum. Here we deduce the potential which results in linear energy levels of the homogeneous two-dimensional quantum harmonic oscillator.

In quantum mechanics, a homogeneous two-dimensional harmonic oscillator is a system described by the potential

$$U(r) = \frac{1}{2} m\omega^2 r^2. \tag{12.43}$$

Solving the Schrödinger equation in potential (12.43) results in the energy spectrum

$$E_\Lambda = \hbar\omega(\Lambda + 1), \quad \Lambda = 0, 1, \dots. \tag{12.44}$$

The simplest way to get this relation is to note that we deal with two uncoupled oscillators (indeed, if x and y are Cartesian coordinates, then (12.43) takes the form $U = \frac{1}{2} m\omega^2(x^2 + y^2)$), each having energy $E_k = \hbar\omega(k + \frac{1}{2})$, then the total energy is just the sum of two such terms and we get (12.44).

Let us first rewrite (12.44) in terms of $n = 1, 2, \dots$; we have $n = \Lambda + 1$ and thus

$$E_n = \alpha n, \quad n = 1, 2, \dots, \tag{12.45}$$

where $\alpha = \hbar\omega$. Our task is to find potential $U = U(r)$ which result in energy levels (12.45). Proceeding as in the previous section we assume that n is continuous and write the differential equation

$$\frac{dE_n}{dn} = \alpha. \tag{12.46}$$

Now, using (12.31) and (12.32) equation (12.46) reduces to

$$2\xi \frac{r_n r_n'}{n^2} - \xi \frac{r_n^2}{n^3} = \alpha, \tag{12.47}$$

where ξ is given by (12.35). The positive solution of (12.47) is given by

$$r_n = \sqrt{\beta n + \frac{\alpha n^3}{2\xi}}, \tag{12.48}$$

where β is a constant due to integration. This leads us to a cubic equation in terms of n, which has one real solution

$$n = \frac{\mathscr{V}(r_n)}{3\alpha} - \frac{2\beta\xi}{\mathscr{V}(r_n)},$$

where $\mathscr{V}(r)$ is given by

$$\mathscr{V}(r) = 3^{1/3}\left(9r^2\alpha^2\xi + \sqrt{3}\sqrt{27r^4\alpha^4\xi^2 + 8\alpha^3\beta^3\xi^3}\right)^{1/3}.$$

We finally get the expression for potential

$$U(r) = \frac{\mathscr{V}(r)}{3} - \frac{2\alpha\beta\xi}{\mathscr{V}(r)} - \frac{9r^2\mathscr{V}(r)^2\alpha^2\xi}{2(\mathscr{V}(r)^2 - 6\alpha\beta\xi)^2}.$$

If we put $\beta = 0$ in (12.48) the potential takes the simple form

$$U(r) = \frac{3}{2}\left(\frac{r^2\alpha^2\xi}{4}\right)^{1/3} \tag{12.49}$$

Substituting (12.49) to (12.31)–(12.32) we get correct energy levels (12.45). It is interesting to then note that, when all constants are substituted, expression (12.49) takes the form

$$U(r) = \frac{3}{2}\left(r\frac{\hbar\omega\pi\sqrt{m}}{\tau}\right)^{2/3}$$

12.9 General Case of Arbitrary Spectrum

One can find explicit relations for the radii of the orbits for an arbitrary spectrum E_n. Indeed, from (12.31) and (12.32) we want to find U as a function of r in terms of a given energy spectrum E_n. As we did above, we assume continuity of the parameter n and take derivatives of both parts of (12.32) in n, which gives us

$$E'_n = r'_n \frac{\xi r_n}{n^2} - \frac{\xi r_n^2}{n^3} + r'_n \frac{\partial U}{\partial r}\bigg|_{r=r_n},$$

where prime denotes the derivative in n. The use of (12.31) allows us to get a differential equation for r_n in terms of only known quantities. We then have

$$2r_n r'_n - \frac{1}{n} r_n^2 = \frac{n^2}{\xi} E'_n.$$

Introducing a new variable $\rho = r^2$ we obtain a linear differential equation which could be rewritten as

$$\frac{d}{dn}\left(\frac{\rho_n}{n}\right) = \frac{1}{\xi} n E'_n$$

which can be integrated to obtain

$$r_n = \sqrt{\frac{1}{\xi} n \left(n E_n - E_1 - \int_1^n E_k dk + \varepsilon \right)}. \tag{12.50}$$

Equation (12.50) expresses the n-th radius in terms of n, i.e. it has the form $r = f(n)$. Now if we invert it we relate n in terms of r, $n = f^{-1}(r)$, which if substituted into (12.32) gives an equation for U in terms of r only (actually in terms of r_n, but we perform interpolation, effectively ignoring the fact that the relation strictly holds only for orbit radii).

Note that in (12.50) we have a constant ε (having units of energy) arising due to the integration, this means that we have a set of potentials resulting in the same energy spectrum. As we see from (12.50) the constant ε could be determined if for example the smallest radius r_1 is known. In Sects. 12.7 and 12.8 the situation was the same resulting in the constant β (see (12.40) and (12.48)). Since ε is more interesting from the point of view of physical interpretation, we provide the expressions for β in terms of ε. For the case of energy levels of the hydrogen atom we have (see (12.40))

$$\beta_{hydr} - \frac{1}{2\xi} \ln(\varepsilon + \gamma)$$

and for the case of a harmonic oscillator we have (see (12.48))

$$\beta_{osc} = \frac{\varepsilon - \frac{1}{2}\alpha}{\xi}.$$

12.10 Energy Spectrum in Various Potentials

As we already seen, for a given potential it is straightforward to compute corresponding energy levels. Indeed, from (12.31) we find $r = r_n$ and upon substitution to (12.32) we get E_n. As in quantum mechanics, potential $U = U(r)$ should be attractive and strong enough to result in finite motion. Below we consider several common central potentials which result in rather simple expressions for an energy spectrum. In what follows the constant ξ is given by (12.35), note that it depends on the time discreteness parameter $1/\tau^2$.

(a) Coulomb potential

$$U(r) = -\frac{\alpha}{r}, \qquad r_n = n^{2/3}\left(\frac{\alpha}{\xi}\right)^{1/3}, \qquad E_n = -\frac{1}{2n^{2/3}}(\alpha^2\xi)^{1/3}.$$

Note that the energy spectrum is different from the $-\gamma/n^2$ spectrum obtained in quantum mechanics for this potential (see section on energy levels of the hydrogen atom for detailed discussion).

(b) Linear potential

$$U(r) = \alpha r, \qquad r_n = \frac{n^2\alpha}{\xi}, \qquad E_n = \frac{3n^2\alpha^2}{2\xi}.$$

(c) Logarithmic potential

$$U(r) = \alpha \ln r, \qquad r_n = n\sqrt{\frac{\alpha}{\xi}}, \qquad E_n = \alpha\left[\frac{1}{2} + \ln\left(n\sqrt{\frac{\alpha}{\xi}}\right)\right].$$

For potentials (a), (b), and (c), $r_n > 0$ if $\alpha > 0$; in all three cases it corresponds to the attraction field.

(d) Polynomial potential

$$U(r) = \alpha r^{\sigma}, \qquad r_n = \left(\frac{n^2\alpha\sigma}{\xi}\right)^{\frac{1}{2-\sigma}}, \qquad E_n = \frac{1}{2}\alpha(2+\sigma)\left(\frac{n^2\alpha\sigma}{\xi}\right)^{\frac{\sigma}{2-\sigma}}. \tag{12.51}$$

This case generalizes cases (a) and (b) described above, although because of importance of these potentials we have provided corresponding expressions explicitly. Note that if we make σ in (12.51) satisfy the equation

$$\frac{2\sigma}{2-\sigma} = 1,$$

i.e. $\sigma = 2/3$, then we get the linear energy spectrum $E_n \sim n$ for the 2D quantum harmonic oscillator (see previous section for detailed discussion).

12.11 Discussion and Conclusion

We have shown that the discrete-time formalism leads to some distinguishable properties of micro-observables that are used to being described by quantum mechanics. In particular, it was shown that finite motion results in a discrete energy spectrum. Of main interest in this chapter are discrete energy levels of the hydrogen atom. We have shown that for the unperturbed hydrogen atom the discrete time formalism is able to give a correct energy spectrum, more precisely we have reconstructed the corresponding "micro"-potential. Here we did not consider Stark or Zeeman effects for the hydrogen atom; it would be interesting to study them also from the point of view of the discrete time formalism.

As we have seen, the discrete time model requires potentials which are different from QM potentials. One may argue this as a disadvantage of the model. We remark that there are no reasons to expect to reproduce QM by using the standard classical potentials. Bohr, Zommerfeld, Heisenberg and many others tried to do this, but they did not succeed. D. Bohm developed a new model (see, e.g., [38, 126] for details) in which quantum mechanics can be reproduced on the classical basis, but, of course, the classical potential could not be preserved—it is perturbed by the quantum potential. And the latter looks not so natural from the classical viewpoint, see, e.g., the quantum potential for the two-slit experiment in [38, 126]. The discrete time model has an analogy with Bohmian mechanics—it tries to reproduce QM by changing potentials. There is, of course, a fundamental difference: the only postulate that is used in the proposed approach is that there exists a quant of time τ.

Another interesting point is that one might expect that our dynamical equations are essentially difference equations which might produce a discrete spectrum. This is not correct—we recall that in our model only time is discrete, but space is still continuous.

There is still an open problem of the quantitative value of the discreteness parameter τ. One might speculate its relation with Planck's time constant—the smallest measurable time interval in ordinary QM and gravity—which is quantitatively given by

$$t_{Pl} = \sqrt{\frac{\hbar G}{c^5}} \approx 5.3910^{-44} \text{ (s)}.$$

Finally, we would like to comment also that there might be a deep interrelation between the energy-time uncertainty relations and Bohr-Sommerfeld quantization rules in quantum mechanics and our discrete time model. In particular one can try to write the Bohr-Sommerfeld semi-classical quantization rules for energy and time as canonical variables. For a system with conserved energy one might get $E_n T_n \sim n\hbar$; this relation holds for example for energy levels E_n and classical periods T_n of the hydrogen atom. On the other hand, the relation (12.29) in discrete mechanics could be treated as the condition for the period to take only discrete values $T_n \sim n\tau$. We can see that, although relations are similar, there is an extra factor E_n in the quantum-mechanical relation. In fact, one may argue that if we make the τ in equations of motion depend on the energy of the system as

$$\tau = \tau_0 \frac{\varepsilon}{E},$$

where τ_0 is the "fundamental" time quantum and ε a "fundamental" energy quantum, we get precisely the semiclassical quantization rules. The question arises of how to treat the energy E here and what will happen with the dynamics.

Chapter 13

Noncommutative Probability in Classical Disordered Systems

As was already mentioned in the previous chapters, our contextual probabilistic investigations demonstrated that, in fact, the gap between classical statistical physics and quantum physics is not so huge as it was commonly believed. This may stimulate not only further thinking about the classical probabilistic derivations of the main distinguishing features of quantum theory, but also investigations in the opposite direction: quantum probabilistic features of classical statistical mechanics. The first step in this direction was taken in [158] by the author and Sergei Kozyrev (Steklov Mathematical Institute of Russian Academy of Science). We investigated the following problem:

"Is it possible to observe correlations of noncommutative observables in a purely classical situation?"

This would mean that the classical system will be described by a noncommutative (or QL) probability model.

In Part II we demonstrated that, by taking into account dependence of probabilities on complexes of experimental physical conditions, *physical contexts,* we can derive the QL-interference. Such a contextual derivation is not directly related to special quantum (e.g. superposition) features of physical systems. We now discuss two examples of *classical statistical mechanical systems* where we obtain the correlation functions in *noncommutative probability space.*

The structure of the present chapter is as follows.

In Sect. 13.1 we investigate the example of noncommutative probability for classical observables as a result of *time averaging.*

In Sect. 13.2 we consider the correlation functions on noncommutative probability space for classical *disordered systems.* The noncommutativity there will be a result of *ensemble averaging.* Also in this section we discuss the considered examples from the point of view of the context dependent interpretation of noncommutative probability.

13.1 Noncommutative Probability and Time Averaging

In the present section we discuss the following problem. Consider the dynamics of a quantum system, described by some Hamiltonian H_0 and the algebra of observables $\mathscr{A}$. Let this (noncommutative) algebra of observables $\mathscr{A}$ contain some (commutative) classical subalgebra $\mathscr{X}$. This classical subalgebra is not conserved by time evolution, but for $X, Y \in \mathscr{C}$ the time evolutions $X(t) = e^{itH_0} X e^{-itH_0}$ and $Y(t) = e^{itH_0} Y e^{-itH_0}$ will commute by definition.

Let us assume that the time evolution defined by Hamiltonian H_0 is very fast, and in experiment we observe some time averaged observables. These time averaged observables, in general, already will not commute, since the classical subalgebra is not conserved by time evolution. This means that, in principle, we might expect that these time averaged operators for classical physical variables $X(t)$, $Y(t)$ from the classical subalgebra $\mathscr{X}$ will have nonclassical correlations.

A natural example of this kind of behavior is observed in the quantum stochastic limit approach. In this approach we consider a quantum system with the Hamiltonian in the form

$$H = H_0 + \lambda H_I$$

where H_0 is called the free Hamiltonian, H_I is called the interaction Hamiltonian, and $\lambda \in \mathbf{R}$ is the coupling constant.

We investigate the dynamics of the system in the new slow time scale of the stochastic limit, taking the van Hove time rescaling

$$t \mapsto t/\lambda^2$$

and considering the limit $\lambda \to 0$. In this limit [4] the free evolutions of the suitable collective operators

$$A(t,k) = e^{itH_0} A(k) e^{-itH_0}$$

will become quantum white noises

$$\lim_{\lambda \to 0} \frac{1}{\lambda} A\left(\frac{t}{\lambda^2}, k\right) = b(t,k).$$

The convergence is understood in the sense of correlators. The $\lambda \to 0$ limit describes the time averaging over infinitesimal intervals of time and allows us to investigate the dynamics on a large time scale, where the effects of interaction with the small coupling constant λ are important. For the details of the procedure see [4].

The collective operators describe joint excitations of different degrees of freedom in systems with interaction, and may have the form of polynomials over creation and annihilation of the field, or may look like combinations of the field and particle operators etc.

For example, for nonrelativistic quantum electrodynamics without the dipole approximation, the collective operator is

$$A_j(k) = e^{ikq} a_j(k) \tag{13.1}$$

where $a_j(k)$ is the annihilation of the electromagnetic (Bose) field with wave vector k and polarization j and $q = (q_1, q_2, q_3)$ is the position operator of a quantum particle (say an electron), $qk = \sum_i q_i k_i$.

The nontrivial fact is that, after the $\lambda \to 0$ limit, depending on the form of the collective operator, the statistics of the noise $b(t,k)$ depends on the form of the collective operator and may be nontrivial.

Consider the following examples.

(1) We may have the possibility

$$[b_i(t,k), b_j^\dagger(t',k')] = 2\pi\delta_{ij}\delta(t-t')\delta(k-k')\delta(\omega(k)-\omega_0), \tag{13.2}$$

which corresponds to the quantum electrodynamics in the dipole approximation, describing the interaction of the electromagnetic field with a two level atom with level spacing (energy difference of the levels) equal to ω_0. Here $\omega(k)$ is the dispersion of the quantum field.

In this case the quantum noise will have the Bose statistics, and different annihilations of the noise will commute:

$$[b_i(t,k), b_j(t',k')] = 0.$$

(2) Another possibility is the relation

$$b_i(t,k)b_j^\dagger(t',k') = 2\pi\delta_{ij}\delta(t-t')\delta(k-k')\delta(\omega(k)+\varepsilon(p)-\varepsilon(p+k)) \tag{13.3}$$

which corresponds to the quantum electrodynamics without the dipole approximation (13.1). Here $\omega(k)$ and $\varepsilon(p)$ are dispersion functions of the field and of the particle correspondingly.

In this case the quantum noise will have the quantum Boltzmann statistics [4], and different annihilations of the noise will not commute:

$$b_i(t,k)b_j(t',k') \neq b_j(t',k')b_i(t,k).$$

The commutation relations of types (13.2), (13.3) are universal in the stochastic limit approach (a lot of systems will have similar relations in the stochastic limit $\lambda \to 0$).

Take two operators $b_i(t,k)$ for the same time and fixed polarization and consider the combinations

$$X(k) = b_i(t,k) + b_i^\dagger(t,k), \qquad X(k') = b_i(t,k') + b_i^\dagger(t,k') \tag{13.4}$$

which correspond to the coordinate operator.

Then for the case of quantum Boltzmann relations (13.3) we have

$$X(k)X(k') \neq X(k')X(k).$$

Of course, we need some regularization of the product of generalized functions.

Operators $X(k)$ before the stochastic limit was performed belonged to the classical subalgebra. More precisely, corresponding combinations $x(k)$ of interacting operators (13.1) would belong to the classical subalgebra

$$x(k) = A_i(k) + A_i^\dagger(k),$$

$$[x(k), x(k')] = 0.$$

We proved that, for the case when after the stochastic limit the statistics of the field become quantum Boltzmannian, operators (13.4) will not commute even if we take them over equal time. This is not a mystery, since in the stochastic limit we work with time averaged observables. But in real experiments we may observe the result of time averaging. The discussed example shows that for quantum electrodynamics beyond the dipole approximation we may observe quantum correlations for time averaged observables in the classical subalgebra.

The situation considered in the present section is similar in some sense to the results of Chap. 12 of this part. There the QL diffraction and interference fringes were observed for discretization of the classical dynamical system. The discretization is an analog, in some sense, of the time averaging procedure. Probably those results might be possible to embed into the frameworks of the approach considered in the present section.

13.2 Noncommutative Probability and Disordered Systems

In the present section we discuss the possibility of using noncommutative probability to describe (classical) disordered systems, following [225, 226]. In these papers the new procedure, called the noncommutative replica procedure, which is an analog of the replica procedure of Edwards and Anderson [88], was proposed to describe the statistical mechanics of quenched disordered systems (for example, spin glasses).

We will not discuss here the standard replica approach, for an introduction to spin glasses and the replica method see [88, 248].

Consider the disordered system with Hamiltonian $H[\sigma, J]$ which depends on the random parameter J which in the most interesting cases (for spin glasses for instance) is the large random $N \times N$ matrix with independent Gaussian matrix elements J_{ij}, considered in the thermodynamic $N \to \infty$ limit.

To describe the system with quenched disorder in [225, 226], it was proposed to consider the state described by the noncommutative replica statistic sum

$$Z^{(p)} = \int \sum_{\{\sigma\}} \exp(-\beta H[\sigma, \Delta J]) \prod_{a=0}^{p-1} \exp\left(-\frac{1}{2} \sum_{i \le j}^{N} J_{ij}^{(a)2}\right) \prod_{i \le j}^{N} dJ_{ij}^{(a)} \tag{13.5}$$

where Δ is the coproduct operation

$$\Delta : J_{ij} \mapsto \frac{1}{\sqrt{p}} \sum_{a=0}^{p-1} J_{ij}^{(a)} \tag{13.6}$$

which maps the matrix element J_{ij} into the linear combination of independent replicas $J_{ij}^{(a)}$, enumerated by the replica index a. This operation was called quenching in [226].

In the large N limit, by the Wigner theorem, see [320], the system of p random matrices with independent variables will give rise to the quantum Boltzmann algebra with p degrees of freedom with the generators A_a, $A_a^\dagger$, $a = 0, \ldots, p-1$ and the relations

$$A_a A_b^\dagger = \delta_{ab}.$$

These operators are the limits of the large random matrices

$$\lim_{N\to\infty} \frac{1}{N} J_{ij}^{(a)} = Q_a = A_a + A_a^\dagger$$

where convergence is understood in the sense of correlators (as in the central limit theorem).

Then in the thermodynamic limit $N \to \infty$ the noncommutative replica procedure (13.6) will take the form of the following map of the quantum Boltzmann algebra with one degree of freedom into a quantum Boltzmann algebra with p degrees of freedom

$$\Delta : Q \mapsto \frac{1}{\sqrt{p}} \sum_{a=0} Q_a.$$

Note that different Q_a do not commute. We see that we again have obtained noncommutative probability in a purely classical system.

Actually the picture is more complicated, compared to the one we discussed above. The correlations of the system in the noncommutative replica approach will be given by

$$\lim_{N\to\infty} \langle (\Delta J)^k \rangle = \left\langle \left(\frac{1}{\sqrt{p}} \sum_{a=0} Q_a \right)^k \right\rangle$$

where the state $\langle \cdot \rangle$ is generated by the noncommutative replica statistic sum (13.5). In principle, it is not clear how to extract noncommutativity from this a set of correlators, since different degrees $(\Delta Q)^k$ commute.

To distinguish noncommutative and commutative systems, we have to consider set of correlation functions which will be large enough. In the present case this set should contain correlations of different linear combinations of Q_a, more general than ΔQ.

This problem may be discussed in the following way. Quenching (13.6) in principle may be related to a particular way of preparation of the disordered system under consideration. If we use different physical preparation of the disordered system, this may result in different quenching procedure. An example of quenching different from (13.6) was discussed in [226]. This example has the form

$$\Delta' : J \mapsto \frac{1}{\sqrt{p}} \sum_{a=0}^{p-1} c_a J_a, \tag{13.7}$$

where c_a are real-valued coefficients, which should satisfy the condition

$$\sum_{a=0}^{p-1} c_a^2 = p.$$

Varying coefficients c_a, we will obtain different quenchings.

Then, using different physical preparations of the system, we measure the correlation functions which will correspond to different quenchings. Afterwards we may (at least in principle) use the Bell inequality to prove that we do (or do we not) really have a noncommutative probability space which describes the behavior of the disordered system under investigation.

Actually the most natural example of this setup is an experiment with spin glasses, where glass transitions with different external magnetic fields were investigated, and nontrivial behavior of magnetization on preparation was observed [248].

We propose to analyze these experiments, taking into account the correlations between systems with different preparations (i.e. frozen in the presence of different external magnetic fields, which in our approach should correspond to different quenchings), and to check the validity of the Bell inequalities.

This would help to check the validity of the noncommutative replica approach itself, since there is no direct way to introduce noncommutativity in the standard replica approach.

The experimental situation discussed here could also be described by using the contextual probabilistic approach. In the contextual framework, probabilities (which are interpreted as conventional ensemble probabilities) depend on physical contexts—complexes of experimental physical conditions. Mathematically this means that in general we could not use one fixed Kolmogorov probability space. We should work with a system of probability spaces depending on physical contexts. In our case various contexts are defined by choosing various external magnetic fields. We recall that by using the contextual probabilistic approach it is possible to obtain the "quantum rule" for the interference of probabilities. Such an interference of probabilities is just another way to describe noncommutativity of probabilities (induced in the conventional formalism by using Hilbert space calculus). Therefore we could expect that, by taking into account contextuality of statistics for disordered systems frozen in the presence of different external magnetic fields, it would be possible to find experimental confirmation of the presence of the noncommutative structure for classical disordered systems (in particular, spin glasses).

Chapter 14

Derivation of Schrödinger's Equation in the Contextual Probabilistic Framework

As we have seen in Part II, the quantum probabilistic calculus can be derived on the basis of a classical (but contextual) probabilistic model. This was an important step in demystification of quantum probabilities. It strongly stimulated investigations on classical (contextual) probabilistic reconstruction of fundamental quantum structures. There are no doubts that the Schrödinger equation plays the most fundamental role in quantum theory. Therefore it would be interesting to investigate the possibility of deriving this equation in our contextual framework. We showed [185, 189] that it is really possible to obtain the Schrödinger equation by starting with classical stochastic dynamics.

We construct a representation in the complex Hilbert space (the space of complex probabilistic amplitudes) of realistic stochastic dynamics for the reference variables. The basic assumption for existence of such a representation is the validity of the *law of conservation of probabilities* for one of the reference observables, e.g., a. In particular, we can consider a as the *energy variable* and b as the *position variable.* However, in general our scheme of the QL representation is more general and it can be applied to all physical quantities with conservation of probabilities. We emphasize that we do not assume conservation of these quantities (e.g., energy) for individual systems. In particular, the law of conservation of energy might be violated for some prequantum realistic dynamics (so we should distinguish the laws of statistical conservation of energy and individual conservation of energy).

Another unexpected feature of our model is that dynamics in the complex Hilbert space (representing prequantum realistic dynamics) can be *nonlinear.* We found conditions of linearity of the Hilbert space image of a prequantum dynamics. We remark that the Hilbert space dynamics is always unitary (both in the linear and nonlinear cases).

We also emphasize that in general the Hilbert space image of prequantum dynamics can be *irreversible*. We found conditions of reversibility. Finally, we found conditions which induce the conventional Schrödinger dynamics: linear reversible unitary dynamics. The Schrödinger dynamics is characterized through dynamics of the interference coefficient (coefficient of supplementarity) which appears in the interference formula of total probability. Dynamics of this coefficient coincides with dynamics for a *harmonic oscillator.*

We recall that the Schrödinger equation can be derived in another realistic framework, namely stochastic electrodynamics, see, e.g., Nelson [252]. There is a crucial difference between my contextual approach and stochastic electrodynamics. In stochastic electrodynamics one should assume a new physical effect—*fluctuations of vacuum.* In the contextual approach one need not make new physical assumptions. The Schrödinger equation is just an image of the classical stochastic dynamics which is created by neglecting selected information. Such an incomplete information representation is induced by representation of physical reality with the aid of two special physical observables—reference observables.

In this chapter it will be always supposed that the matrix $\mathbf{P}^{b|a}$ for the (dichotomous) reference observables, see Part II, is double stochastic.

14.1 Representation of Contextual Probabilistic Dynamics in the Complex Hilbert Space

Let us assume that the reference observables, see Part II, a and b evolve with time

$$a = a(t, \omega), \qquad b = b(t, \omega).$$

To simplify matters, we consider evolutions which do not change ranges of values of the reference observables $X_a = \{\alpha_1, \alpha_2\}$ and $X_b = \{\beta_1, \beta_2\}$ do not depend on time. Thus, for any t, $a(t, \omega) \in X_a$ and $b = b(t, \omega) \in X_b$. These are random walks with two-point state spaces Y and X. Since our main aim is the contextual probabilistic realistic reconstruction of QM, we should restrict our considerations to evolutions with the trigonometric interference. We proceed under the following assumption.

(CTRB) (Conservation of trigonometric behavior) *The set of trigonometric contexts does not depend on time*

$$\mathcal{C}^{\text{tr}}_{a(t)|b(t)} = \mathcal{C}^{\text{tr}}_{a(t_0)|b(t_0)}.$$

By (CTB) if a context $C \in \mathcal{C}^{\text{tr}}_{a(t_0)|b(t_0)}$, i.e., at the initial instant of time the coefficients of supplementarity $|\lambda(b(t_0) = x|a(t_0), C)| \leq 1$, then the coefficients $\lambda(b(t) = x|a(t), C)$ will always fluctuate in the segment $[0, 1]$.[1]

For each instant of time t, we can use the formalism of contextual quantization, see Part II: a context C can be represented by the complex probability amplitude

$$\psi(t, x) \equiv \psi_C^{b(t)|a(t)}(x)$$
$$= \sqrt{p_C^{a(t)}(\alpha_1) p^{b(t)|a(t)}(x|\alpha_1)} + e^{i\theta_C^{b(t)|a(t)}(x)} \sqrt{p_C^{a(t)}(\alpha_2) p^{b(t)|a(t)}(x|\alpha_2)}.$$

[1] Of course, there can be considered more general dynamics in which the trigonometric probabilistic behavior can be transformed into the hyperbolic one and vice versa. But we shall not try to study the most general dynamics. Our aim is to show that the conventional Schrödinger dynamics can be easily found among contextual dynamics in the complex Hilbert space.

We remark that the observable $a(t)$ is represented by the self-adjoint operator $\hat{a}(t)$ defined by its eigenvectors

$$e_1^a(t) = \begin{pmatrix} \sqrt{p(t;\beta_1|\alpha_1)} \\ \sqrt{p(t;\beta_2|\alpha_1)} \end{pmatrix}, \qquad e_2^a(t) = e^{i\theta_C(t)} \begin{pmatrix} \sqrt{p(t;\beta_1|\alpha_2)} \\ -\sqrt{p(t;\beta_2|\alpha_1)} \end{pmatrix},$$

where

$$p(t;x|y) = p^{b(t)|a(t)}(x|y), \qquad \theta_C(t) = \theta_C^{b(t)|a(t)}(\beta_1)$$

and where we set $e_j^a(t) \equiv e_j^{a(t)}$. We recall that

$$\theta_C^{b(t)|a(t)}(\beta_2) = \theta_C^{b(t)|a(t)}(\beta_1) + \pi,$$

since the matrix of transition probabilities is assumed to be double stochastic for all instances of time, see Part II.

We shall describe dynamics of the wave function $\psi(t,x)$ starting with assumptions (CP) and (CTP). Then these assumptions will be completed by the set (a)–(b) of mathematical assumptions which will imply the conventional Schrödinger evolution.

(CP) (Conservation of a-probabilities) *The probability distribution of the a-observable is preserved in the process of evolution*

$$p_C^{a(t)}(y) = p_C^{a(t_0)}(y), \quad y \in X_a, \tag{14.1}$$

for any context $C \in \mathscr{C}^{\mathrm{tr}}_{a(t_0)|b(t_0)}$. This statistical conservation of the a-quantity will have very important dynamical consequences. We also assume that the law of conservation of transition probabilities holds.

(CTP) (Conservation of transition probabilities) *Probabilities $p(t;x|y)$ are conserved in the process of evolution*

$$p(t;x|y) = p(t_0;x|y) \equiv p(x|y). \tag{14.2}$$

Under the latter assumption we have

$$e_1^a(t) \equiv e_1^a(t_0), \qquad e_2^a(t) = e^{i[\theta_C(t)-\theta_C(t_0)]} e_2^a(t_0). \tag{14.3}$$

Remark 14.1. If the $a(t)$-basis evolves according to (14.3), then $\hat{a}(t) = \hat{a}(t_0) = \hat{a}$. Hence the whole stochastic process $a(t, \omega)$ is represented by one fixed self-adjoint operator. We emphasize that random variables $a(t_1, \omega)$ and $a(t_2, \omega)$, $t_1 \neq t_2$, can differ essentially as functions of the random parameter ω. Nevertheless, they are represented by the same quantum operator.

Thus under assumptions (CTRB), (CP) and (CTP) we have

$$\psi(t) = u_1^a e_1^a(t) + u_2^a e_2^a(t) = u_1^a e_1^a(t_0) + e^{i\xi_C(t,t_0)} u_2^a e_2^a(t_0),$$

where $u_j^a = \sqrt{p_C^{a(t_0)}(\alpha_j)}$, $j = 1, 2$, and

$$\xi_C(t, t_0) = \theta_C(t) - \theta_C(t_0).$$

Let us consider the unitary operator $\hat{U}(t, t_0) : \mathscr{H} \to \mathscr{H}$ defined by this transformation of basis: $e^a(t_0) \to e^a(t)$. In the basis $e^a(t_0) = \{e_1^a(t_0), e_2^a(t_0)\}$ the $\hat{U}(t, t_0)$ can be represented by the matrix

$$\hat{U}(t, t_0) = \begin{pmatrix} 1 & 0 \\ 0 & e^{i\xi_C(t,t_0)} \end{pmatrix}.$$

We obtained the following dynamics in the Hilbert space $\mathscr{H}$

$$\psi(t) = \hat{U}(t, t_0)\psi(t_0). \tag{14.4}$$

This dynamics looks very similar to the Schrödinger dynamics in Hilbert space. However, the dynamics (14.4) is essentially more general than Schrödinger's dynamics. In fact, the unitary operator $\hat{U}(t, t_0) = \hat{U}(t, t_0, C)$ depends on the context C, i.e., on the initial state $\psi(t_0)$

$$\hat{U}(t, t_0) \equiv \hat{U}(t, t_0, \psi(t_0)).$$

In fact, we derived the following dynamical equation

$$\psi(t) = \hat{U}(t, t_0, \varphi_0)\varphi_0, \tag{14.5}$$

where, for any φ_0, $\hat{U}(t, t_0, \varphi_0)$ is a family of unitary operators.

The conditions (CTRB), (CP) and (CTP) are natural from the physical viewpoint (if the a-observable is considered as an analog of energy, see further considerations). But these conditions do not imply that the Hilbert space image of the contextual realistic dynamics is a linear unitary dynamics.

In general the Hilbert space projection of realistic prequantum dynamics is non-linear.

To obtain a linear dynamics, we should make the following assumption:

(CI) (Context independence of the increment of the probabilistic phase) *The phase difference*

$$\xi_C(t, t_0) = \theta_C(t) - \theta_C(t_0)$$

does not depend on C.

Under this assumption the unitary operator $\hat{U}(t, t_0)$ does not depend on C.

$$\hat{U}(t, t_0) = \begin{pmatrix} 1 & 0 \\ 0 & e^{i\xi(t,t_0)} \end{pmatrix}. \tag{14.6}$$

Thus (14.4) is the equation of linear unitary evolution. The main problem in these considerations is to find a physical basis of the condition (CI): the increment of statistical disturbance should be the same for all contexts, see Sect. 14.2 for the detailed analysis. The linear unitary evolution (14.4) is still essentially more general than the conventional Schrödinger dynamics. To obtain the Schrödinger evolution, we need a few standard mathematical assumptions:

(a) Dynamics is continuous: the map $(t, t_0) \to \hat{U}(t, t_0)$ is continuous.[2]

(b) Dynamics is deterministic.

[2] Recall that we are considering the finite-dimensional case. Thus there is no problem of the choice of topology.

(c) Dynamics is invariant with respect to time-shifts; $\hat{U}(t, t_0)$ depends only on $t - t_0 : \hat{U}(t, t_0) \equiv \hat{U}(t - t_0)$.

The assumption of determinism can be described by the following relation

$$\psi(t; t_0, \psi_0) = \psi(t; t_1, \psi(t_1; t_0, \psi_0)), \quad t_0 \leq t_1 \leq t,$$

where $\psi(t; t_0, \psi_0) = \hat{U}(t, t_0)\psi_0$.

It is well known that under the assumptions (a), (b), (c) the family of (linear) unitary operators $\hat{U}(t, t_0)$ corresponds to the one-parameter group of unitary operators

$$\hat{U}(t) = e^{\frac{i}{h}\hat{H}t}, \tag{14.7}$$

where $\hat{H} : \mathcal{H} \to \mathcal{H}$ is a self-adjoint operator. Here $h > 0$ is a scaling factor (e.g., the Planck constant). We have

$$\hat{H} = \begin{pmatrix} 0 & 0 \\ 0 & E \end{pmatrix}, \tag{14.8}$$

where

$$E = -h\left[\frac{\theta_C(t) - \theta_C(t_0)}{t - t_0}\right].$$

Hence Schrödinger evolution in the complex Hilbert space corresponds to contextual probabilistic dynamics with linear evolution of the probabilistic phase

$$\theta_C(t) = \theta_C(t_0) - \frac{E}{h}(t - t_0). \tag{14.9}$$

Let us consider a stochastic process (rescaling of the process $a(t, \omega)$)

$$H(t, \omega) = \begin{cases} 0, & a(t, \omega) = \alpha_1, \\ E, & a(t, \omega) = \alpha_2. \end{cases} \tag{14.10}$$

Since the probability distributions of the processes $a(t, \omega)$) and $H(t, \omega)$ coincide (up to rescaling of ranges of values), we have

$$p_C^{H(t)}(0) = p_C^{a(t)}(\alpha_1) \equiv p_C^{a(t_0)}(\alpha_1) = p_C^{H(t_0)}(0), \tag{14.11}$$

$$p_C^{H(t)}(E) = p_C^{a(t)}(\alpha_2) \equiv p_C^{a(t_0)}(\alpha_2) = p_C^{H(t_0)}(E). \tag{14.12}$$

If $E > 0$ we can interpret $H(t, \omega)$ as the energy observable and the operator $\hat{H}$ as its Hilbert space image. We emphasize, see Remark 14.1, that the whole "energy process" $H(t, \omega)$ is represented by a single self-adjoint nonnegative operator $\hat{H}$ in the Hilbert space. This operator, "quantum Hamiltonian", is the Hilbert space projection of the energy process which is defined on the "prespace" Ω. In principle, random variables $H(t_1, \omega), H(t_2, \omega), t_1 \neq t_2$, can be very different (as functions of ω). We have only the law of *statistical conservation of energy*

$$p_C^{H(t)}(z) \equiv p_C^{H(t_0)}(z), \quad z = 0, E. \tag{14.13}$$

In general (depending on dynamics of the coefficient of supplementarity) the eigenvalue E need not be positive. So in general we have a dynamical equation corresponding to some statistically conserved quantity. Of course, the representation (14.7) is equivalent to the Schrödinger equation

$$ih\frac{d\hat{U}}{dt}(t) = \hat{H}\hat{U}(t), \quad \hat{U}(0) = I,$$

where I is the unit operator.

14.2 Schrödinger Dynamics and Coefficients of Supplementarity

We discuss here consequences of the condition (CI) for the measurable quantity, namely, the coefficient of supplementarity

$$\lambda_C(t) \equiv \lambda(b(t) = \beta_1 | a(t), C) = \cos\theta_C(t).$$

By (CI) we have $\theta_C(t) - \theta_C(t_0) = q(t, t_0)$, where q does not depend on C. Thus $\theta'_C(t) = q'(t, t_0) \equiv f(t, t_0)$ does not depend on C. We remark that in principle $f(t, t_0)$ can depend on t_0, see Remark 14.3 below for details. Let us investigate an interesting special case:

(DT) *The function f does not depend on t_0.*

Here $f = f(t)$. Suppose that $f(t)$ is a continuous function. Hence

$$\theta_C(t) = \theta_C(t_0) + \int_{t_0}^{t} f(s)\,ds$$

and

$$\lambda_C(t) = \cos\left[\theta_C(t_0) + \int_{t_0}^{t} f(s)\,ds\right].$$

We have the following differential equation for the coefficient $\lambda_C(t)$

$$\lambda'_C(t) = \pm f(t)\sqrt{1 - \lambda_C^2(t)}. \tag{14.14}$$

To find the coefficient of supplementarity (under above assumptions), one should solve the Cauchy problem for the differential equation (14.14) with the initial condition

$$\lambda_C(t_0) = \lambda(b(t_0) = \beta_1 | a(t_0), C) \equiv \cos\theta_C(t_0). \tag{14.15}$$

In this case the assumption (CI) can be written in the form of the Cauchy problem (14.14), (14.15). The evolution family has the form

$$\hat{U}(t, t_0) = \begin{pmatrix} 1 & 0 \\ 0 & e^{i\int_{t_0}^{t} f(s)ds} \end{pmatrix}. \tag{14.16}$$

This evolution is continuous and deterministic. To prove that the condition of determinism (b) holds, we use the additivity of the integral. In general such evolutions are not invariant with respect to time-shifts; they correspond to Schrödinger evolutions with time dependent generators

$$ih\frac{d\hat{U}(t, t_0)}{dt} = \hat{H}(t)\hat{U}(t, t_0), \quad \hat{U}(t_0, t_0) = I,$$

where

$$\hat{H}(t) = \begin{pmatrix} 0 & 0 \\ 0 & E(t) \end{pmatrix}, \tag{14.17}$$

where $E(t) = -hf(t)$. Let us consider two very simple, but illustrative examples.

Example 14.1 (Schrödinger dynamics). Let $f(t) = -E/h = \text{Const}$. Then $\theta_C(t) = \theta_0 - E(t - t_0)/h$. Here

$$\hat{U}(t) = \begin{pmatrix} 1 & 0 \\ 0 & e^{-i\frac{Et}{h}} \end{pmatrix}.$$

This is the Schrödinger dynamics; so here the conditions (a)–(c) are automatically satisfied. We remark that in this case the coefficient of supplementarity satisfies the second-order differential equation, namely, the equation for *harmonic oscillations*

$$\frac{d^2\lambda}{dt^2}(t) + \omega^2\lambda(t) = 0, \tag{14.18}$$

where $\omega = E/h$. This is the direct consequence of the representation

$$\lambda(t) = \cos[\theta_0 - E(t - t_0)/h].$$

Hence:

The Schrödinger dynamics is characterized by the harmonic fluctuations of the coefficient of supplementarity.

Example 14.2. Let $f(t) = -Et/h$, $E > 0$. Here $\theta_C(t) = \theta_C(t_0) - E(t^2 - t_0^2)/2h$ and

$$\hat{U}(t, t_0) = \begin{pmatrix} 1 & 0 \\ 0 & e^{-iE(t^2 - t_0^2)/2h} \end{pmatrix}.$$

This is a linear unitary deterministic and continuous dynamics, but it is not invariant with respect to time-shifts. This is the Schrödinger dynamics with time-dependent generator of evolution. The $\hat{H}(t)$ is positively defined and it can be considered as time-dependent Hamiltonian.

Remark 14.2 (Approximate reversibility of the Hilbert space evolution). If the function $f = f(t, t_0)$ depends nontrivially on t_0, then the evolution of the wave function is not deterministic. It is *irreversible*. We remark that if determinism of the evolution of the probabilistic phase is violated, then $f(t, t_0)$ nontrivially depends on t_0 (and vice versa). The violation of determinism for the phase evolution means that $\theta_C(t)$

is not uniquely determined by $\theta_C(t_0)$; so the following condition of determinism

$$\theta_C(t;\theta_C(t_1;\theta_C(t_0))) = \theta_C(t;\theta_C(t_0))$$

should be violated, $t_0 \leq t_1 \leq t$. Here $\theta_C(t;\theta_C(t_0))$ is the probabilistic phase at the instant of time t under the condition that it was equal to $\theta_C(t_0)$ at the initial instant of time t_0. We remark that reversible Hilbert space evolutions could appear as approximations of irreversible Hilbert space evolutions. Suppose that

$$f(t,t_0) = f(t) + \varepsilon f_1(t,t_0),$$

where ε is negligibly small. Then by neglecting terms of the ε-magnitude we can approximately describe the Hilbert space evolution as reversible. Of course, there should be chosen some scale. It is natural to use the scale based on the Planck constant h. Thus if $\varepsilon \ll h$, then the irreversible evolution

$$\hat{U}(t,t_0) = \begin{pmatrix} 1 & 0 \\ 0 & e^{i[\int_{t_0}^t f(s)ds + \varepsilon \int_{t_0}^t f_1(t,t_0)ds]} \end{pmatrix} \tag{14.19}$$

can be approximately considered as the reversible evolution (14.7).

Remark 14.3 (Approximate linearity of the Hilbert space evolution). Arguments which are similar to the arguments of the previous Remark can be applied to the problem of linearization of general nonlinear dynamics in the complex Hilbert space. Let $f = f(t)$ (so dynamics is deterministic) and let $f(t)$ be an analytic function

$$f(t) = \sum_{n=0}^{\infty} f_n t^n \equiv f_0 + f_1(t).$$

Suppose that

$$\varepsilon = \max_{0 \leq t \leq T} |f_1(t)| \ll h.$$

If $\varepsilon \ll h$, then the Hilbert space dynamics could be approximately considered as a linear dynamics.

We also can combine arguments of both remarks.

Conclusion. The contextual realistic dynamics can be represented (under assumptions (CTRB), (CP), and (CTP)) as unitary dynamics in complex Hilbert space. In general such dynamics are nonlinear and irreversible. Dynamics are linear iff the condition (CI) holds. The contextual dynamics in Hilbert space are reduced to the conventional Schrödinger evolution under the additional assumptions (a)–(c). In particular, the assumption (b) implies reversibility. The Schrödinger dynamics is the Hilbert space projection of the realistic dynamics with harmonic oscillations of the coefficient of supplementarity, see (14.18). The reversible and linear Schrödinger dynamics can be considered as an approximation of irreversible and nonlinear dynamics in Hilbert space.

Part V
Hyperbolic Quantum Mechanics

This part is devoted to hyperbolic quantum mechanics which arises naturally in the contextual probabilistic (Växjö) model. Not any context (or better to say statistical data about context) can be represented by a complex-valued probabilistic amplitude. Some contexts are represented by amplitudes taking values in a so-called hyperbolic algebra. One of the main distinguishing features of hyperbolic quantization is violation of the principle of superposition. We remark that the classical limit of hyperbolic quantum mechanics (when the Planck constant is considered as infinitely small) coincides with the limit of conventional quantum mechanics, namely, classical mechanics on the phase space. We prove this result by using the theory of pseudo-differential operators with hyperbolic-valued symbols.

Chapter 15

Representation of Contextual Statistical Model by Hyperbolic Amplitudes

In this chapter we continue development of the contextual statistical model—the Växjö model. As we know from Part II, besides contexts producing the conventional trigonometric cos-interference, there exist contexts producing the hyperbolic cosh-interference. We start with the corresponding formula of total probability with the interference term, see Part II. Then (in the same way as in the trigonometric case) we represent such contexts by hyperbolic probabilistic amplitudes or in the abstract formalism by normalized vectors of a hyperbolic analogue of Hilbert space. A hyperbolic analogue of Born's rule is obtained. Probabilistically conjugate, see Part II, Chap. 3, observables a and b are represented by noncommutative operators $\hat{a}$ and $\hat{b}$.

This chapter can be considered as the first step towards hyperbolic quantum probability.

As in the trigonometric case, each hyperbolic QL-representation is based on a fixed pair of observables a and *b—reference observables*—which produce the contextual image of the Växjö model in hyperbolic Hilbert space.

In Part II, Chap. 4 we introduced a class $\mathcal{C}^{\mathrm{tr}}$ of contexts ("trigonometric contexts") which can be represented by complex probabilistic amplitudes inducing a representation in the complex Hilbert space. The $\mathcal{C}^{\mathrm{tr}}$ consists of context producing the conventional trigonometric cos-interference. However, in general the set of contexts is not reduced to the class of trigonometric contexts $\mathscr{C}^{\rm tr}$. There exist contexts producing the hyperbolic cosh-interference. The set of hyperbolic contexts is denoted by the symbol $\mathscr{C}^{\rm hyp}$.

In this chapter we show that it is possible to represent contexts belonging to $\mathscr{C}^{\rm hyp}$ by *hyperbolic amplitudes*—by using a hyperbolic version of QLRA. Such amplitudes take values in the set of "hyperbolic numbers" (two-dimensional Clifford algebra). It will be demonstrated that in the hyperbolic framework we can proceed quite far in the same directions as in the trigonometric framework. The crucial difference between two cases is that in *the hyperbolic case the principle of superposition is violated.*

15.1 Hyperbolic Algebra

Instead of the field of complex numbers **C**, we shall use so-called *hyperbolic numbers,* namely, the two-dimensional Clifford algebra, **G**. We call this algebra *hyperbolic algebra.*[1] Denote by the symbol j the generator of the algebra **G** of hyperbolic numbers

$$j^2 = 1.$$

[1] Of course, it is rather dangerous to invent one's own name for a notion established almost as firmly as complex numbers. We use a new name, hyperbolic algebra, for the well-known algebraic object, the two-dimensional Clifford algebra, for the following reasons. First we explain why we dislike using the standard notion of Clifford algebra in this particular case. The standard Clifford machinery was developed around noncommutative features of general Clifford algebras. The two-dimensional Clifford algebra, hyperbolic algebra in our terminology, is commutative. Commutativity of **G** is very important in our considerations. We now explain why we propose the name hyperbolic algebra. Hyperbolic functions are naturally related to the algebraic structure of **G** through a hyperbolic generalization of Euler's formula for the complex numbers. This is the crucial point of our considerations—the possibility to use this algebraic structure to represent some special transformations for hyperbolic functions.

The algebra $\mathbf{G}$ is the two-dimensional real (commutative) algebra with basis $e_0 = 1$ and $e_1 = j$. Elements of $\mathbf{G}$ have the form

$$z = x + jy, \quad x, y \in \mathbf{R}.$$

We have $z_1+z_2 = (x_1+x_2)+j(y_1+y_2)$ and $z_1 z_2 = (x_1 x_2+y_1 y_2)+j(x_1 y_2+x_2 y_1)$. This algebra is commutative. It is not a field—not every element has an inverse. We introduce an involution in $\mathbf{G}$ by setting

$$\bar{z} = x - jy$$

and set

$$|z|^2 = z\bar{z} = x^2 - y^2.$$

We remark that

$$|z| = \sqrt{x^2 - y^2}$$

is not well defined for an arbitrary $z \in \mathbf{G}$. We set

$$\mathbf{G}_+ = \{z \in \mathbf{G} : |z|^2 \geq 0\}.$$

We remark that $\mathbf{G}_+$ is a multiplicative semigroup as follows from the equality

$$|z_1 z_2|^2 = |z_1|^2 |z_2|^2.$$

Thus, for $z_1, z_2 \in \mathbf{G}_+$, we have that $|z_1 z_2|$ is well defined and $|z_1 z_2| = |z_1||z_2|$. We define a hyperbolic exponential function by using a hyperbolic analogue of Euler's formula

$$e^{j\theta} = \cosh\theta + j\sinh\theta, \quad \theta \in \mathbf{R}.$$

We remark that

$$e^{j\theta_1} e^{j\theta_2} = e^{j(\theta_1+\theta_2)}, \qquad \overline{e^{j\theta}} = e^{-j\theta}, \qquad |e^{j\theta}|^2 = \cosh^2\theta - \sinh^2\theta = 1.$$

Hence,

$$z = \pm e^{j\theta}$$

always belongs to $\mathbf{G}_+$. We also have

$$\cosh\theta = \frac{e^{j\theta} + e^{-j\theta}}{2}, \qquad \sinh\theta = \frac{e^{j\theta} - e^{-j\theta}}{2j}.$$

We set

$$\mathbf{G}_+^* = \{z \in \mathbf{G}_+ : |z|^2 > 0\}.$$

Let $z \in \mathbf{G}_+^*$. We have

$$z = |z|\left(\frac{x}{|z|} + j\frac{y}{|z|}\right) = \operatorname{sign} x|z|\left(\frac{x \operatorname{sign} x}{|z|} + j\frac{y \operatorname{sign} x}{|z|}\right).$$

As

$$\frac{x^2}{|z|^2} - \frac{y^2}{|z|^2} = 1,$$

we can represent

$$x \operatorname{sign} x = \cosh\theta$$

and

$$y \operatorname{sign} x = \sinh\theta,$$

where the phase θ is uniquely defined. We can represent each $z \in \mathbf{G}_+^*$ as

$$z = \operatorname{sign} x|z|e^{j\theta}.$$

By using this representation we can easily prove that $\mathbf{G}_+^*$ is a multiplicative group. Here

$$\frac{1}{z} = \frac{\operatorname{sign} x}{|z|}e^{-j\theta}.$$

The unit circle in $\mathbf{G}$ is defined as

$$S_1 = \{z \in \mathbf{G} : |z|^2 = 1\} = \{z = \pm e^{j\theta}, \theta \in (-\infty, +\infty)\}.$$

It is a multiplicative subgroup of $\mathbf{G}_+^*$.

To construct a $\mathbf{G}$-linear representation of the set $\mathscr{C}^{\mathrm{hyp}}$ of hyperbolic contexts, we shall use the following elementary formula

$$D = A + B \pm 2AB \cosh\theta = |\sqrt{A} \pm e^{j\theta}\sqrt{B}|^2, \qquad (15.1)$$

for real coefficients $A, B > 0$.

15.2 Hyperbolic Version of Quantum-Like Representation Algorithm

Everywhere below we study contexts producing the *hyperbolic interference* for probabilistically conjugate dichotomous random variables $a = \alpha_1, \alpha_2, b = \beta_1, \beta_2$. This pair of variables will be fixed—the *reference variables.* We set

$$X_a = \{\alpha_1, \alpha_2\}, \qquad X_b = \{\beta_1, \beta_2\}$$

("spectra" of observables a and b). For each pair a, b of reference variables we construct a representation of the set of contexts $\mathscr{C}^{\mathrm{hyp}}$, see Part II: Chap. 4, in hyperbolic Hilbert space, QL-representation.

15.2.1 Hyperbolic Probability Amplitude, Hyperbolic Born's Rule

Let $C \in \mathscr{C}^{\mathrm{hyp}}$. We set

$$p_C^a(y) = \mathbf{P}(a = y|C), \qquad p_C^b(x) = \mathbf{P}(b = x|C), \qquad p(x|y) = \mathbf{P}(b = x|a = y),$$

$x \in X_b$, $y \in X_a$. The interference formula of total probability, Part II, Chap. 4, can be written in the following form

$$p_C^b(x) = \sum_{y \in X_a} p_C^a(y) p(x|y) \pm 2 \cosh\theta_C(x) \sqrt{\Pi_{y \in X_a} p_C^a(y) p(x|y)}, \qquad (15.2)$$

where

$$\theta_C(x) = \theta(b = x|a, C) = \pm \operatorname{arccosh} |\lambda(b = x|a, C)|, \quad x \in X_b, \ C \in \mathscr{C}^{\text{hyp}}.$$

Here the coefficient of supplementarity λ is defined by

$$\lambda(b = x|a, C) = \frac{p_C^b(x) - \sum_{y \in X_a} p_C^a(y) p(x|y)}{2\sqrt{\Pi_{y \in X_a} p_C^a(y) p(x|y)}}. \tag{15.3}$$

By using (15.1) we can represent the probability $p_C^b(x)$ as the square of the hyperbolic amplitude

$$p_C^b(x) = |\psi_C(x)|^2, \tag{15.4}$$

where

$$\psi(x) \equiv \psi_C(x) = \sqrt{p_C^a(\alpha_1) p(x|\alpha_1)} + \varepsilon_C(x) e^{j\theta_C(x)} \sqrt{p_C^a(\alpha_2) p(x|\alpha_2)}. \tag{15.5}$$

Here $\varepsilon_C(x) = \operatorname{sign} \lambda(x|a, C)$. We remark that

$$\sum_{x \in X_b} \varepsilon_C(x) = 0. \tag{15.6}$$

Thus we have a *hyperbolic generalization of Born's rule* for the b-variable.

The formula (15.5) gives the hyperbolic version of the QL representation algorithm—QLRA. For any hyperbolic context C, by starting with the probabilistic data—$p_C^b(x)$, $p_C^a(y)$, $p(x|y)$—QLRA produces the hyperbolic amplitude ψ_C. This algorithm can be used in any domain of science to create the QL-representation of probabilistic data (for a special class of contexts). We point out that QLRA contains the reference observables a and b as parameters. Hence the hyperbolic amplitude given by (15.5) depends on a, b

$$\psi_C \equiv \psi_C^{b|a}.$$

Definition 15.1. *The hyperbolic amplitude ψ_C produced by QLRA is called a QL wave function (of the complex of physical conditions, context C) or a QL state.*

15.2.2 Hyperbolic Hilbert Space

Hyperbolic Hilbert space $\mathscr{H}$ is a $\mathbf{G}$-linear space (or more precisely: a module over the commutative algebra) endowed with a $\mathbf{G}$-linear scalar product—a map

$$\langle \cdot, \cdot \rangle : \mathscr{H} \times \mathscr{H} \to \mathbf{G}$$

such that

(1) it is linear with respect to the first argument

$$\langle az + bw, u \rangle = a\langle z,\ u \rangle + b\langle w, u \rangle, \quad a, b \in \mathbf{G},\ z, w, u \in \mathscr{H};$$

(2) it is conjugate-symmetric

$$\langle z, u \rangle = \overline{\langle u, z \rangle};$$

(3) it is nondegenerate $\langle z, u \rangle = 0$ for all $u \in \mathscr{H}$ iff $z = 0$.

We shall use the "square-norm" defined as

$$\|\psi\|^2 = \langle \psi, \psi \rangle.$$

It is clear that in general the norm

$$\|\psi\| = \sqrt{\langle \psi, \psi \rangle}$$

is not well defined.

We do not consider any topology on hyperbolic Hilbert space $\mathscr{H}$, because we shall restrict our considerations to the finite-dimensional case, see [138] for the general case.

Remark 15.1. If we consider $\mathscr{H}$ as just an $\mathbf{R}$-linear space, then $\langle \cdot, \cdot \rangle$ is a bilinear form which is not positively defined. In particular, in the two dimensional case we have the signature $(+, -, +, -)$.

15.2.3 Hyperbolic Hilbert Space Representation

We introduce the space $E_b = \Phi(X_b, \mathbf{G})$ of functions $\psi : X_b \to \mathbf{G}$. Since $X_b = \{\beta_1, \beta_2\}$, the E_b is the two-dimensional $\mathbf{G}$-module. We define the $\mathbf{G}$-scalar product by setting

$$\langle \varphi, \psi \rangle = \sum_{x \in X_b} \varphi(x)\overline{\psi(x)},$$

where $z \to \bar{z}$ is conjugation in the algebra $\mathbf{G}$. The system of functions

$$\{e_x^b = \delta(\cdot - x)\}_{x \in X_b}$$

is an orthonormal basis in the hyperbolic Hilbert space $\mathscr{H} = (E_b, \langle \cdot, \cdot \rangle)$. Here $\delta(\cdot - x)$ is the x-shift of the Dirac δ-function.

We have the hyperbolic analogue of Born's rule in $\mathscr{H}$

$$p_C^b(x) = |\langle \psi_C, e_x^b \rangle|^2. \tag{15.7}$$

Let $X_b \subset \mathbf{R}$. By (15.7) we obtain hyperbolic Hilbert space representation of the expectation of the observable b

$$\begin{aligned} E(b|C) &= \sum_{x \in X_b} x p_C^b(x) = \sum_{x \in X_b} x |\psi_C(x)|^2 = \sum_{x \in X_b} x \langle \psi_C, e_x^b \rangle \overline{\langle \psi_C, e_x^b \rangle} \\ &= \langle \hat{b} \psi_C, \psi_C \rangle, \end{aligned} \tag{15.8}$$

where the (self-adjoint) operator $\hat{b} : \mathscr{H} \to \mathscr{H}$ is determined by its eigenvectors $\hat{b} e_x^b = x e_x^b, x \in X_b$. This is the multiplication operator in the space of $\mathbf{G}$-valued functions $\Phi(X_b, \mathbf{G})$

$$\hat{b}\psi(x) = x\psi(x). \tag{15.9}$$

By (15.8) the conditional expectation of the observable b is represented with the aid of the self-adjoint operator $\hat{b}$. We emphasize that representations (15.8), (15.9) are just convenient mathematical representations.

Thus we constructed a **G**-linear representation of hyperbolic contexts of the contextual statistical model

$$J^{b|a} : \mathscr{C}^{\text{hyp}} \to \mathscr{H}.$$

We set

$$S_{\mathscr{C}^{\text{hyp}}} = J^{b|a}(\mathscr{C}^{\text{hyp}}).$$

This is a subset of the unit sphere

$$S = \{\psi \in \mathscr{H} : \|\psi\|^2 = 1\}$$

of hyperbolic Hilbert space $\mathscr{H}$. We introduce the coefficients

$$\begin{aligned} &u_j^a = \sqrt{p_C^a(\alpha_j)}, \qquad u_j^b = \sqrt{p_C^b(\beta_j)}, \qquad p_{ij} = p(\beta_j|\alpha_i), \\ &u_{ij} = \sqrt{p_{ij}}, \qquad \theta_j = \theta_C(\beta_j) \end{aligned} \tag{15.10}$$

and

$$\varepsilon_i = \varepsilon(\beta_i).$$

We remark that the coefficients u_j^a, u_j^b depend on a context C; so

$$u_j^a = u_j^a(C), \qquad u_j^b = u_j^b(C).$$

We also consider the *matrix of transition probabilities* $\mathbf{P}^{b|a} = (p_{ij})$. It is always a *stochastic matrix:* $p_{i1} + p_{i2} = 1$, $i = 1, 2$. In further considerations we shall also consider *double stochastic* matrices: $p_{1j} + p_{2j} = 1$, $j = 1, 2$, cf. Part II.

We represent a state ψ_C by

$$\psi_C = v_1^b e_{\alpha_1}^b + v_2^b e_2^b,$$

where

$$v_i^b = u_1^a u_{1i} + \varepsilon_i u_2^a u_{2i} e^{j\theta_i}.$$

So

$$p_C^b(\beta_i) = |v_i^b|^2 = |u_1^a u_{1i} + \varepsilon_i u_2^a u_{2i} e^{j\theta_i}|^2.$$

This is the **G**-linear representation of the hyperbolic interference of probabilities. This formula can also be derived in the formalism of hyperbolic Hilbert space, see Sect. 15.3. We remark that here the **G**-linear combination

$$u_1^a u_{1i} + \varepsilon_i u_2^a u_{2i} e^{j\theta_i}$$

belongs to $\mathbf{G}_+^*$.

Thus for any context $C_0 \in \mathscr{C}^{\text{hyp}}$ we can represent ψ_{C_0} in the form

$$\psi_{C_0} = u_1^a e_{\alpha_1}^a + u_2^a e_{\alpha_2}^a,$$

where

$$e_{\alpha_1}^a = \begin{pmatrix} u_{11} \\ u_{12} \end{pmatrix}, \qquad e_{\alpha_2}^a = \begin{pmatrix} \varepsilon_1 e^{j\theta_1} u_{21} \\ \varepsilon_2 e^{j\theta_2} u_{22} \end{pmatrix}. \tag{15.11}$$

As in the **C**-case, we introduce the matrix

$$V = (v_{ij}) = \begin{pmatrix} \langle e_{\alpha_1}^a, e_{\beta_1}^b \rangle & \langle e_{\alpha_1}^a, e_{\beta_2}^b \rangle \\ \langle e_{\alpha_2}^a, e_{\beta_1}^b \rangle & \langle e_{\alpha_2}^a, e_{\beta_2}^b \rangle \end{pmatrix}$$

$$= \begin{pmatrix} \sqrt{p(\beta_1|\alpha_1)} & \sqrt{p(\beta_2|\alpha_1)} \\ \varepsilon_1 e^{j\theta_1}\sqrt{p(\beta_1|\alpha_2)} & \varepsilon_2 e^{j\theta_2}\sqrt{p(\beta_2|\alpha_2)} \end{pmatrix}. \tag{15.12}$$

We remark that here coefficients $v_{ij} \in \mathbf{G}_+^*$. In the same way as in the complex case, Born's rule

$$p_{C_0}^a(\alpha_i) = |(\psi_{C_0}, e_\alpha^a)|^2 \tag{15.13}$$

holds for the a-basis $\{e_\alpha^a\}$ iff it is an orthonormal basis in $\mathscr{H}$. The latter is equivalent to the **G**-unitary of the matrix V,

$$\overline{V}^* V = I,$$

or

$$\bar{v}_{11} v_{11} + \bar{v}_{21} v_{21} = 1, \qquad \bar{v}_{12} v_{12} + \bar{v}_{22} v_{22} = 1, \tag{15.14}$$

$$\bar{v}_{11} v_{12} + \bar{v}_{21} v_{22} = 0. \tag{15.15}$$

Thus $1 = u_{11}^2 + u_{21}^2 = p(\beta_1|\alpha_1) + p(\beta_1|\alpha_2)$ and $1 = u_{12}^2 + u_{22}^2 = p(\beta_2|\alpha_1) + p(\beta_2|\alpha_2)$. Thus the first two equations of the **G**-unitary are equivalent to the double stochasticity of $\mathbf{P}^{b|a}$ (as in the **C**-case). We remark that (15.14) can be written as

$$|v_{11}|^2 + |v_{21}|^2 = 1, \qquad |v_{12}|^2 + |v_{22}|^2 = 1, \tag{15.16}$$

cf. Sect. 15.3. The third unitarity-equation (15.15) can be written as

$$u_{11}u_{12} + \varepsilon_1\varepsilon_2 e^{j(\theta_2-\theta_1)} u_{21}u_{22} = 0. \tag{15.17}$$

By using double stochasticity of $\mathbf{P}^{a|b}$ we obtain $e^{j\theta_1} = e^{j\theta_2}$. Thus

$$\theta_1 = \theta_2. \tag{15.18}$$

Lemma 15.1. *Let a and b be incompatible observables and let* $\mathbf{P}^{b|a}$ *be double stochastic. Then*

$$\cosh\theta_C(\beta_2) = \cosh\theta_C(\beta_1) \tag{15.19}$$

for any context $C \in \mathscr{C}^{\text{hyp}}$.

Proof. By Lemma 3.1, Part II: Chap. 3, we have

$$\sum_x \varepsilon(x)\cosh\theta_C(x)\sqrt{\Pi_y p_C^a(y) p(x|y)} = 0.$$

Double stochasticity of $\mathbf{P}^{b|a}$ implies (15.19).

The constraint (15.19) induced by double stochasticity can be written as the constraint to phases

$$\theta_C(\beta_2) = \pm\theta_C(\beta_1). \tag{15.20}$$

To obtain unitarity of the matrix V of transition $\{e_\beta^b\} \to \{e_\alpha^a\}$ we should choose phases according to (15.18). And by (15.20) we can always do this for a double stochastic matrix of transition probabilities.

By choosing such a representation we obtain the hyperbolic generalization of Born's rule for the a-variable

$$p_C^a(\alpha) = |(\psi, e_\alpha^a)|^2. \tag{15.21}$$

We now investigate the possibility of using one fixed basis $\{e_\alpha^a \equiv e_\alpha^a(C_0)\}$, $C_0 \in \mathscr{C}^{\text{hyp}}$, for all states $\psi_C, C \in \mathscr{C}^{\text{hyp}}$. For any $C \in \mathscr{C}^{\text{hyp}}$ we would like to have the representation

$$\phi_C = v_1^a(C)e_{\alpha_1}^a(C_0) + v_2^a(C)e_{\alpha_2}^a(C_0), \quad \text{where } |v_j^a(C)|^2 = p_C^a(\alpha_j). \tag{15.22}$$

We have

$$\psi_C(\beta_1) = u_1^a(C)v_{11}(C_0) + \varepsilon_C(\beta_1)\varepsilon_{C_0}(\beta_1)e^{j[\theta_C(\beta_1)-\theta_{C_0}(\beta_1)]}u_2^a(C)v_{12}(C_0),$$

$$\psi_C(\beta_2) = u_1^a(C)v_{21}(C_0) + \varepsilon_C(\beta_2)\varepsilon_{C_0}(\beta_2)e^{j[\theta_C(\beta_2)-\theta_{C_0}(\beta_2)]}u_2^a(C)v_{22}(C_0).$$

Thus to obtain (15.22) we should have

$$\varepsilon_C(\beta_1)\varepsilon_{C_0}(\beta_1)e^{j[\theta_C(\beta_1)-\theta_{C_0}(\beta_1)]} = \varepsilon_C(\beta_2)\varepsilon_{C_0}(\beta_2)e^{j[\theta_C(\beta_2)-\theta_{C_0}(\beta_2)]}.$$

Thus

$$\theta_C(\beta_1) - \theta_{C_0}(\beta_1) = \theta_C(\beta_2) - \theta_{C_0}(\beta_2), \quad \text{or}$$

$$\theta_C(\beta_1) - \theta_C(\beta_2) = \theta_{C_0}(\beta_1) - \theta_{C_0}(\beta_2).$$

By choosing the representation with (15.18) we satisfy the above condition.

Theorem 15.1. *We can construct the QL (hyperbolic Hilbert space) representation of the set of hyperbolic contexts $\mathscr{C}^{\text{hyp}}$ such that the hyperbolic analogue of Born's rule holds for both reference observables a, b (which are assumed to be probabilistically conjugate) iff the matrix $\mathbf{P}^{b|a}$ is double stochastic.*

We remark that in the Kolmogorov framework, basic contexts $C_\beta = \{\omega \in \Omega : b(\omega) = \beta\}$, $\beta \in X_b$, always belong to $\mathscr{C}^{\text{hyp}}$, see Part II, Chap. 4, details of the proof of Theorem 4.3. Thus $\psi_{C_\beta} \in \mathscr{H}$; and $C_\beta \in \mathscr{C}^{\text{tr}} \cap \mathscr{C}^{\text{hyp}}$ iff the observables a and b are symmetrically conditioned.

15.3 Hyperbolic Quantization

As in the ordinary quantum formalism, we represent physical states by normalized vectors of hyperbolic Hilbert space $\mathcal{H} : \psi \in S_1$. We shall consider only dichotomous physical variables and quantum states belonging to two-dimensional Hilbert space. Thus everywhere below $\mathcal{H}$ denotes a two-dimensional space. Let $a = \alpha_1, \alpha_2$ and $b = \beta_1, \beta_2$ be two physical variables. We represent them (in Dirac's notation) by **G**-linear operators

$$\hat{a} = \alpha_1 |\alpha_1\rangle \otimes \langle\alpha_1| + \alpha_2 |\alpha_2\rangle \otimes \langle\alpha_2|$$

and

$$\hat{b} = \beta_1 |\beta_1\rangle \otimes \langle\beta_1| + \beta_2 |\beta_2\rangle \otimes \langle\beta_2|,$$

where $\{|\alpha_i\rangle\}_{i=1,2}$ and $\{|\beta_i\rangle\}_{i=1,2}$ are two orthonormal bases in $\mathcal{H}$. The latter condition plays a fundamental role in hyperbolic quantum mechanics. This is an analogue of the representation of physical observables by self-adjoint operators in the conventional quantum mechanics (in the complex Hilbert space).

Let ψ be a state (normalized vector belonging to $\mathcal{H}$). We can perform the following operation (which is well defined from the mathematical point of view). We expand the vector ψ with respect to the basis $\{|\beta_i\rangle\}_{i=1,2}$

$$\psi = v_1^b |\beta_1\rangle + v_2^b |\beta_2\rangle, \tag{15.23}$$

where the coefficients (coordinates) v_i^b belong to **G**. We remark that we consider the two-dimensional **G**-Hilbert space. There exists (by definition) a basis consisting of two vectors. As the basis $\{|\beta_i\rangle\}_{i=1,2}$ is orthonormal, we have (as in the complex case) that

$$|v_1^b|^2 + |v_2^b|^2 = 1. \tag{15.24}$$

However, we could not automatically use Born's probabilistic interpretation for normalized vectors in hyperbolic Hilbert space: it may be that $v_i^b \notin \mathbf{G}_+$ and hence $|v_i^b|^2 < 0$ (in fact, in the complex case we have $\mathbf{C} = \mathbf{C}_+$; thus there is no problem with positivity). Since we do not want to consider negative probabilities, in such a case we cannot use the hyperbolic version of Born's probability interpretation.

Definition 15.2. *A state ψ is decomposable with respect to the system of states $\{|\beta_i\rangle\}_{i=1,2}$ (b-decomposable) if*

$$v_i^b \in \mathbf{G}_+. \tag{15.25}$$

In such a case we can use a generalization of Born's probabilistic interpretation for hyperbolic Hilbert space. Numbers

$$p_\psi^b(\beta_i) = |v_i^b|^2, \quad i = 1, 2,$$

are interpreted as probabilities for values $b = \beta_i$ for the state ψ.

We remark that in this framework (here we started with hyperbolic Hilbert space and not with a contextual statistical model, cf. Sect. 15.3) hyperbolic generalization of Born's rule is a postulate!

Thus decomposability is not a mathematical notion. This is not just linear algebraic decomposition of a vector with respect to a basis. This is a physical notion describing the possibility of probability interpretation of a measurement over a state. As was already mentioned, in hyperbolic quantum mechanics a state $\psi \in \mathscr{H}$ is not always decomposable. Thus for an observable b there can exist a state ψ such that the probabilities $p_\psi^b(\beta_i)$ are not well defined.

One of the reasons for this could be the impossibility of performing the b-measurement for systems in the state ψ. Such a situation is quite natural from the experimental viewpoint. Moreover, it is surprising that in ordinary quantum (as well as classical) theory one can measure any observable in any state. I think that it is just a consequence of the fact that there was fixed the set of states corresponding to a rather special class of physical observables. Thus in the hyperbolic quantum formalism for each state $\psi \in \mathcal{H}$, there exists its own set of observables $\mathcal{O}(\psi)$. And in general

$$\mathcal{O}(\psi) \neq \mathcal{O}(\phi).$$

We cannot exclude another possibility. The set of observables $\mathcal{O}$ does not depend on a state ψ. And the result of an individual measurement of any $b \in \mathcal{O}$ is well defined for any state ψ. But relative frequencies of realizations of the value $b = \beta_k$ do not converge to any limit. Therefore probabilities are not well defined. Thus the principle of statistical stabilization should be violated.

Remark 15.2. We now consider a contextual statistical model. We apply the QLRA to hyperbolic contexts of this model. Let $\psi \in S_{\mathcal{C}^{\text{hyp}}}$. Thus $\psi = \psi_C$ for some context $C \in \mathcal{C}^{\text{hyp}}$. Let the matrix of transition probabilities $\mathbf{P}^{b|a}$ be double stochastic. Then ψ is decomposable with respect to both reference variables b and a. Moreover, basis vectors $e_\beta^b = |\beta\rangle$ are a-decomposable and vice versa.

We now start the derivation of the hyperbolic probabilistic rule by using the hyperbolic Hilbert space formalism. Suppose that a state $\psi \in \mathcal{H}$ is a-decomposable

$$\psi = v_1^a|\alpha_1\rangle + v_2^a|\alpha_2\rangle$$

and the coefficients $v_i^a \in \mathbf{G}_+$.

We also suppose that each state $|\alpha_i\rangle$ is decomposable with respect to the system of states $\{|\beta_i\rangle\}_{i=1,2}$. We have

$$|\alpha_1\rangle = v_{11}|\beta_1\rangle + v_{12}|\beta_2\rangle, \qquad |\alpha_2\rangle = v_{21}|\beta_1\rangle + v_{22}|\beta_2\rangle, \tag{15.26}$$

where the coefficients v_{ik} belong to $\mathbf{G}_+$. We have (since both bases are orthonormal)

$$|v_{11}|^2 + |v_{12}|^2 = 1, \qquad |v_{21}|^2 + |v_{22}|^2 = 1, \tag{15.27}$$

cf. (15.16). We can use the probabilistic interpretation of numbers $p_{ik} = |v_{ik}|^2$, namely $p_{ik} = p_{|\alpha_i\rangle}(\beta_k)$ is the probability for $b = \beta_k$ in the state $|\alpha_i\rangle$.

Let us consider matrix $V = (v_{ik})$. As in the complex case, the matrix V is unitary, since the vectors $|\alpha_1\rangle = (v_{11}, v_{12})$ and $|\alpha_2\rangle = (v_{21}, v_{22})$ are orthonormal. Hence we have normalization conditions (15.27) and the orthogonality condition

$$v_{11}\bar{v}_{21} + v_{12}\bar{v}_{22} = 0, \tag{15.28}$$

cf. (15.15). It must be noticed that in general unitarity does not imply that $v_{ik} \in \mathbf{G}_+$. The latter condition is an additional constraint on the unitary matrix V. Let us consider the matrix $\mathbf{P}^{b|a} = (p_{ik})$. This matrix is double stochastic (since V is unitary).

By using the $\mathbf{G}$-linear space calculation (the change of basis) we get $\psi = v_1^b|\beta_1\rangle + v_2^b|\beta_2\rangle$, where $v_1^b = v_1^a v_{11} + v_2^a v_{21}$ and $v_2^b = v_1^a v_{12} + v_2^a v_{22}$.

We remark that *decomposability is not transitive.* In principle ψ may be not decomposable with respect to $\{|\beta_i\rangle\}_{i=1,2}$, despite the decomposability of ψ with respect to $\{|\alpha_i\rangle\}_{i=1,2}$ and the decomposability of the latter system with respect to $\{|\beta_i\rangle\}_{i=1,2}$.

The possibility of decomposability is based on two (totally different) conditions: (15.24), normalization, and (15.25), positivity. Any $\mathbf{G}$-unitary transformation preserves the normalization condition. Thus we get automatically that $|v_1^b|^2 + |v_2^b|^2 = 1$. However, the condition of positivity in general is not preserved: it can be that $v_i^b \notin \mathbf{G}_+$ even if we have $v_i^a \in \mathbf{G}_+$ and the matrix V is $\mathbf{G}$-unitary.

Finally, suppose that ψ is decomposable with respect to $\{|\beta_i\rangle\}_{i=1,2}$. Thus $v_k^b \in \mathbf{G}_+$. Therefore coefficients $p_\psi^b(\beta_i) = |v_i^b|^2$ can be interpreted as probabilities for $b = \beta_k$ for the $\mathbf{G}$-quantum state ψ.

Let us consider states such that coefficients v_i^a, v_{ik} belong to $\mathbf{G}_+^*$. We can uniquely represent them as

$$v_i^a = \pm\sqrt{p_\psi^a(\alpha_i)}e^{j\xi_i}, \qquad v_{ik} = \pm\sqrt{p_{ik}}e^{j\gamma_{ik}}, \quad i, k, = 1, 2.$$

We find that

$$p^b_\psi(\beta_1) = p^a_\psi(\alpha_1)p_{11} + p^a_\psi(\alpha_2)p_{21} + 2\varepsilon_1 \cosh\theta_1\sqrt{p^a_\psi(\alpha_1)p_{11}p^a_\psi(\alpha_2)p_{21}}, \tag{15.29}$$

$$p^b_\psi(\beta_2) = p^a_\psi(\alpha_1)p_{12} + p^a_\psi(\alpha_2)p_{22} + 2\varepsilon_2 \cosh\theta_2\sqrt{p^a_\psi(\alpha_1)p_{12}p^a_\psi(\alpha_2)p_{22}}, \tag{15.30}$$

where $\theta_i = \eta + \gamma_i$ and $\eta = \xi_1 - \xi_2, \gamma_1 = \gamma_{11} - \gamma_{21}, \gamma_1 = \gamma_{12} - \gamma_{22}$ and $\varepsilon_i = \pm$. To find the right relation between signs of the last terms in (15.29), (15.30), we use the normalization condition

$$|v^b_2|^2 + |v^b_2|^2 = 1 \tag{15.31}$$

(which is a consequence of the normalization of ψ and orthonormality of the system $\{|\beta_i\rangle\}_{i=1,2}$).

Remark 15.3. We remark that the normalization condition (15.31) can be reduced to relations between coefficients of the transition matrix V. So it does not depend on the original a-decomposition of ψ, namely coefficients v^a_i. A condition of positivity, $|v^b_i|^2 \geq 0$, could not be written by using only coefficients of V. We also need to use coefficients v^a_i. Therefore it seems to be impossible to find such a class of linear transformations V that would preserve a condition of positivity, "decomposition-group" of operators.

Equation (15.31) is equivalent to the equation

$$\sqrt{p_{12}p_{22}}\cosh\theta_2 \pm \sqrt{p_{11}p_{21}}\cosh\theta_2 = 0. \tag{15.32}$$

Thus we have to choose opposite signs in (15.29), (15.30). Unitarity of V also implies that $\theta_1 - \theta_2 = 0$, so $\gamma_1 = \gamma_2$. We recall that in the ordinary quantum mechanics we have similar conditions, but trigonometric functions are used instead of hyperbolic and phases γ_1 and γ_2 are such that $\gamma_1 - \gamma_2 - \pi$.

Finally, we get that unitary linear transformations in the **G**-Hilbert space (in the domain of decomposable states) represent the following transformation of probabilities

$$p_\psi^b(\beta_1) = p_\psi^a(\alpha_1)p_{11} + p_\psi^a(\alpha_2)p_{21} \pm 2\varepsilon_1 \cosh\theta_1 \sqrt{p_\psi^a(\alpha_1)p_{11}p_\psi^a(\alpha_2)p_{21}}, \tag{15.33}$$

$$p_\psi^b(\beta_2) = p_\psi^a(\alpha_1)p_{12} + p_\psi^a(\alpha_2)p_{22} \mp 2\varepsilon_2 \cosh\theta_2 \sqrt{p_\psi^a(\alpha_1)p_{12}p_\psi^a(\alpha_2)p_{22}}. \tag{15.34}$$

This is the hyperbolic interference. In Part II it was derived from the contextual statistical model. Then in Sect. 15.2 of this chapter, by using the interference formulas, we obtained the hyperbolic Hilbert space representation for contexts. In this section we started directly from hyperbolic Hilbert space representation and derived the hyperbolic interference of probabilities.

We remark that the state dynamics is described by the hyperbolic version of the Schrödinger equation

$$j\frac{\partial \psi}{\partial t} = \hat{H}\psi.$$

Here the state ψ belongs to hyperbolic Hilbert space $\mathcal{H}$, the evolution parameter t is real, $\hat{H} : \mathcal{H} \to \mathcal{H}$ is a symmetric **G**-linear operator having basis consisting of its eigenvectors.

Chapter 16

Hyperbolic Quantum Mechanics as Deformation of Conventional Classical Mechanics

In this chapter we develop Fourier analysis over the hyperbolic algebra. We demonstrate that classical mechanics has, besides the well-known quantum deformation over complex numbers, another deformation, namely, hyperbolic quantum mechanics. The classical Poisson bracket can be obtained as the limit $h \to 0$ not only of the ordinary Moyal bracket, but also a hyperbolic analogue of the Moyal bracket.

16.1 On the Classical Limit of Hyperbolic Quantum Mechanics

In the previous chapter we derived the hyperbolic interference starting with our general contextual model—the Växjö model. Then we derived such an interference by using hyperbolic quantum mechanics. In hyperbolic quantum mechanics observables are represented by self-adjoint operators in hyperbolic Hilbert space—a Hilbert module over the hyperbolic algebra. The crucial role is played not by linear superposition of state vectors (which loses their physical meaning and should be considered as a purely mathematical operation), but by decomposition of a state with respect to an observable. As we have seen, the conventional principle of superposition, see Part I: Chap. 1, is violated.

We remark that hyperbolic quantization also appears naturally in relativistic quantum physics. The hyperbolic numbers offer the possibility of representing the four-component Dirac spinor as a two-component hyperbolic spinor. Hucks has shown [128] that the Lorentz group is equivalent to the hyperbolic unitary group. Poteous [273] proved the unitarity of a special linear group with the help of the double field, which corresponds to the null basis representation of the hyperbolic numbers. S. Ulrych investigated the hyperbolic representation of Poincare mass [301–303]. He also studied symmetries in the hyperbolic Hilbert space [301–303]. Applications of hyperbolic numbers in general relativity can be found in the paper [227] of G. Kunstatter et al. These intensive applications of hyperbolic numbers in quantum physics induce a natural question

What is a classical limit of the hyperbolic QM?

We study this problem in this chapter, see also [211]. We recall that in conventional quantum mechanics this problem was solved by using the *deformation quantization* framework, see particularly the work of Moyal [249] (and, e.g., [138] for a modern presentation and various generalizations; in particular, to functional superanalysis). In this framework it was proved that when $h \to 0$ the Moyal bracket on the space $E(Q \times P)$ of symbols $a(q, p)$ of pseudo-differential operators (representing quantum observables) was transformed into the Poisson bracket.

We shall use the same approach in the hyperbolic case. We develop *Fourier analysis and calculus of pseudo-differential operators over the hyperbolic algebra* **G** and we found the limit of the hyperbolic Moyal bracket. Surprisingly we obtain the standard Poisson bracket. Thus

The classical limit of hyperbolic quantum mechanics is ordinary classical mechanics.

Classical mechanics can be deformed in two ways by using complex and hyperbolic representations. These deformations describe two different types of interference of probabilities: the trigonometric interference and the hyperbolic interference.

We also point to a recent publication of R. Hudson [129] in that he obtained a similar result, but by using a completely different approach (without consideration of hyperbolic numbers).

As always, the hyperbolic algebra is denoted by the symbol **G**. We remark that for any $y \in \mathbf{R}$ the map

$$\mathbf{R} \to \mathbf{G}, \qquad x \to \chi_y(x) = e^{jyx},$$

is an additive **G**-valued character

$$\chi_y(x_1 + x_2) = \chi_y(x_1)\chi_y(x_2), \quad x_1, x_2 \in \mathbf{R},$$

$$|\chi_y(x)| = 1.$$

We shall use these **G**-valued characters on **R** to define an analogue of the Fourier transform and pseudo-differential operators. We also introduce on **G** the positive norm $\|z\| = \sqrt{x^2 + y^2}$. which will be used in analysis over **G**.

16.2 Ultra-Distributions and Pseudo-Differential Operators over the Hyperbolic Algebra

We recall that for a function $\varphi : \mathbf{R} \to \mathbf{C}$ the Fourier transform is defined by

$$\tilde{\varphi}(p) = \frac{1}{2\pi h} \int_{-\infty}^{+\infty} e^{\frac{-ipq}{h}} \varphi(q)\, dq$$

and the inverse Fourier transform given by:

$$\varphi(q) = \int_{-\infty}^{+\infty} e^{\frac{ipq}{h}} \tilde{\varphi}(p)\, dp. \tag{16.1}$$

These formulas are well defined for, e.g., functions $\varphi \in \mathscr{S}$, where $\mathscr{S}$ is the space of Schwartz test functions. A pseudo-differential operator $\hat{a}$ with the symbol $a(q, p)$ is defined by

$$\hat{a}(\varphi)(q) = \int_{-\infty}^{+\infty} a(q, p) e^{\frac{iqp}{h}} \tilde{\varphi}(p)\, dp. \tag{16.2}$$

We would like to use the analogous definitions in the case of functions $\varphi : \mathbf{R} \to \mathbf{G}$, and $a : \mathbf{R} \times \mathbf{R} \to \mathbf{G}$ by using instead of additive $\mathbf{C}$-valued characters $x \to e^{iyx}$ additive $\mathbf{G}$-valued characters $x \to e^{jyx}$. The only problem is that the latter exponent is not bounded and, e.g., the class of functions $\mathscr{S}$ cannot be used as the base of the hyperbolic Fourier calculus. Even if we chose the space D of test functions with compact supports, then, for $\varphi \in \mathscr{D}$, the inverse Fourier transform (16.1) is in general not well defined.

One of the ways to proceed in such a case is to use the theory of analytic generalized functions, ultradistributions, cf. [138]. Let us consider the space $\mathscr{A}(\mathbf{R}, \mathbf{G})$ of analytic functions

$$f(x) = \sum_{n=0}^{\infty} f_n x^n, \quad f_n \in \mathbf{G},$$

and $\|f\|_R = \sum_{n=0}^{\infty} \|f_n\| R^n < \infty,\ \forall R > 0$. The $\mathbf{G}$-module $\mathscr{A}(\mathbf{R}, \mathbf{G})$ can be endowed with the topology given by the system of norms $\{\|\cdot\|_R\}$. This is a complete metrizable $\mathbf{G}$-module (Frechet module). We denote by the symbol $\mathscr{A}'(\mathbf{R}, \mathbf{G})$ the space of continuous $\mathbf{G}$-linear functionals

$$\lambda : \mathscr{A}(\mathbf{R}, \mathbf{G}) \to \mathbf{G}.$$

Functions $\varphi \in \mathscr{A}(\mathbf{R}, \mathbf{G})$ are called analytic test functions, functionals $\varphi \in \mathscr{A}'(\mathbf{R}, \mathbf{G})$ are called ($\mathbf{G}$-valued) ultradistributions. As usual in the theory of distributions, we define the derivative of $\lambda \in \mathscr{A}'(\mathbf{R}, \mathbf{G})$ by $(\frac{d\lambda}{dx}, \varphi) = -(\lambda, \frac{d\varphi}{dx})$. This operation is well defined in the space $\mathscr{A}'(\mathbf{R}, \mathbf{G})$. The Fourier transform of an ultradistribution $\lambda \in \mathscr{A}'(\mathbf{R}, \mathbf{G})$ is the function

$$\hat{\lambda}(y) \equiv \mathscr{F}(\lambda)(y) = (\lambda(x), e^{jyx}), \quad y \in \mathbf{R}.$$

Properties of the Fourier transform are collected in the following proposition and theorem.

Proposition 16.1. *For any ultradistribution* $\lambda \in \mathscr{A}'(\mathbf{R}, \mathbf{G})$ *its Fourier transform is infinitely differentiable. We have*

$$\frac{d^n}{dy^n}\mathscr{F}(\lambda)(y) = j^n \mathscr{F}(x^n \lambda(y)),$$

$$\mathscr{F}\left(\frac{d^n \lambda}{dx^n}\right)(y) = (-jy)^n \mathscr{F}(\lambda)(y).$$

We denote the Fourier-image of the space of ultradistributions by the symbol $E(\mathbf{R}, \mathbf{G})$.

We remark that the Dirac δ-function $\delta(x)$ belongs to $\mathscr{A}'(\mathbf{R}, \mathbf{G})$ and as always, we have $\mathscr{F}(\delta^{(n)}) = (-jy)^n$. Thus, in particular, the space $E(\mathbf{R}, \mathbf{G})$ contains all polynomials with coefficients belonging to $\mathbf{G}$. The description of the space $E(\mathbf{R}, \mathbf{G})$ is given by the following theorem.

Theorem 16.1 (Paley-Wiener). *The Fourier-image* $E(\mathbf{R}, \mathbf{G})$ *is equal to the space*

$$\left\{\varphi \in \mathscr{A}(\mathbf{R}, \mathbf{G}) : \left\| \frac{d^n \varphi}{dy^n}(0) \right\| \leq C_\varphi R_\varphi^n \right\}.$$

Thus the Fourier-image consists of $\mathbf{G}$-valued analytic functions which have exponentially growing derivatives. The proof of this theorem is rather long and we do not present it here. This proof is similar to the proof of the analogous theorem in superanalysis, see [138].

To proceed to the theory of $\mathbf{G}$-valued pseudo-differential operators, we chose the space of symbols $a(q, p) \in E(Q \times P, \mathbf{G})$, where $Q \times P = \mathbf{R}^2$ is the (ordinary) phase space. We can easily generalize all previous constructions to the multi-dimensional case.

The map

$$\mathscr{F} : \mathscr{A}'(\mathbf{R}, \mathbf{G}) \to E(\mathbf{R}, \mathbf{G})$$

is one-to-one. Thus, for any $v \in E(\mathbf{R}, \mathbf{G})$, there exists the unique ultradistribution $\lambda \in \mathscr{A}'(\mathbf{R}, \mathbf{G}) : \mathscr{F}(\lambda) = v$. We denote this λ by the symbol $\tilde{v}$. We shall also use (as people do in physics) the integral symbol to denote the action of an ultradistribution λ to a test function

$$f : (\lambda, f) \equiv \int f(x)\lambda(dx).$$

In particular,

$$\mathscr{F}(\lambda)(y) \equiv \int e^{jyx}\lambda(dx),$$

and, for a symbol $a \in E(Q \times P, \mathbf{G})$, we have

$$a(q, p) = \int e^{j(qp_1 + pq_1)}\tilde{a}(dp_1 dq_1). \tag{16.3}$$

To introduce into the model the Planck parameter $h > 0$, we modify the definition of the Fourier transform for functions φ from the domain of definition of a pseudo-differential operator

$$\varphi(q) = \int e^{\frac{jqp}{h}}\tilde{\varphi}(dp),$$

where $\lambda = \tilde{\varphi} \in \mathscr{A}'$. At the same time we preserve the definition (16.3) of the Fourier transform for symbols. We define the pseudo-differential operator $\hat{a}$ with the symbol $a \in E(Q \times P, \mathbf{G})$ by the natural generalization of (16.1)

$$\hat{a}(\varphi)(q) = \int a(q, p)e^{\frac{jpq}{h}}\tilde{\varphi}(dp). \tag{16.4}$$

We remark that $E(P, \mathbf{G}) \subset \mathscr{A}(P, \mathbf{G})$. Thus the function $f(p) \equiv a(q, p)e^{\frac{jpq}{h}} \in \mathscr{A}(P, \mathbf{G})$ for any $q \in \mathbf{R}$. Hence we can apply $\lambda \equiv \tilde{\varphi} \in \mathscr{A}'(P, \mathbf{G})$ to the analytic test function f. In principle, the formula (16.4) can be used to define a pseudo-differential operator $\hat{a}$ with a symbol $a \in \mathscr{A}(Q \times P, \mathbf{G})$. However, I do not know how to prove the correspondence principle for this larger class of symbols.

Let $a(q, p) = q$. Then

$$\hat{a}(\varphi)(q) = \int qe^{\frac{jpq}{h}}\tilde{\varphi}(dp) = q\varphi(q).$$

Let $a(q, p) = p$. Then

$$\hat{a}(\varphi)(q) = \int p e^{\frac{jpq}{h}} \tilde{\varphi}(dp) = \frac{h}{j} \frac{d}{dq} \int e \frac{jpq}{h} \tilde{\varphi}(dp) = \frac{h}{j} \frac{d}{dq} \varphi(q).$$

The first operator $\hat{q}$ is the position operator and the second operator $\hat{p}$ is the momentum operator. This is the hyperbolic Schrödinger representation

$$\hat{q} = q, \qquad \hat{p} = \frac{h}{j} \frac{d}{dq}.$$

We have the hyperbolic canonical commutation relation

$$[\hat{q}, \hat{p}] = \hat{q}\hat{p} - \hat{p}\hat{q} = -hj.$$

Proposition 16.2. *Any symbol $a \in E(Q \times P, \mathbf{G})$ defines the operator*

$$\hat{a} : E(Q, \mathbf{G}) \to E(Q, \mathbf{G}).$$

Proof. As always, we define the direct product of distributions $\lambda_1, \lambda_2 \in \mathscr{A}(\mathbf{R}, \mathbf{G})$

$$(\lambda_1 \otimes \lambda_2(x_1, x_2), \varphi(x_1, x_2)) = (\lambda_1(x_1), (\lambda_2(x_2), \varphi(x_1, x_2)))$$

for $\varphi \in \mathscr{A}(\mathbf{R}^2, \mathbf{G})$. This operation $\mathscr{A}(\mathbf{R}, \mathbf{G}) \times \mathscr{A}(\mathbf{R}, \mathbf{G}) \to \mathscr{A}(\mathbf{R}^2, \mathbf{G})$ is well defined. We have

$$\begin{aligned} \hat{a}(\varphi)(q) &= \int \left[\int e^{j(p_1 q + q_1 p)} \tilde{a}(dp_1 dq_1) \right] e^{\frac{jpq}{h}} \tilde{\varphi}(dp) \\ &= \int e^{j(p_1 q + q_1 p) + \frac{jpq}{h}} \tilde{a} \otimes \tilde{\varphi}(dp_1 dq_1 dp). \end{aligned}$$

Let us consider the $\mathbf{G}$-linear continuous operator

$$\mathbf{S} : \mathscr{A}(P, \mathbf{G}) \to \mathscr{A}(P \times Q \times P, \mathbf{G}),\ S(f)(p_1, q_1, p) = f(p + p_1 h) e^{j q_1 p}.$$

Then we have

$$\hat{a}(\varphi)(q) = \int e^{\frac{jqp}{h}} \tilde{a} \otimes \tilde{\varphi} \circ \mathbf{S}(dp).$$

Thus

$$\hat{a}(\varphi) = \mathscr{F}(\lambda), \lambda \in \mathscr{A}'(P, \mathbf{G}).$$

We have

$$(\tilde{a} \otimes \tilde{\varphi} \circ \mathbf{S}, f) = (\tilde{a} \otimes \tilde{\varphi}, \mathbf{S}(f)),$$

and, since $\mathbf{S}$ is continuous,

$$\lambda = \tilde{a} \otimes \tilde{\varphi} \circ \mathbf{S} \in \mathscr{A}'.$$

In fact, any pseudo-differential operator $\hat{a} : E \to E$ is continuous in a natural topology of inductive limit on E. However, we shall not use this fact.

Proposition 16.3. *Any pseudo-differential operator can be represented in the form*

$$\hat{a}(\varphi)(q) = \int e^{jqp_1} \varphi(q + hq_1) \tilde{a}(dp_1 dq_1). \tag{16.5}$$

Proof. We have

$$\hat{a}(\varphi)(q) = \int \Big[\int e^{\frac{jp(q+hq_1)}{h}} \tilde{\varphi}(dp) \Big] e^{jqp_1} \tilde{a}(dp_1 dq_1).$$

Theorem 16.2 (The formula of composition). *For any two pseudo-differential operators $\hat{a}_1, \hat{a}_2 : E(Q, \mathbf{G}) \to E(Q, \mathbf{G})$ with symbols $a_1, a_2 \in E(Q \times P, \mathbf{G})$, the composition $\hat{a} = \hat{a}_1 \circ \hat{a}_2$ is again a pseudo-differential operator with the symbol $a \in E(Q \times P, \mathbf{G})$ and*

$$a(q, p) = a_1 * a_2(q, p) = \int e^{jq(p_1+p_2)+jp(q_1+q_2)+jhq_1 p_2} \tilde{a}_1 \otimes \tilde{a}_2(dp_1 dq_1 dp_2 dq_2). \tag{16.6}$$

Proof. By (16.5) we have

$$\begin{aligned}
\hat{a}_1(\hat{a}_2(\varphi))(q) &= \int e^{jqp_1} a_2(\varphi)(q + hq_1) \tilde{a}_1(dp_1 dq_1) \\
&= \int e^{jqp_1} \Big[\int e^{j(q+hq_1)p_2} \varphi(q + hq_1 + hq_2) \tilde{a}_2(dp_2 dq_2) \Big] \tilde{a}_1(dp_1 dq_1) \\
&= \int e^{jq(p_1+p_2)} e^{jhq_1 p_2} \varphi(q + h(q_1 + q_2)) \tilde{a}_1 \otimes \tilde{a}_2(dp_1 dq_1 dp_2 dq_2).
\end{aligned}$$

We introduce a **G**-linear continuous operator

$$B : \mathscr{A}(P \times Q, \mathbf{G}) \to \mathscr{A}(P \times Q \times P \times Q, \mathbf{G}),$$

$$B(f)(p_1, q_1, p_2, q_2) = e^{jhq_1 p_2} f(p_1 + p_2, q_1 + q_2).$$

We can write

$$\hat{a}_1(\hat{a}_2(\varphi))(q) = \int e^{jqp_1} \varphi(q + hq_1) \tilde{a}_1 \otimes \tilde{a}_2 \circ B(dp_1 dq_1).$$

Since B is a continuous operator, $\lambda = a_1 \otimes a_2 \circ B \in \mathscr{A}'$. Thus $\hat{a}_1 \circ \hat{a}_2$ is also a pseudo-differential operator and its symbol

$$\begin{aligned} a(q, p) = \mathscr{F}(\lambda)(q, p) &= \int e^{j(qp_1 + pq_1)} \tilde{a}_1 \otimes \tilde{a}_2 \circ B(dp_1 dq_1) \\ &= \int e^{j(q(p_1+p_2)+p(q_1+q_2))} e^{jhq_1 p_2} \tilde{a}_1 \otimes \tilde{a}_2 (dp_1 dq_1 dp_2 dq_2). \end{aligned}$$

We now introduce on the space $E(Q \times P, \mathbf{G})$ of symbols the hyperbolic Moyal bracket

$$\{a_1, a_2\}_*(q, p) = a_1 * a_2(q, p) - a_2 * a_1(q, p),$$

where the operation $*$ is defined by (16.6). We remark that $* = *(h)$ depends on the Planck parameter $h > 0$. Thus the Moyal bracket also depends on $h : \{a_1, a_2\}_{*(h)}$. On the space of smooth functions $f : Q \times P \to \mathbf{G}$ we introduce the Poisson bracket

$$\{a_1, a_2\}(q, p) = \frac{\partial a_1}{\partial p}(q, p) \frac{\partial a_2}{\partial q}(q, p) - \frac{\partial a_1}{\partial q}(q, p) \frac{\partial a_2}{\partial p}(q, p).$$

The space $(E(Q \times P, \mathbf{G}), \{\cdot,\cdot\})$ is a Lie algebra. It contains the Lie algebra of classical mechanics, $E(Q \times \mathbf{P}, \mathbf{R})\{\cdot,\cdot\})$.

Theorem 16.3. *Let* $a_1, a_2 \in E(Q \times P, \mathbf{G})$. *Then*

$$\lim_{h \to 0} \frac{j}{h} \{a_1, a_2\}_{*(h)}(q, p) = \{a_1, a_2\}(q, p), \quad (q, p) \in Q \times P. \tag{16.7}$$

Proof. We have

$$\begin{aligned}\{a_1, a_2\}_*(q, q) &= \int e^{jq(p_1+p_2)+jp(q_1+q_2)}[e^{jhq_1p_2} - e^{jhq_2p_1}]\tilde{a}_1 \\ &\quad \otimes \tilde{a}_2(dp_1dq_1dp_2dq_2) \\ &= jh \int e^{jq(p_1+p_2)+jp(q_1+q_2)}[q_1p_2 - q_2p_1]\tilde{a}_1 \\ &\quad \otimes \tilde{a}_2(dp_1dq_1dp_2dq_2) + 0(h).\end{aligned}$$

We also have

$$\begin{aligned}\frac{\partial a_1}{\partial p}(q, p)\frac{\partial a_2}{\partial q}(q, p) &= \frac{\partial}{\partial p}\int e^{j(qp_2+pq_1)}\tilde{a}_1(dp_1dq_1)\frac{\partial}{\partial q}\int e^{j(qp_2+pq_2)}\tilde{a}_2(dp_2dq_2) \\ &= \int j^2q_1p_2e^{jq(p_1+p_2)+jp(q_1+q_2)}\tilde{a}_1 \otimes \tilde{a}_2(dp_1dq_1dp_2dq_2).\end{aligned}$$

Thus we obtain the following hyperbolic Fourier representation of the Poisson bracket

$$\{a_1, a_2\}(q, p) = \int [q_1p_2 - q_2p_1]e^{jq(p_1+p_2)+jp(q_1+q_2)}\tilde{a}_1 \otimes \tilde{a}_2(dp_1dq_1dp_2dq_2).$$

Hence we have proved (16.7).

Conclusion. The hyperbolic quantum mechanics in the limit $h \to 0$ coincides with the classical mechanics.

16.3 The Classical Limit of the Hyperbolic Quantum Field Theory

The classical limit for quantum systems with an infinite number of degrees of freedom was investigated (on a rigorous mathematical level) in [132–135, 138]. I used the theory of ultradistributions on infinite-dimensional spaces to build the calculus of infinite-dimensional pseudo-differential operators and introduce the Moyal deformation of the Poisson bracket on the infinite-dimensional case, see [132–135, 138] for detail. We can do the same in the hyperbolic case.

Let X be an infinite-dimensional real topological vector (locally convex) space, e.g., the space $\mathscr{S}(\mathbf{R}^n)$ of Schwartz test functions or the space $\mathscr{S}^*(\mathbf{R}^n)$ of Schwartz distributions. Denote by the symbol Y the $\mathbf{R}$-dual space of X—the space of $\mathbf{R}$-linear continuous functionals $y : X \to \mathbf{R}$). As always, we use the notation $(y, x) = y(x)$.

Denote the space ($\mathbf{G}$-module) of analytic functions $f : X \to \mathbf{G}$ by the symbol $\mathscr{A}(X, \mathbf{G})$ and the space ($\mathbf{G}$-module) of continuous $\mathbf{G}$-linear functionals $\lambda : \mathscr{A} \to \mathbf{G}$ by the symbol $\mathscr{A}'(X, \mathbf{G})$.

We choose $\mathscr{A}(X, \mathbf{G})$ as the space of (analytic) $\mathbf{G}$-valued test functions and $\mathscr{A}'(X, \mathbf{G})$ as the space of $\mathbf{G}$-valued (ultra) distributions. $\mathbf{G}$-valued additive characters[1] $(y \in Y)$ on X,

$$x \to e^{j(y,x)}$$

belong to the space of $\mathbf{G}$-valued analytic functions. We define the Fourier transform of an ultradistribution $\lambda \in \mathscr{A}'(X, \mathbf{G})$ by

$$\hat{\lambda}(y) = \mathscr{F}(\lambda)(y) = (\lambda, e^{j(y,\cdot)}), \quad y \in Y.$$

This is an analytic function on the dual space $Y = X^*$ (endowed with the strong topology). We denote the Fourier image of the space $\mathscr{A}'(X, \mathbf{G})$ of ultradistributions by the symbol $E(Y, \mathbf{G})$. By using methods developed in [138] we can try to obtain an internal description of this $\mathbf{G}$-module, Paley-Wiener theorem. However, this is not a trivial problem.

It is important for us that $E(Y, \mathbf{G})$ contains cylindrical polynomials (as well as "nuclear polynomials"). Under some topological restrictions on X the Fourier transform

$$\mathscr{F} : \mathscr{A}(X, \mathbf{G}) \to E(Y, \mathbf{G})$$

[1] We recall that $|e^{j(y,x)}| = 1, x \in X$.

is a one-to-one map. We consider such a class of infinite-dimensional spaces, e.g., $X = \mathscr{S}(\mathbf{R}^n)$, $Y = \mathscr{S}^*(\mathbf{R}^n)$, or vice versa. Thus, for any $v \in E(Y, \mathbf{G})$ there exists the unique ultradistribution $\tilde{v} \in \mathscr{A}'(X, \mathbf{G})$ such that

$$v(y) = \mathscr{F}(\tilde{v})(y) \equiv \int e^{j(y,x)} \tilde{v}(dy)$$

(as always, we use the integral symbol to denote the pairing between an ultradistribution and a test function). In the same way as in the finite-dimensional case we introduce pseudo-differential operators with symbols $a \in E(Q \times P, \mathbf{G})$. Here the infinite-dimensional phase-space is introduced in the following way.

Let Q be an $\mathbf{R}$-linear locally convex space which is reflexive. Thus dual space $Q^* = P$ of Q (endowed with the strong topology) has the dual space $P^* = Q$. The space $Q \times P$ is the phase-space. We remark that $(Q \times P)^* = Q^* \times P^* = P \times Q$. In the above scheme we put $X = P \times Q$ and $Y = Q \times P$ and proceed

$$\varphi(q) = \int e^{\frac{j(p,q)}{h}} \tilde{\varphi}(dp), \quad \varphi \in E(Q, \mathbf{G}),$$

$$a(q,p) = \int e^{j(p_1,q)+j(p,q_1} \tilde{a}(dp_1 dq_1), \quad a \in E(Q \times P, \mathbf{G});$$

$$\hat{a}(\varphi)(q) = \int a(q,p) e^{\frac{j(p,q)}{h}} \tilde{\varphi}(dp).$$

By analogy with one-dimensional case we prove:

Theorem 16.4. *For any symbol $a \in E(Q \times P, \mathbf{G})$ the pseudo-differential operator $\hat{a} : E(Q, \mathbf{G}) \to E(Q, \mathbf{G})$. For $a_1, a_2 \in E(Q \times P, \mathbf{G})$, the operator $\hat{a} = \hat{a}_1 \circ \hat{a}_2$ is again a pseudo-differential operator with the symbol*

$$a(q,p) = a_1 * a_2(q,p)$$
$$= \int e^{j(p_1+p_2,q)+j(p,q_1+q_2)+jh(p_2,q_1)} \tilde{a}_1 \otimes \tilde{a}_2(dp_1 dq_1 dp_2 q_2).$$

We have the correspondence principle (16.7) *where $\{\cdot,\cdot\}_{*(h)}$ and $\{\cdot,\cdot\}$ are the Moyal and Poisson brackets, respectively.*

Conclusion. The classical limit of the hyperbolic quantum theory with an infinite number of degrees of freedom coincides with ordinary classical mechanics on the infinite-dimensional phase-space.

Thus we have two deformations of classical field theory: complex second quantization and hyperbolic second quantization.

16.4 Hyperbolic Fermions and Hyperbolic Supersymmetry

Let B be an algebra over a field K and let A be a ring which is also a B-module. Let operations of ring and module be connected in the natural way (in the same way as operations of a ring and a linear space are related in the case of an ordinary algebra). Such an algebraic structure A will be called a B-algebra.

The standard example is some space A of functions $f : \mathbf{R}^m \to \mathbf{G}$. They are $\mathbf{G}$-algebras.

Let us consider a supercommutative Banach $\mathbf{G}$-superalgebra $\Lambda = \Lambda_0 \oplus \Lambda_1$, see, e.g., [138].

For example, Λ can be a Grassmann $\mathbf{G}$-algebra with n-generators $\theta_1, \ldots, \theta_n$

$$\mathbf{G}_n = \left\{ u = \sum_\alpha c_\alpha \theta^\alpha : c_\alpha \in \mathbf{G} \right\}$$

and $\alpha = (\alpha_1, \ldots, \alpha_n)$, $\alpha_j = 0, 1$, $\theta^\alpha = \theta_1^{\alpha_1} \cdots \theta_n^{\alpha_n}$ and $\theta_i \theta_j = -\theta_j \theta_i$.

In the same way as in [138] we should consider $\mathbf{G}$-superalgebras $\Lambda = \Lambda_0 \oplus \Lambda_1$ with trivial Λ_1-annihilators

$${}^{\perp}\Lambda_1 = \{u \in \Lambda : u\lambda = 0,\ \forall \lambda \in \Lambda_1\} = \{0\}.$$

All Grassmann **G**-algebras with a finite number of generators have nontrivial Λ_1-annihilators. As an example of a supercommutative Banach **G**-superalgebra with trivial Λ_1-annihilator we can consider an infinite-dimensional Banach-Grassmann **G**-superalgebra, see [138].

We consider the superspace over $\mathbf{G} : \mathbf{R}^{k,l} = \Lambda_0^k \times \Lambda_1^l$ and construct the hyperbolic calculus of super pseudo-differential operators by combining results of Sect. 16.2 and [138]. We obtain the following result.

Theorem 16.5. *A hyperbolic Moyal super bracket is a deformation of the ordinary Poisson bracket on a superspace.*

We have seen that in some aspects hyperbolic quantum mechanics is similar to conventional complex quantum mechanics, e.g., quantization can be performed in the same way as it was done in the complex case—by using the calculus of pseudo-differential operators, the same is valid for the correspondence principle. Moreover, both theories have the same classical limit. However, there are differences in the probability interpretation of states. Therefore we need the notion of decomposability (the conventional superposition principle loses its probabilistic counterpart). Recently I and my post-doc Gavriel Segre found that even von Neumann's theorem about equivalence of different representations of canonical commutation relations is not valid in the hyperbolic case [202].

References

1. Aaronson, S.: Is quantum mechanics an island in theory space?. In: Khrennikov, A.Yu. (ed.) Quantum Theory: Reconsideration of Foundations, vol. 2. Ser. Math. Model., vol. 10, pp. 15–28. Växjö University Press, Växjö (2003); electronic volume: http://www.vxu.se/msi/forskn/publications.html
2. Accardi, L.: The probabilistic roots of the quantum mechanical paradoxes. In: Diner, S., Fargue, D., Lochak, G., Selleri, F. (eds.) The Wave–Particle Dualism. A Tribute to Louis de Broglie on his 90th Birthday, pp. 47–55. Reidel, Dordrecht (1970)
3. Accardi, L.: Urne e Camaleoni: Dialogo sulla Realta, le Leggi del Caso e la Teoria Quantistica. Il Saggiatore, Rome (1997)
4. Accardi, L., Lu, Y.G., Volovich, I.V.: Quantum Theory and its Stochastic Limit. Texts and Monographs in Physics. Springer, Berlin (2002)
5. Accardi, L.: Some loopholes to save quantum nonlocality. In: Adenier, G., Khrennikov, A.Yu. (eds.) Foundations of Probability and Physics, vol. 3. Ser. Conference Proceedings, vol. 750, pp. 1–20. American Institute of Physics, Melville (2005)
6. Accardi, L.: Could one now convince Einstein? In: Adenier, G., Khrennikov, A.Yu., Nieuwenhuizen, Th.M. (eds.) Quantum Theory: Reconsideration of Foundations, vol. 3. Ser. Conference Proceedings, vol. 810, pp. 3–18. American Institute of Physics, Melville (2006)
7. Accardi, L., Khrennikov, A.Yu.: Chameleon effect, the range of values hypothesis, and reproducing the EPR-Bohm correlations. In: Adenier, G., Fuchs, C., Khrennikov, A.Yu. (eds.) Foundations of Probability and Physics, vol. 3. Ser. Conference Proceedings, vol. 889, pp. 21–29. American Institute of Physics, Melville (2007)

8. Adenier, G., Khrennikov, A.Yu., Nieuwenhuizen, Th.M. (eds.): Quantum Theory: Reconsideration of Foundations, vol. 3. Ser. Conference Proceedings, vol. 810. American Institute of Physics, Melville (2006)
9. Adenier, G., Khrennikov, A.Yu.: Anomalies in EPR-Bell experiments. In: Adenier, G., Khrennikov, A.Yu., Nieuwenhuizen, Th.M. (eds.) Quantum Theory: Reconsideration of Foundations, vol. 3. Ser. Conference Proceedings, vol. 810, pp. 283–293. American Institute of Physics, Melville (2006)
10. Adenier, G., Fuchs, C., Khrennikov, A.Yu. (eds.): Foundations of Probability and Physics, vol. 3. Ser. Conference Proceedings, vol. 889. American Institute of Physics, Melville (2007)
11. Adenier, G., Khrennikov, A.Yu.: Is the fair sampling assumption supported by EPR experiments?. Phys. B: At. Mol. Opt. Phys. **40**(1), 131–141 (2007)
12. Aerts, D., Czachor, M., Pawlovski, M.: Security in quantum cryptography vs. nonlocal hidden variables. In: Adenier, G., Fuchs, C., Khrennikov, A.Yu. (eds.) Foundations of Probability and Physics, vol. 4. Ser. Conference Proceedings, vol. 89, pp. 71–78. American Institute of Physics, Melville (2007)
13. Aichele, T., Herzog, U., Scholz, M., Benson, O.: Single photon generation and simultaneous observation of wave and particle properties. In: Adenier, G., Khrennikov, A.Yu. (eds.) Foundations of Probability and Physics, vol. 3. Ser. Conference Proceedings, vol. 750, pp. 35–41. American Institute of Physics, Melville (2005)
14. Albeverio, S., Hoegh-Krohn, R.: Phys. Lett. B **177**, 175 (1989)
15. Albeverio, S., Yasue, K., Zambrini, J.C.: Euclidean quantum mechanics: analytical approach. Ann. Inst. H. Poincare, Phys. Theor. **50**, 259–308 (1989)
16. Allahverdyan, A.E., Balian, R., Nieuwenhuizen, Th.M.: The quantum measurement process in an exactly solvable model. In: Khrennikov, A.Yu. (ed.) Foundations of Probability and Physics, vol. 3. Ser. Conference Proceedings, vol. 750, pp. 16–24. American Institute of Physics, Melville (2005)
17. Allahverdyan, A., Khrennikov, A.Yu., Nieuwenhuizen, Th.M.: Brownian entanglement. Phys. Rev. A **71**, 0321021–03210214 (2005)
18. Andreev, V.A., Man'ko, V.I.: Two-particle spin states and generalized Bell's inequalities. JETP Lett. **72**, 93–96 (2000)
19. Andreev, V.A., Man'ko, V.I.: The classification of two-particle spin states and generalized Bell inequalities. Phys. Lett. A **281**, 278–288 (2001)

20. Andreev, V.A., Man'ko, V.I.: Bell's inequality for two-particle mixed spin states. Theor. Math. Phys. **140**, 1135–1145 (2004)
21. Andreev, V.A.: Reduction of the two-spin state density matrix and evaluation of the left-hand side of the Bell-CHSH inequality. J. Russ. Laser Res. **27**, 327–331 (2006)
22. Andreev, V.A., Man'ko, V.I., Man'ko, O.V., et al.: Tomography of spin states, the entanglement criterion, and Bell's inequalities. Theor. Math. Phys. **146**, 140–151 (2006)
23. Andreev, V.A., Man'ko, V.I.: Quantum tomography and verification of generalized Bell-CHSH inequalities. In: Adenier, G., Khrennikov, A.Yu. (eds.) Foundations of Probability and Physics, vol. 3. Ser. Conference Proceedings, vol. 750, pp. 42–48. American Institute of Physics, Melville (2005)
24. Appleby, D.M.: Concerning dice and divinity. In: Adenier, G., Fuchs, C., Khrennikov, A.Yu. (eds.) Foundations of Probability and Physics, vol. 3. Ser. Conference Proceedings, vol. 889, pp. 30–39. American Institute of Physics, Melville (2007)
25. Ashwanden, M., Philipp, W., Hess, K., Adenier, G.: Local time dependent instruction-set model for the experiment of Pan et al. In: Adenier, G., Khrennikov, A.Yu., Nieuwenhuizen, Th.M. (eds.) Quantum Theory: Reconsideration of Foundations, vol. 3. Ser. Conference Proceedings, vol. 810, pp. 437–446. American Institute of Physics, Melville (2006)
26. Aspect, A., Dalibard, J., Roger, G.: Experimental test of bell's inequalities using time-varying analyzers. Phys. Rev. Lett. **49**, 1804 (1982)
27. Atmanspacher, H., Primas, H.: Epistemic and ontic quantum realities. In: Adenier, G., Khrennikov, A.Yu. (eds.) Foundations of Probability and Physics, vol. 3. Ser. Conference Proceedings, vol. 750, pp. 49–62. American Institute of Physics, Melville (2005)
28. Bacciagaluppi, G.: Classical extensions, classical representations and Bayesian updating in quantum mechanics. In: Khrennikov, A.Yu. (ed.) Quantum Theory: Reconsideration of Foundations, vol. 2. Ser. Math. Model., vol. 10, pp. 55–70. Växjö University Press, Växjö (2003); electronic volume: http://www.vxu.se/msi/forskn/publications.html
29. Ballentine, L.E.: The statistical interpretation of quantum mechanics. Rev. Mod. Phys. **42**, 358–381 (1989)
30. Ballentine, L.E.: Quantum Mechanics. Prentice Hall, Englewood Cliffs (1989)
31. Ballentine, L.E.: Interpretations of probability and quantum theory. In: Khrennikov, A.Yu. (ed.) Foundations of Probability and Physics. Quantum Probability and White Noise Analysis, vol. 13, pp. 71–84. WSP, Singapore (2001)

32. Ballentine, L.E.: The classical limit of quantum mechanics and its implications for the foundations of quantum mechanics. In: Khrennikov, A.Yu. (ed.) Quantum Theory: Reconsideration of Foundations, vol. 2. Ser. Math. Model., vol. 10, pp. 71–82. Växjö University Press, Växjö (2003); electronic volume: http://www.vxu.se/msi/forskn/publications.html
33. Belavkin, V.P.: Visible information dynamics of hidden quantum mechanics. In: Khrennikov, A.Yu. (ed.) Foundations of Probability and Physics, vol. 2. Ser. Math. Model., vol. 5, pp. 69–98. Växjö University Press, Växjö (2003); electronic volume: http://www.vxu.se/msi/forskn/publications.html
34. Bell, J.: Speakable and Unspeakable in Quantum Mechanics. Cambridge University Press, Cambridge (1987)
35. Beltrametti, E.G., Cassinelli, G.: The logic of quantum mechanics. In: Rota, G.-C. (ed.) Encyclopedia of Mathematics and its Applications. Cambridge University Press, Cambridge (1984)
36. Bernasconi, J., Gustafson, K.: Contextual quick-learning and generalization by humans and machines. Netw. Comput. Neural Syst. **85**, 106 (1998)
37. Bialynicki-Birula, I.: Rényi entropy and the uncertainty relations. In: Adenier, G., Fuchs, C., Khrennikov, A.Yu. (eds.) Foundations of Probability and Physics, vol. 3. Ser. Conference Proceedings, vol. 889, pp. 52–61. American Institute of Physics, Melville (2007)
38. Bohm, D., Hiley, B.: The Undivided Universe: an Ontological Interpretation of Quantum Mechanics. Routledge and Kegan Paul, London (1993)
39. Bohr, N.: Can quantum-mechanical description of physical reality be considered complete? Phys. Rev. **48**, 696–702 (1935)
40. Bohr, N.: The philosophical writings of Niels Bohr, 3 vols. Ox Bow Press, Woodbridge (1987)
41. Boyer, T.H.: A brief survey of stochastic electrodynamics. In: Barut, A.O. (ed.) Foundations of Radiation Theory and Quantum Electrodynamics, pp. 141–162. Plenum, New York (1980)
42. Brida, G., Genovese, M., Gramegna, M., Novero, C., Predazzi, E.: A first test of Wigner function local realistic model. Phys. Lett. A **299**, 121 (2002)
43. Bulinski, A.V., Khrennikov, A.Yu.: Nonclassical total probability formula and quantum interference probabilities. Stat. Probab. Lett. **70**, 49–58 (2004)
44. Busch, P., Grabowski, M., Lahti, P.: Operational Quantum Physics. Springer, Berlin (1995)
45. Busch, P.: Less (precision) is more (information): quantum information in fuzzy probability theory. In: Khrennikov, A.Yu. (ed.) Quantum Theory: Reconsideration of Foundations, vol. 2.

Ser. Math. Model., vol. 10, pp. 113–148. Växjö University Press, Växjö (2003); electronic volume: http://www.vxu.se/msi/forskn/publications.html
46. Busemeyer, J.B., Wang, Z., Townsend, J.T.: Quantum dynamics of human decision making. J. Math. Psychol. **50**, 220–241 (2006)
47. Busemeyer, J.B., Wang, Z.: Quantum information processing explanation for interactions between inferences and decisions. In: Bruza, P.D., Lawless, W., van Rijsbergen, K., Sofge, D.A. (eds.) Quantum Interaction. AAAI Spring Symposium, pp. 91–97. AAAI Press, Menlo Park (2007). Technical Report SS-07-08
48. Casado, A., Marshal, T., Santos, E.: Parametric downconversion experiments in the Wigner representation. J. Opt. Soc. Am. B **14**, 494–502 (1997)
49. Caticha, A.: From objective amplitudes to Bayesian probability. In: Adenier, G., Fuchs, C., Khrennikov, A.Yu. (eds.) Foundations of Probability and Physics, vol. 3. Ser. Conference Proceedings, vol. 889, pp. 62–70. American Institute of Physics, Melville (2007)
50. Chaitin, G.J.: On the length of programs for computing finite binary sequences: statistical considerations. J. ACM **16**, 145–159 (1969)
51. Chaitin, G.J.: A theory of program size formally identical to information theory. J. ACM **22**, 329–340 (1975)
52. Choustova, O.A.: Pilot wave quantum model for the stock market. In: Khrennikov, A.Yu. (ed.) Quantum Theory: Reconsideration of Foundations. Ser. Math. Model., vol. 2, pp. 41–58. Växjö University Press, Växjö (2002); electronic volume: http://www.vxu.se/msi/forskn/publications.html
53. Choustova, O.: Bohmian mechanics for financial processes. J. Mod. Opt. **51**, 1111 (2004)
54. Choustova, O.: Price dynamics of shares and Bohmian mechanics: deterministic or stochastic model? In: Adenier, G., Khrennikov, A.Yu. (eds.) Foundations of Probability and Physics, vol. 3. Ser. Conference Proceedings, vol. 750, pp. 274–282. American Institute of Physics, Melville (2005)
55. Choustova, O.: Information field model for agents of financial market. In: Petitjean, M. (ed.) Proceedings of FIS2005, The Third Conference on the Foundations of Information Science. MDPI Publishers, Basel (2005); electronic volume: http://www.mdpi.org/fis2005/proceedings.html
56. Choustova, O.: Quantum Bohmian model for financial market. Physica A **374**, 304–314 (2006)

57. Choustova, O.: Quantum modeling of nonlinear dynamics of stock prices: Bohmian approach. Theor. Math. Phys. **152**, 405–416 (2007)

58. Choustova, O.: Quantum Bohmian model for financial market. Licentiat Thesis. University of Växjö Press, Växjö (2007)

59. Clauser, J.F., Horne, M.A., Shimony, A., Holt, R.A.: Phys. Rev. Lett. **49**, 1804 (1969)

60. Clauser, J.F., Shimony, A.: Bell's theorem. Experimental tests and implications. Rep. Prog. Phys. **41**, 1881 (1978)

61. Cole, D.: Simulation results related to stochastic electrodynamics. In: Adenier, G., Khrennikov, A.Yu., Nieuwenhuizen, Th.M. (eds.) Quantum Theory: Reconsideration of Foundations, vol. 3. Ser. Conference Proceedings, vol. 810, pp. 99–113. American Institute of Physics, Melville (2006)

62. Conte, E., Todarello, O., Federici, A., Vitiello, F., Lopane, M., Khrennikov, A., Zbilut, J.P.: Some remarks on an experiment suggesting quantum-like behavior of cognitive entities and formulation of an abstract quantum mechanical formalism to describe cognitive entity and its dynamics. Chaos Solitons Fractals **31**, 1076–1088 (2006)

63. D Ariano, G.M.: Operational axioms for quantum mechanics. In: Adenier, G., Fuchs, C., Khrennikov, A.Yu. (eds.) Foundations of Probability and Physics, vol. 3. Ser. Conference Proceedings, vol. 889, pp. 79–105. American Institute of Physics, Melville (2007)

64. Davidson, M.P.: Stochastic models of quantum mechanics—a perspective. In: Adenier, G., Fuchs, C., Khrennikov, A.Yu. (eds.) Foundations of Probability and Physics, vol. 3. Ser. Conference Proceedings, vol. 889, pp. 106–119. American Institute of Physics, Melville (2007)

65. Davies, E.B., Lewis, J.T.: Commun. Math. Phys. **17**, 239–260 (1970)

66. De Baere, W.: Lett. Nuovo Cimento **39**, 234–238 (1984)

67. De Baere, W.: Lett. Nuovo Cimento **40**, 488–498 (1984)

68. De Baere, W.: Adv. Electron. Electron Phys. **68**, 245–280 (1986)

69. De Baere, W.: Subquantum nonreproducibility and the complete local description of physical reality. In: Khrennikov, A.Yu. (ed.) Quantum Theory: Reconsideration of Foundations. Ser. Math. Model., vol. 2, pp. 59–74. Växjö University Press, Växjö (2002); electronic volume: http://www.vxu.se/msi/forskn/publications.html

70. De Broglie, L.: The Current Interpretation of Wave Mechanics. A Critical Study. Elsevier, Amsterdam (1964)

71. De la Pena, L., Cetto, A.M.: The Quantum Dice: An Introduction to Stochastic Electrodynamics. Kluwer Academic, Dordrecht (1996)

72. De la Pena, L.: Does quantum mechanics accept a stochastic support? Found. Phys. **12**, 1017 (1982)
73. De la Pena, L.: New formulation of stochastic theory and quantum mechanics. J. Math. Phys. **10**, 1620–1630 (1969)
74. De la Pena, L., Cetto, A.M.: Phys. Rev. D **3**, 795 (1971)
75. De la Pena, L., Cetto, A.M.: Recent development in linear stochastic electrodynamics. In: Adenier, G., Khrennikov, A.Yu., Nieuwenhuizen, Th.M. (eds.) Quantum Theory: Reconsideration of Foundations, vol. 3. Ser. Conference Proceedings, vol. 810, pp. 3–18, 131–140. American Institute of Physics, Melville (2006)
76. De Muynck, W.M., De Baere, W.: Ann. Israel Phys. Soc. **12**, 1–22 (1996)
77. De Muynck, W.M., De Baere, W., Marten, H.: Interpretations of quantum mechanics, joint measurement of incompatible observables, and counterfactual definiteness. Found. Phys. **24**, 1589–1663 (1994)
78. De Muynck, W.M., Stekelenborg, J.T.: On the significance of the bell inequalities for the locality problem in different realistic interpretations of quantum mechanics. Ann. Phys. **45**, 222–234 (1988)
79. De Muynck, W.M.: Interpretations of quantum mechanics, and interpretations of violations of Bell's inequality. In: Khrennikov, A.Yu. (ed.) Foundations of Probability and Physics. Series PQ-QP: Quantum Probability and White Noise Analysis, vol. 13, pp. 95–104. WSP, Singapore (2001)
80. De Muynck, W.M.: Foundations of Quantum Mechanics, an Empiricists Approach. Kluwer Academic, Dordrecht (2002)
81. Deutsch, D.: The fabric of reality. How Much Can Our Four Deepest Theories of the World Explain? Penguin, Allen Lane (1997)
82. d'Espagnat, B.: Veiled Reality. An Analysis of Present-day Quantum Mechanical Concepts. Addison-Wesley, Reading (1995)
83. d'Espagnat, B.: Conceptual Foundations of Quantum Mechanics. Perseus Books, Reading (1999)
84. Dirac, P.A.M.: The Principles of Quantum Mechanics. Clarendon, Oxford (1995)
85. Eberhard, P.H.: Bell's theorem without hidden variables. Nuovo Cimento B **38**, 75–80 (1977)
86. Eberhard, P.H.: Phys. Rev. Lett. **49**, 1474 (1982)
87. Eberhard, P.H.: Nuovo Cimento B **46**, 392 (1978)
88. Edwards, S.F., Anderson, P.W.: J. Phys. F **5**, 965 (1975)

89. Einstein, A., Podolsky, B., Rosen, N.: Can quantum-mechanical description of physical reality be considered complete? Phys. Rev. **47**, 777–780 (1935)
90. Ezhov, A.A., Khrennikov, A.Yu.: Agents with left and right dominant hemispheres and quantum statistics. Phys. Rev. E **71**, 016138 (2005)
91. Feynman, R., Hibbs, A.: Quantum Mechanics and Path Integrals. McGraw-Hill, New York (1965)
92. Feynman, R.P.: Negative probability. In: Hiley, B.J., Peat, F.D. (eds.) Quantum Implications, Essays in Honour of David Bohm, pp. 235–246. Routledge and Kegan Paul, London (1987)
93. Fine, A.: Hidden variables, joint probability, and the bell inequalities. Phys. Rev. Lett. **48**, 291–295 (1982)
94. Folse, H.J.: Bohr's conception of the quantum mechanical state of a system and its role in the framework of complementarity. In: Khrennikov, A.Yu. (ed.) Quantum Theory: Reconsideration of Foundations. Ser. Math. Model., vol. 2, pp. 83–98. Växjö University Press, Växjö (2002); electronic volume: http://www.vxu.se/msi/forskn/publications.html
95. Fuchs, C.A.: The anti-Växjö interpretation of quantum mechanics. In: Khrennikov, A.Yu. (ed.) Quantum Theory: Reconsideration of Foundations. Ser. Math. Model., vol. 2, pp. 99–116. Växjö University Press, Växjö (2002); electronic volume: http://www.vxu.se/msi/forskn/publications.html
96. Fuchs, C.: Delirium quantum (or, where I will take quantum mechanics if it will let me). In: Adenier, G., Fuchs, C., Khrennikov, A.Yu. (eds.) Foundations of Probability and Physics, vol. 3. Ser. Conference Proceedings, vol. 889, pp. 438–462. American Institute of Physics, Melville (2007)
97. Genovese, M.: Research on hidden variable theories: a review of recent progresses. Phys. Rep. **413**, 319 (2005)
98. Gill, R.: Time, finite statistics and Bell's fifth position. In: Khrennikov, A.Yu. (ed.) Foundations of Probability and Physics, vol. 2. Ser. Math. Model., vol. 5, pp. 179–206. Växjö University Press, Växjö (2003); electronic volume: http://www.vxu.se/msi/forskn/publications.html
99. Gisin, N., Gisin, B.: Bell inequality for arbitrary many settings of the analyzers. Phys. Lett. A **260**, 323 (1999)
100. Gnedenko, B.: Theory of Probability. CRC, Boca Raton (1998)

101. Greenberger, D., Horne, M., Zeilinger, A.: Going beyond Bell's theorem. In: Kafatos, M. (ed.) Bell's Theorem, Quantum Theory, and Conceptions of the Universe, pp. 73–76. Kluwer Academic, Dordrecht (1989)
102. Grib, A.A., Rodrigues, W.A.: Nonlocality in Quantum Physics. Kluwer Academic/Plenum, New York (1999)
103. Grib, A.A., Parfionov, G.N.: Can the game be quantum? In: Notes of Scient. Sem. of the Petersburg's Branch of the Mathematical Institute of the Russian Academy of Sciences, vol. 291, pp. 1–24. St. Petersburg (2002)
104. Grib, A., Khrennikov, A., Starkov, K.: Probability amplitude in quantum-like games. In: Khrennikov, A.Yu. (ed.) Quantum Theory: Reconsideration of Foundations. Ser. Math. Model., vol. 10, pp. 703–722. Växjö University Press, Växjö (2004); electronic volume: http://www.vxu.se/msi/forskn/publications.html
105. Grib, A., Khrennikov, A., Parfionov, G., Starkov, K.: Quantum equilibria for macroscopic systems. J. Phys. A: Math. Gen. **39**, 8461–8475 (2006)
106. Gudder, S.P.: Trans. AMS **119**, 428–442 (1965)
107. Gudder, S.P.: Axiomatic Quantum Mechanics and Generalized Probability Theory. Academic Press, New York (1970)
108. Gudder, S.P.: An approach to quantum probability. In: Khrennikov, A.Yu. (ed.) Foundations of Probability and Physics. Quantum Probability and White Noise Analysis, vol. 13, pp. 147–160. WSP, Singapore (2001)
109. Hardy, L.: Quantum theory from intuitively reasonable axioms. In: Khrennikov, A.Yu. (ed.) Quantum Theory: Reconsideration of Foundations. Ser. Math. Model., vol. 2, pp. 117–130. Växjö University Press, Växjö (2002); electronic volume: http://www.vxu.se/msi/forskn/publications.html
110. Harkavy, A.A.: Quantum mechanical information is ubiquitous. In: Kafatos, M. (ed.) Bell's Theorem, Quantum Theory and Conceptions of the Universe, pp. 11–24. Kluwer Academic, Dordrecht (1989)
111. Haven, E.: A Discussion on embedding the Black-Scholes option pricing model in a quantum physics setting. Physica A **304**, 507–524 (2002)
112. Haven, E.: Bohmian mechanics in a macroscopic quantum system. In: Adenier, G., Khrennikov, A.Yu., Nieuwenhuizen, Th.M. (eds.) Foundations of Probability and Physics, vol. 3. Ser. Conference Proceedings, vol. 810, pp. 330–340. American Institute of Physics, Melville (2006)

113. Heisenberg, W.: The Physical Principles of the Quantum Theory. Dover, London (1989)

114. Heisenberg, W.: Physics and Philosophy. Harper and Row. Harper Torchbooks, New York (1958)

115. Helland, I.S.: Quantum theory as a statistical theory. In: Khrennikov, A.Yu. (ed.) Quantum Theory: Reconsideration of Foundations. Ser. Math. Model., vol. 2, pp. 219–242. Växjö University Press, Växjö (2002); electronic volume: http://www.vxu.se/msi/forskn/publications.html

116. Hess, K., Philipp, W.: A possible loophole in the theorem of Bell. Proc. Natl. Acad. Sci. **98**, 14224 (2001)

117. Hess, K., Philipp, W.: Exclusion of time in Mermin's proof of Bell-type inequalities. In: Khrennikov, A.Yu. (ed.) Quantum Theory: Reconsideration of Foundations, vol. 2. Ser. Math. Model., vol. 10, pp. 243–254. Växjö University Press, Växjö (2003); electronic volume: http://www.vxu.se/msi/forskn/publications.html

118. Hess, K., Philipp, W.: Europhys. Lett. **57**, 775 (2002)

119. Hess, K., Philipp, W.: Bell's theorem: critique of proofs with and without inequalities. In: Adenier, G., Khrennikov, A.Yu. (eds.) Foundations of Probability and Physics, vol. 3. Ser. Conference Proceedings, vol. 750, pp. 150–155. American Institute of Physics, Melville (2005)

120. Hess, K.: In memoriam Walter Philipp. In: Adenier, G., Fuchs, C., Khrennikov, A.Yu. (eds.) Foundations of Probability and Physics, vol. 3. Ser. Conference Proceedings, vol. 889, pp. 3–6. American Institute of Physics, Melville (2007)

121. Hilbert, D., von Neumann, J., Nordheim, L.: Uber die Grundlagen der Quantenmechanik. Math. Ann. **98**, 1–30 (1927)

122. Hiley, B.: Phase space distributions of quantum phenomena. In: Khrennikov, A.Yu. (ed.) Quantum Theory: Reconsideration of Foundations, vol. 2. Ser. Math. Model., vol. 10, pp. 267–286. Växjö University Press, Växjö (2003); electronic volume: http://www.vxu.se/msi/forskn/publications.html

123. Hiesmayer, B.C.: Bell inequalities for neutral kaon system. In: Khrennikov, A.Yu. (ed.) Quantum Theory: Reconsideration of Foundations, vol. 2. Ser. Math. Model., vol. 10, pp. 255–266. Växjö University Press, Växjö (2003); electronic volume: http://www.vxu.se/msi/forskn/publications.html

124. Holevo, A.S.: Probabilistic and Statistical Aspects of Quantum Theory. North-Holland, Amsterdam (1982)

125. Holevo, A.S.: Statistical Structure of Quantum Theory. Springer, Berlin (2001)

126. Holland, P.: The Quantum Theory of Motion. Cambridge University Press, Cambridge (1993)

127. Home, D., Selleri, F.: Nuovo Cimento Riv. **14**, 2 (1991)

128. Hucks, J.: Hyperbolic complex structures in physics. J. Math. Phys. **34**, 5986–6008 (1993)

129. Hudson, R.: Translation invariant phase space mechanics. In: Khrennikov, A.Yu. (ed.) Quantum Theory: Reconsideration of Foundations, vol. 2. Ser. Math. Model., vol. 10, pp. 301–314. Växjö University Press, Växjö (2004); electronic volume: http://www.vxu.se/msi/forskn/publications.html

130. Jauch, J.M.: Foundations of Quantum Mechanics. Addison-Wesley, Reading (1968)

131. Khrennikov, A.Yu.: P-adic quantum mechanics with p-adic valued functions. J. Math. Phys. **32**, 932–937 (1984)

132. Khrennikov, A.Yu.: Second quantization and pseudo-differential operators. Theor. Math. Phys. **66**, 339–349 (1985)

133. Khrennikov, A.Yu.: The Feynman measure on the phase space and symbols of infinite-dimensional pseudo-differential operators. Math. Notes **37**, 734–742 (1985)

134. Khrennikov, A.Yu.: Infinite-dimensional pseudo-differential operators. Izv. Akad. Nauk USSR, Ser. Math. **51**, 46–68 (1987)

135. Khrennikov, A.Yu.: The principle of correspondence in quantum theories of field and relativistic bosonic string. Mat. Sb. **180**, 763–786 (1989)

136. Khrennikov, A.Yu.: P-adic Valued Distributions and their Applications to the Mathematical Physics. Kluwer Academic, Dordrecht (1994)

137. Khrennikov, A.Yu.: Non-Archimedean Analysis: Quantum Paradoxes, Dynamical Systems and Biological Models. Kluwer Academic, Dordrecht (1997)

138. Khrennikov, A.Yu.: Superanalysis. Kluwer Academic, Dordrecht (1999)

139. Khrennikov, A.Yu.: Interpretations of Probability. VSP International Science Publishers, Utrecht (1999)

140. Khrennikov, A.Yu.: Statistical measure of ensemble nonreproducibility and correction to Bell's inequality. Nuovo Cimento B **115**, 179–184 (1999)

141. Khrennikov, A.Yu.: Non-Kolmogorov probability models and modified Bell's inequality. J. Math. Phys. **41**, 1768–1777 (2000)

142. Khrennikov, A.Yu.: A perturbation of CHSH inequality induced by fluctuations of ensemble distributions. J. Math. Phys. **41**, 5934–5944 (2000)

143. Khrennikov, A.Yu., Petersson, H.: Theorem of Paley-Wiener for generalized analytic functions on infinite-dimensional spaces. Izv. Akad. Nauk, Ser. Math. **65**, 201–224 (2001)

144. Khrennikov, A.Yu.: Contextualist viewpoint to Greenberger-Horne-Zeilinger paradox. Phys. Lett. A **278**, 307–314 (2001)

145. Khrennikov, A.Yu.: Linear representations of probabilistic transformations induced by context transitions. J. Phys. A: Math. Gen. **34**, 9965–9981 (2001)

146. Khrennikov, A.Yu. (ed.): Foundations of Probability and Physics. Series PQ-QP: Quantum Probability and White Noise Analysis, vol. 13. WSP, Singapore (2001)

147. Khrennikov, A.Yu.: In: Classical and Quantum Mental Models and Freud's Theory of Unconscious Mind. Ser. Math. Model, vol. 1, Växjö Univ. Press, Växjö (2002); electronic volume: http://www.vxu.se/msi/forskn/publications.html

148. Khrennikov, A.Yu.: Ensemble fluctuations and the origin of quantum probabilistic rule. J. Math. Phys. **43**, 789–802 (2002)

149. Khrennikov, A.Yu.: Quantum statistics via perturbation effects of preparation procedures. Nuovo Cimento B **117**, 267–281 (2002)

150. Khrennikov, A.Yu.: Frequency analysis of the EPR-Bell argumentation. Found. Phys. **32**, 1159–1174 (2002)

151. Khrennikov, A.Yu. (ed.): Quantum Theory: Reconsideration of Foundations. Ser. Math. Model., vol. 2. Växjö University Press, Växjö (2002); electronic volume: http://www.vxu.se/msi/forskn/publications.html

152. Khrennikov, A.Yu., Volovich, Ja.I.: Discrete time leads to quantum–like interference of deterministic particles. In: Khrennikov, A.Yu. (ed.) Quantum Theory: Reconsideration of Foundations. Ser. Math. Model., vol. 2, pp. 441–454. Växjö University Press, Växjö (2002); electronic volume: http://www.vxu.se/msi/forskn/publications.html

153. Khrennikov, A.Yu., Volovich, Ja.I.: Interference effect for probability distributions of deterministic particles. In: Khrennikov, A.Yu. (ed.) Quantum Theory: Reconsideration of Foundations. Ser. Math. Model, vol. 2, pp. 455–462. Växjö University Press, Växjö (2002); electronic volume: http://www.vxu.se/msi/forskn/publications.html

154. Khrennikov, A.Yu., Volovich, I.V.: Local realism, contextualism and loopholes in Bell's experiments. In: Khrennikov, A.Yu. (ed.) Foundations of Probability and Physics, vol. 2. Ser. Math. Model., vol. 5, pp. 325–344. Växjö University Press, Växjö (2002); electronic volume: http://www.vxu.se/msi/forskn/publications.html

155. Khrennikov, A.Yu., Volovich, I.V.: Quantum nonlocality, EPR model, and Bell's theorem. In: Proceedings of the 3rd Sakharov Conference on Physics, vol. 2, pp. 269–276. WSP, Singapore (2003)
156. Khrennikov, A.Yu. (ed.): Foundations of Probability and Physics, vol. 2. Ser. Math. Model. vol. 5. Växjö University Press, Växjö (2003); electronic volume: http://www.vxu.se/msi/forskn/publications.html
157. Khrennikov, A.Yu., Smolyanov, O.G.: Nonclassical Kolmogorovian type models describing quantum experiments. Dokl. Acad. Nauk **388**, 27–32 (2003); English translation: Dokl. Math. **67**, 93–97 (2003)
158. Khrennikov, A.Yu., Kozyrev, S.V.: Noncommutative probability in classical disordered systems. Physica A **326**, 456–463 (2003)
159. Khrennikov, A.Yu.: Contextual viewpoint to quantum stochastics. J. Math. Phys. **44**, 2471–2478 (2003)
160. Khrennikov, A.Yu.: Representation of the Kolmogorov model having all distinguishing features of quantum probabilistic model. Phys. Lett. A **316**, 279–296 (2003)
161. Khrennikov, A.Yu.: Hyperbolic quantum mechanics. Adv. Appl. Clifford Algebras **13**(1), 1–9 (2003)
162. Khrennikov, A.Yu.: Quantum-like formalism for cognitive measurements. Biosystems **70**, 211–233 (2003)
163. Khrennikov, A.Yu.: Interference of probabilities and number field structure of quantum models. Ann. Phys. **12**, 575–585 (2003)
164. Khrennikov, A.Yu.: Event-independence, collective-independence, EPR-Bohm experiment and incompleteness of quantum mechanics. Phys. Essays **16**, 26–32 (2003)
165. Khrennikov, A.Yu., Smolyanov, O.G., Truman, A.: Kolmogorov probability spaces describing quantum correlations. Dokl. Acad. Nauk **393**, 28–32 (2003); English translation: Dokl. Math. **68**, 452–456 (2003)
166. Khrennikov, A.Yu.: Quantum-psychological model of the stock market. Probl. Perspect. Manag. **1**, 137–148 (2003)
167. Khrennikov, A.Yu.: Information Dynamics in Cognitive, Psychological, Social, and Anomalous Phenomena. Kluwer Academic, Dordrecht (2004)
168. Khrennikov, A.Yu.: Växjö interpretation-2003: Realism of contexts. In: Khrennikov, A.Yu. (ed.) Quantum Theory: Reconsideration of Foundations, vol. 2. Ser. Math.

Model., vol. 10, pp. 323–338. Växjö University Press, Växjö (2004); electronic volume: http://www.vxu.se/msi/forskn/publications.html

169. Khrennikov, A.Yu., Loubenets, E.: On relation between probabilities in quantum and classical experiments. Found. Phys. **34**, 689–704 (2004)

170. Khrennikov, A.Yu.: Contextual approach to quantum mechanics and the theory of the fundamental prespace. J. Math. Phys. **45**, 902–921 (2004)

171. Khrennikov, A.Yu.: EPR-Bohm experiment and interference of probabilities. Found. Phys. Lett. **17**, 691–700 (2004)

172. Khrennikov, A.Yu., Volovich, Ja.I.: Discrete time dynamical models and their quantum-like context-dependent properties. J. Mod. Opt. **51**, 113–114 (2004)

173. Khrennikov, A.Yu.: On unification of classical and quantum probability theories. J. Mod. Opt. **51**, 109 (2004)

174. Khrennikov, A.Yu.: Frequency derivation of the EPR-Bohm correlations. Nuovo Cimento **119**, 131–147 (2004)

175. Khrennikov, A.Yu.: On quantum-like probabilistic structure of mental information. Open Syst. Inf. Dyn. **11**(3), 267–275 (2004)

176. Khrennikov, A.Yu.: Accentuate the negative. Wilmott (Serving the Quantative Finance Community), September, 30–33 (2004)

177. Khrennikov, A.Yu. (ed.): Quantum Theory: Reconsideration of Foundations, vol. 2. Ser. Math. Model., vol. 10. Växjö University Press, Växjö (2004); electronic volume: http://www.vxu.se/msi/forskn/publications.html

178. Khrennikov, A.Yu. (ed.): Foundations of Probability and Physics, vol. 3. Ser. Conference Proceedings, vol. 750, American Institute of Physics, Melville (2005)

179. Khrennikov, A.Yu., Smolyanov, O.G.: Probabilistic measurement models with noncommuting and commuting observables. Dokl. Acad. Nauk **402**, 748–753 (2005). English Translation: Dokl. Math. 71 (2005) 461–465

180. Khrennikov, A.Yu.: The principle of supplementarity: a contextual probabilistic viewpoint to complementarity, the interference of probabilities, and the incompatibility of variables in quantum mechanics. Found. Phys. **35**, 1655–1693 (2005)

181. Khrennikov, A.Yu.: Interference in the classical probabilistic model and its representation in complex Hilbert space. Physica E **29**, 226–236 (2005)

182. Khrennikov, A.Yu.: Interference in the classical probabilistic framework. Fuzzy Sets Syst. **155**, 4–17 (2005)

183. Khrennikov, A.Yu.: Hyperbolic quantum mechanics. Dokl. Acad. Nauk **402**, 170–172 (2005). English Translation: Dokl. Math. **71**, 363–365 (2005)
184. Khrennikov, A.Yu., Smolyanov, O.G., Truman, A.: Kolmogorov probability spaces describing Accardi models for quantum correlations. Open Syst. Inf. Dyn. **12**, 371–384 (2005)
185. Khrennikov, A.Yu.: On the representation of contextual probabilistic dynamics in the complex Hilbert space: linear and nonlinear evolutions, Schrödinger dynamics. Nuovo Cimento **120**, 353–366 (2005)
186. Khrennikov, A.Yu.: A pre-quantum classical statistical model with infinite-dimensional phase space. J. Phys. A: Math. Gen. **38**, 9051–9073 (2005)
187. Khrennikov, A.Yu., Kozyrev, S.V.: Contextual quantization and the principle of complementarity of Probabilities. Open Syst. Inf. Dyn. **12**, 303–318 (2005)
188. Khrennikov, A.Yu., Volovich, I.V.: Local realistic representation for correlations in the original EPR-model for position and momentum. Soft Comput. A: Fusion Found. Methodol. Appl. **10**, 521–529 (2005)
189. Khrennikov, A.Yu.: Schrödinger dynamics as the Hilbert space projection of a realistic contextual probabilistic dynamics. Europhys. Lett. **69**(5), 678–684 (2005)
190. Khrennikov, A.Yu.: From classical statistical model to quantum model through ignorance of information. In: Petitjean, M. (ed.) Proceedings of FIS2005, the Third Conference on the Foundations of Information Science. MDPI Publishers, Basel (2005); electronic volume: http://www.mdpi.org/fis2005/proceedings.html
191. Khrennikov, A.Yu.: To quantum mechanics through projection of classical statistical mechanics on prespace. In: Buccheri, R., Elitzur, A.C., Saniga, M. (eds.) Endophysics Time, Quantum and the Subjective. WSP, Singapore (2005)
192. Khrennikov, A.Yu.: Reconstruction of quantum theory on the basis of the formula of total probability. In: Adenier, G., Khrennikov, A.Yu. (eds.) Foundations of probability and physics, vol. 3. Ser. Conference Proceedings, vol. 750, pp. 187–218. American Institute of Physics, Melville (2005)
193. Khrennikov, A.Yu.: Generalizations of quantum mechanics induced by classical statistical field theory. Found. Phys. Lett. **18**, 637–650 (2006)
194. Khrennikov, A.Yu.: Linear and nonlinear analogues of the Schrödinger equation in the contextual approach to quantum mechanics. Dokl. Akad. Nauk **404**, 33–36 (2005), Dokl. Math. **72**, 791–794 (2005)

195. Khrennikov, A.Yu.: Representation of the contextual statistical model by hyperbolic amplitudes. J. Math. Phys. **46**, 062111–062124 (2005)
196. Khrennikov, A.Yu.: Distributions and pseudo-differential operators on infinite-dimensional spaces and applications in quantum physics. Oper. Theory: Adv. Appl. **164**, 161–172 (2006)
197. Khrennikov, A.Yu., Smolyanov, O.G.: Nonstationary Kolmogorov probability models describing quantum measurements. Dokl. Math. **73**, 286–291 (2006)
198. Khrennikov, A.Yu., Volovich, Ya.I.: Energy levels of hydrogen atom in discrete time dynamics. Open Syst. Inf. Dyn. **13**, 119–132 (2006)
199. Khrennikov, A.Yu.: Quantum-like brain: "interference of minds". Biosystems **84**, 225–241 (2006)
200. Khrennikov, A.Yu.: Nonlinear Schrödinger equations from prequantum classical statistical field theory. Phys. Lett. A **357**, 171–176 (2006)
201. Khrennikov, A.Yu.: Prequantum classical statistical field theory: complex representation, Hamilton-Schrödinger equation, and interpretation of stationary states. Found. Phys. Lett. **19**, 299–319 (2006)
202. Khrennikov, A.Yu., Segre, G.: Von Neumann uniqueness theorem doesn't hold in hyperbolic quantum mechanics. Int. J. Theor. Phys. **45**, 1869–1890 (2006)
203. Khrennikov, A.Yu.: Classical representation of quantum means and quantum dynamics. Dokl. Math. **73**, 445–448 (2006)
204. Khrennikov, A.Yu.: Quantum mechanics as an asymptotic projection of statistical mechanics of classical fields. In: Adenier, G., Khrennikov, A.Yu., Nieuwenhuizen, Th.M. (eds.) Quantum Theory: Reconsideration of Foundations, vol. 3. Ser. Conference Proceedings, vol. 810, pp. 179–197. American Institute of Physics, Melville (2006)
205. Khrennikov, A.Yu.: Representation of quantum field theory as classical statistical mechanics for field functionals. Dokl. Math. 74 (2006)
206. Khrennikov, A.Yu.: On the problem of hidden variables for quantum field theory. Nuovo Cimento B **121**(5), 505–521 (2006)
207. Khrennikov, A.Yu.: The formula of total probability with the interference term and the representation of the Kolmogorov model in the Hilbert space. Theory Probab. Appl. **51**, 1–18 (2006)
208. Khrennikov, A.Yu.: To quantum averages through asymptotic expansion of classical averages on infinite-dimensional space. J. Math. Phys. **48**(1), 013512 (2007)

209. Khrennikov, A.Yu.: A mathematician's viewpoint to Bell's theorem: in memory of Walter Philipp. In: Adenier, G., Fuchs, C., Khrennikov, A.Yu. (eds.) Foundations of Probability and Physics, vol. 3. Ser. Conference Proceedings, vol. 889, pp. 7–20. American Institute of Physics, Melville (2007)
210. Khrennikov, A.Yu.: Quantum mechanics for military officers. In: Adenier, G., Fuchs, C., Khrennikov, A.Yu. (eds.) Foundations of Probability and Physics, vol. 3. Ser. Conference Proceedings, vol. 889, pp. 137–151. American Institute of Physics, Melville (2007)
211. Khrennikov, A.Yu.: Fourier analysis over hyperbolic algebra, pseudo-differential operators, and hyperbolic deformation of classical mechanics. Infin. Dimens. Anal. Quantum Probab. Relat. Top. **10**, 421–438 (2007)
212. Kim, Y.S., Noz, M.E.: Can one do quantum mechanics without Einstein?. In: Adenier, G., Fuchs, C., Khrennikov, A.Yu. (eds.) Foundations of Probability and Physics, vol. 3. Ser. Conference Proceedings, vol. 889, pp. 152–161. American Institute of Physics, Melville (2007)
213. Kirkpatrick, K.A.: "Quantal" behavior in classical probability. Found. Phys. Lett. **16**(3), 199–224 (2003)
214. Klyshko, D.N.: The Bell and GHZ theorems: a possible three-photon interference experiment and the question of nonlocality. Phys. Lett. A **172**, 399–403 (1993)
215. Klyshko, D.N.: Phys. Lett. A **176**, 415–420 (1993)
216. Klyshko, D.N.: Phys. Lett. A **218**, 119–127 (1996)
217. Klyshko, D.N.: Phys. Lett. A **247**, 261–266 (1998)
218. Klyshko, D.N.: Laser Phys. **6**, 1056–1076 (1996)
219. Klyshko, D.N.: Ann. N.Y. Acad. Sci. **755**, 13–27 (1995)
220. Klyshko, D.N.: Usp. Fiz. Nauk **168**, 975–1015 (1998)
221. Kochen, S., Specker, E.: The problem of hidden variables in quantum mechanical systems. J. Math. Mech. **17**, 59–87 (1967)
222. Kolmogoroff, A.N.: Grundbegriffe der Wahrscheinlichkeitsrechnung. Springer, Berlin (1933); English translation: Kolmogorov, A.N.: Foundations of Theory of Probability. Chelsea Publishing Company, New York (1956)
223. Kolmogorov, A.N.: Prob. Inf. Transf. **1**, 1–7 (1965)
224. Kolmogorov, A.N.: Russ. Math. Surv. **38**, 27–36 (1983)
225. Kozyrev, S.V.: The noncommutative replica approach. http://xxx.lanl.gov/abs/cond-mat/0110238

226. Kozyrev, S.V.: The noncommutative replica procedure. http://xxx.lanl.gov/abs/cond-mat/0210196
227. Kustatter, G., Moffat, J.W., Malzan, J.: J. Math. Phys. **24**, 886–894 (1983)
228. Lamb, W.E.: The Interpretation of Quantum Mechanics (Edited and annotated by Mehra, Ja.). Rinton Press, Princeton (2001)
229. Lande, A.: Foundations of Quantum Theory. Yale University Press, Yale (1955)
230. Lande, A.: New Foundations of Quantum Mechanics. Cambridge University Press, Cambridge (1965)
231. Larsson, J.-Å.: Quantum Paradoxes, Probability Theory, and Change of Ensemble. Linköping University Press, Linköping (2000)
232. Larsson, J.-Å.: Bell inequalities for position measurements. In: Khrennikov, A.Yu. (ed.) Quantum Theory: Reconsideration of Foundations. Ser. Math. Model., vol. 10, pp. 353–364. Växjö University Press, Växjö (2003); electronic volume: http://www.vxu.se/msi/forskn/publications.html
233. Larsson, J.-Å., Gill, R.: Bell's inequality and the coincidence time loophole. In: Adenier, G., Khrennikov, A.Yu. (eds.) Foundations of Probability and Physics, vol. 3. Ser. Conference Proceedings, vol. 750, pp. 228–235. American Institute of Physics, Melville (2005)
234. Li, M., Vitanyi, P.: An Introduction to Kolmogorov Complexity. Springer, Berlin (1993)
235. Light, P., Butterworth, G.: Context and Cognition. Erlbaum, Hillsdale (1993)
236. Loubenets, E.R.: General framework for the probabilistic description of experiments. In: Khrennikov, A.Yu. (ed.) Quantum Theory: Reconsideration of Foundations, vol. 2. Ser. Math. Model., vol. 10, pp. 385–387. Växjö University Press, Växjö (2004); electronic volume: http://www.vxu.se/msi/forskn/publications.html
237. Ludwig, G.: Foundations of Quantum Mechanics. Springer, Berlin (1983)
238. Lüders, G.: Ann. Phys., Lpz. **8**, 322 (1951)
239. Mackey, G.W.: Mathematical Foundations of Quantum Mechanics. Benjamin, New York (1963)
240. Manko, V.I.: Classical formulation of quantum mechanics. J. Russ. Laser Res. **17**, 579–584 (1996)
241. Manko, O.V., Manko, V.I.: Classical mechanics is not $h \to 0$ limit of quantum mechanics. J. Russ. Laser Res. **25**, 477–492 (2004)
242. Manko, V.I., Shchukin, E.V.: A charged particle in an electric field in the probability representation of quantum mechanics. J. Russ. Laser Res. **22**, 545–560 (2001)

243. Manko, M.A., Manko, V.I., Mendes, V.I.: A probabilistic operator symbol framework for quantum information. J. Russ. Laser Res. **27**, 507–532 (2006)

244. Martin-Löf, P.: The definition of random sequences. Inf. Control **9**, 602–619 (1966)

245. Martin-Löf, P.: Complexity oscillations in infinite binary sequences. Z. Warsch. Verw. Geb. **19**, 225–230 (1971)

246. Mermin, N.D.: Is the moon there when nobody looks?. Reality and quantum theory. Phys. Today **April**, 38–41 (1985)

247. Mermin, N.D.: Whose knowledge?. In: Khrennikov, A.Yu. (ed.) Quantum Theory: Reconsideration of Foundations. Ser. Math. Model., vol. 2, pp. 261–270. Växjö University Press, Växjö (2002); electronic volume: http://www.vxu.se/msi/forskn/publications.html

248. Mezard, M., Parisi, G., Virasoro, M.: Spin-Glass Theory and Beyond. World Scientific, Singapore (1987)

249. Moyal, J.E.: Quantum mechanics as a statistical theory. Proc. Camb. Philos. Soc. **45**, 99–124 (1949)

250. Muckenheim, W.: A review of extended probability. Phys. Rep. **133**, 338–401 (1986)

251. Nánásiová, O., Khrennikov, A.Yu.: Representation theorem of observables on a quantum system. Int. J. Theor. Phys. **45**, 469–482 (2006)

252. Nelson, E.: Quantum Fluctuations. Princeton University Press, Princeton (1985)

253. Nieuwenhuizen, Th.M.: Classical phase space density for relativistic electron. In: Adenier, G., Khrennikov, A.Yu., Nieuwenhuizen, Th.M. (eds.) Quantum Theory: Reconsideration of Foundations, vol. 3. Ser. Conference Proceedings, vol. 810, pp. 198–210. American Institute of Physics, Melville (2006)

254. Nieuwenhuizen, T.M., Mehmani, B., Spicka, V., Aghdami, M.J., Khrennikov, A.Yu. (eds): Beyond the Quantum. WSP, Singapore (2007)

255. Pearle, P.: Hidden-variable example based upon data rejection. Phys. Rev. D **2**, 1418 (1970)

256. Peres, A.: Unperformed experiments have no results. Am. J. Phys. **46**, 745 (1978)

257. Peres, A.: Quantum Theory: Concepts and Methods. Kluwer Academic, Dordrecht (1994)

258. Perez-Suarez, M., Santos, D.J.: Quantum mechanics as an information theory: some further musings on a Fuchsian proposal. In: Khrennikov, A.Yu. (ed.) Quantum Theory: Reconsideration of Foundations. Ser. Math. Model., vol. 2, pp. 469–478. Växjö University Press, Växjö (2002); electronic volume: http://www.vxu.se/msi/forskn/publications.html

259. Pitowsky, I.: Resolution of the Einstein-Podolsky-Rosen and Bell paradoxes. Phys. Rev. Lett. **48**, 1299–1302 (1982)

260. Pitowsky, I.: Deterministic model of spin and statistics. Phys. Rev. D **27**, 2316–2326 (1983)
261. Piotrowski, E.W., Sladkowski, J.: Quantum-like approach to financial risk: quantum anthropic principle. Acta Phys. Pol. B **32**, 3873–3879 (2001)
262. Piotrowski, E.W., Sladkowski, J.: Quantum market games. Physica A **312**, 208–216 (2002)
263. Piotrowski, E.W., Sladkowski, J., Syska, J.: Interference of quantum market strategies. Physica A **318**, 516–528 (2003)
264. Piotrowski, E.W., Sladkowski, J.: An invitation to quantum game theory. Int. J. Theor. Phys. **42**, 1089 (2003)
265. Piotrowski, E.W.: Fixed point theorem for simple quantum strategies in quantum market games. Physica A **324**, 196–200 (2003)
266. Piotrowski, E.W., Sladkowski, J.: Quantum games in finance. Quant. Finance **4**, C61–C67 (2004)
267. Piotrowski, E.W., Sladkowski, J.: Quantum diffusion of prices and profits. Physica A **345**, 185–195 (2005)
268. Piotrowski, E.W., Schroeder, M., Zambrzycka, A.: Quantum extension of European option pricing based on the Ornstein-Uhlenbeck process. Physica A **368**, 176–182 (2006)
269. Plotnitsky, A.: Reading Bohr: Complementarity, Epistemology, Entanglement, and Decoherence. In: Gonis, A., Turchi, P. (eds.) NATO Workshop Decoherence and Its Implications for Quantum Computations, pp. 3–37. IOS, Amsterdam (2001)
270. Plotnitsky, A.: The Knowable and Unknowable: Modern Science, Nonclassical Thought, and the “Two Cultures”. University Michigan Press, Michigan (2002)
271. Plotnitsky, A.: This is an extremely funny thing, something must be hidden behind that: Quantum waves and quantum probability with Erwin Schrödinger. In: Adenier, G., Khrennikov, A.Yu. (eds.) Foundations of Probability and Physics, vol. 3. Ser. Conference Proceedings, vol. 750, pp. 388–408. American Institute of Physics, Melville (2005)
272. Plotnitsky, A.: Reading Bohr: Physics and Philosophy. Springer, Dordrecht (2006)
273. Porteous, I.: Clifford Algebras and the Classical Groups. Cambridge University Press, Cambridge (1995)
274. Rastal, P.: Found. Phys. **13**, 555 (1983)
275. Renyi, A.: Probability Theory. North-Holland, Amsterdam (1970)
276. Santos, E.: Found. Phys. **21**, 221 (1991)
277. Santos, E.: Phys. Rev. Lett. **66**, 1388 (1991)

278. Schack, R.: Bayesian versus frequentist predictions in quantum tomography. In: Adenier, G., Fuchs, C., Khrennikov, A.Yu. (eds.) Foundations of Probability and Physics, vol. 3. Ser. Conference Proceedings, vol. 889, pp. 230–234. American Institute of Physics, Melville (2007)
279. Schrödinger, E.: Die gegenwärtige Situation in der Quantenmechanik. Naturwissenschaften **23**, 807–812, 823–828; 844-849 (1935)
280. Schrödinger, E.: In: Bitbol, M., Darrigol, O. (eds.) Philosophy and the Birth of Quantum Mechanics. Editions Frontieres, Gif-sur-Yvette (1992)
281. Segal, W., Segal, I.E.: The Black-Scholes pricing formula in the quantum context. Proc. Natl. Acad. Sci. USA **95**, 4072–4075 (1998)
282. Serdobolskii, V.I.: Constructive foundations of randomness. In: Khrennikov, A.Yu. (ed.) Foundations of Probability and Physics. Quantum Probability and White Noise Analysis, vol. 13, pp. 335–349. WSP, Singapore (2001)
283. Shimony, A.: Search for a Naturalistic World View. Cambridge University Press, Cambridge (1993)
284. Shimony, A.: On mentality, quantum mechanics and the actualization of potentialities. In: Penrose, R., Longair, M. (eds.) The Large, the Small and the Human Mind. Cambridge University Press, New York (1997)
285. Shiryaev, A.N.: Kolmogorov: life and creative activities. Ann. Probab. **17**, 866–944 (1989)
286. Shiryaev, A.N.: Probability. Springer, New York (1984)
287. Smolyanov, O.G.: Infinite-dimensional pseudo-differential operators and Schrödinger quantization. Dokl. Akad. Nauk USSR **263**, 558–561 (1982)
288. Solomonoff, R.J.: In: Progress of Symposia in Applied Math., AMS 43 (1962)
289. Solomonoff, R.J.: A formal theory of inductive inference, Part I + A formal theory of inductive inference, Part II. Inf. Contr. **7**, 1–2 (1964)
290. Solomonoff, R.J.: IEEE Trans. Inform. Theory **4/5**, 662–664 (1968)
291. Stapp, H.P.: S-matrix interpretation of quantum theory. Phys. Rev. D **3**, 1303 (1971)
292. Stapp, H.P.: Mind, Matter and Quantum Mechanics. Springer, Berlin (1993)
293. Stenholm, S.: Two dogmas of quantum theory. In: Khrennikov, A.Yu. (ed.) Quantum Theory: Reconsideration of Foundations, vol. 2. Ser. Math. Model., vol. 10, pp. 519–532. Växjö University Press, Växjö (2003); electronic volume: http://www.vxu.se/msi/forskn/publications.html
294. Svozil, K.: Quantum Logic. Springer, Berlin (1998)
295. Svozil, K.: Randomness and Undecidesibility in Physics. WSP, Singapore (1993)

296. Svozil, K.: On counterfactuals and contextuality. In: Adenier, G., Khrennikov, A.Yu. (eds.) Foundations of Probability and Physics, vol. 3. Ser. Conference Proceedings, vol. 750, pp. 351–360. American Institute of Physics, Melville (2005)
297. 't Hooft, G.: Quantum gravity as a dissipative deterministic system. http://xxx.lanl.gov/abs/gr-qc/9903084 (1999)
298. 't Hooft, G.: The mathematical basis for deterministic quantum mechanics. http://arxiv.org/abs/quant-ph/0604008 (2006)
299. 't Hooft, G.: The free-will postulate in quantum mechanics. http://xxx.lanl.gov/abs/quant-ph/0701097v1 (2007)
300. Tornier, E.: Wahrscheinlichkeitsrechnunug und allgemeine Integrationstheorie. Leipzig University Press, Leipzig (1936)
301. Ulrych, S.: The Poincaré mass operator in terms of a hyperbolic algebra. Phys. Lett. B **612**, 89–91 (2005)
302. Ulrych, S.: Symmetries in the hyperbolic Hilbert space. Phys. Lett. B **618**, 233–236 (2005)
303. Ulrych, S.: Relativistic quantum physics with hyperbolic numbers. Preprint Zürich, Switzerland (2005)
304. van Lambalgen, M.: Von Mises' definition of random sequences reconsidered. J. Symb. Log. **52**, 725–755 (1987)
305. Ville, J.: Etude Critique de la Notion de Collective. Gauthier–Villars, Paris (1939)
306. Vitiello, G.: My Double Unveiled—the Dissipative Quantum Model of Brain. Benjamins, Amsterdam (2001)
307. Vladimirov, V.S., Volovich, I.V., Zelenov, E.I.: P-Adic Analysis and Mathematical Physics. WSP, Singapore (1993)
308. Volovich, I.V.: Quantum cryptography in space and Bell's theorem. In: Khrennikov, A.Yu. (ed.) Foundations of Probability and Physics. QP–PQ: Quantum Prob. White Noise Anal., vol. 13, pp. 364–372. WSP, River Edge (2001)
309. Von Mises, R.: Grundlagen der Wahrscheinlichkeitsrechnung. Math. Z. **5**, 52–99 (1919)
310. Von Mises, R.: Probability, Statistics and Truth. Macmillan, London (1957)
311. Von Mises, R.: The Mathematical Theory of Probability and Statistics. Academic, London (1964)
312. Von Neumann, J., Morgenstern, O.: Theory of Games and Economic Behavior. Princeton University Press, Princeton (1947)

313. Von Neuman, J.: Mathematical Foundations of Quantum Mechanics. Princeton University Press, Princeton (1955)
314. Vorob'ev, N.N.: Consistent families of measures and their extensions. Theory Probab. Appl. **7**, 147–162 (1962)
315. Wald, A.: Die Widerspruchsfreiheit des Kollektivbegriffs in der Wahrscheinlichkeitsrechnung. Ergeb. Math. Kolloqu. **8**, 38–72 (1938)
316. Weihs, G.: A test of Bell's inequality with spacelike separation. In: Adenier, G., Fuchs, C., Khrennikov, A.Yu. (eds.) Foundations of Probability and Physics, vol. 4. Ser. Conference Proceedings, vol. 889, pp. 250–262. American Institute of Physics, Melville (2007)
317. Weyl, H.: The Theory of Groups and Quantum Mechanics. Dover, New York (1950)
318. Wightman, A.S.: Hilbert's sixth problem: mathematical treatment of the axioms of physics. Proc. Symp. Pure Math. **28**, 147–233 (1976)
319. Wigner, E.P.: The problem of measurement. Am. J. Phys. **31**, 6–46 (1963)
320. Wigner, E.P.: Characteristic vectors of bordered matrices with infinite dimensions. Ann. Math. **62**, 548–564 (1955)
321. Zheng, Z., Xuegang, Y.: Hyperbolic Schrödinger equation. Adv. Appl. Clifford Anal. **14**(2), 207–220 (2004)
322. Zvonkin, A.K., Levin, L.A.: A complexity of finite objects and justification of the notions of information and randomness on the basis of algorithms. Usp. Mat. Nauk **25**, 85–127 (1970)

Index

MIX
Papier | Fördert
gute Waldnutzung
FSC® C083411

Kolibri 360 GmbH
Ferdinand-Jühlke-Straße 7
99095 Erfurt, Deutschland
produktsicherheit@kolibri360.de